Advances in Experimental Medicine and Biology

Volume 1122

More information about this series at http://www.springer.com/series/5584

Alexander Birbrair
Editor

Pericyte Biology in Different Organs

Springer

Editor
Alexander Birbrair
Department of Radiology
Columbia University Medical Center
New York, NY, USA

Department of Pathology
Federal University of Minas Gerais
Belo Horizonte, MG, Brazil

ISSN 0065-2598 ISSN 2214-8019 (electronic)
Advances in Experimental Medicine and Biology
ISBN 978-3-030-11092-5 ISBN 978-3-030-11093-2 (eBook)
https://doi.org/10.1007/978-3-030-11093-2

Library of Congress Control Number: 2019934955

This Springer imprint is published by the registered company Springer Nature Switzerland AG
The registered company address is: Gewerbestrasse 11, 6330 Cham, Switzerland

Preface

This book's initial title was "Pericyte Biology: Development, Homeostasis and Disease." However, due to the current great interest in this topic, we were able to assemble more chapters than would fit in one book, covering pericyte biology under distinct circumstances. Therefore, the book was subdivided into three volumes entitled *Pericyte Biology - Novel Concepts*, *Pericyte Biology in Different Organs*, and *Pericyte Biology in Disease*.

This book *Pericyte Biology in Different Organs* presents contributions by expert researchers and clinicians in the multidisciplinary areas of medical and biological research. The chapters provide timely detailed overviews of recent advances in the field. This book describes the major contributions of pericytes to different organs' biology in physiological and pathological conditions. Further insights into the biology of pericytes will have important implications for our understanding of organ development, homeostasis, and disease. The authors focus on the modern methodologies and the leading-edge concepts in the field of cell biology. In recent years, remarkable progress has been made in the identification and characterization of pericytes in several tissues using state-of-the-art techniques. These advantages facilitated the identification of pericyte subpopulations and definition of the molecular basis of pericytes' role within different organs. Thus, the present book is an attempt to describe the most recent developments in the area of pericyte behavior which is one of the emergent hot topics in the field of molecular and cellular biology today. Here, we present a selected collection of detailed chapters on what we know so far about the pericytes in various tissues and under distinct pathophysiological conditions. Thirteen chapters written by experts in the field summarize the present knowledge about the roles of pericytes in different organs.

Herbert A. Reitsamer and colleagues from Paracelsus Medical University/SALK discuss the role of pericytes in the retina. Limor Landsman from Tel Aviv University describes pericytes in the pancreas. Lynn M. Schnapp and colleagues from the Medical University of South Carolina compile our understanding of pericyte biology in the lung. Jyoti Gautam and Yao Yao from the University of Georgia update us with what we know about skeletal muscle pericytes. Mercedes Fernandez and colleagues from the University of Barcelona summarize current knowledge on gut pericytes. Yuya Kunisaki from Kyushu University Hospital addresses the importance of pericytes in the bone marrow. Martin Canis and Mattis Bertlich from the University Hospital Munich focus on cochlear pericytes. Maria Angelica Miglino and colleagues from the University of S*ã*o Paulo introduce our current knowledge about placental pericytes. Enis Kostallari and Vijay H. Shah from the Mayo Clinic discuss the roles of pericytes in the liver. Motohiro Komaki from Kanagawa Dental University introduces what we know about pericytes in the periodontal ligament. Linda L. Lee and Vishnu Chintalgattu from Amgen Inc. talk about pericytes in the heart. Clifford L. Librach and colleagues from the University of Toronto focus on umbilical cord pericytes. Finally, Michail S. Davidoff from the *University Medical Center Hamburg-Eppendorf* gives an overview of pericytes in the testis.

It is hoped that the articles published in this book will become a source of reference and inspiration for future research ideas. I would like to express my deep gratitude to my wife Veranika Ushakova and Mr. Murugesan Tamilsevan from Springer, who helped at every step of the execution of this project.

This book is dedicated to the memory of my grandfather Pavel Sobolevsky, PhD, a renowned mathematician, who passed away during the creation of this piece.

My grandfather Pavel Sobolevsky z"l, PhD (March 26, 1930–August 16, 2018)

New York, NY, USA
Belo Horizonte, MG, Brazil

Alexander Birbrair

Contents

Contributors

Rodrigo S. N. Barreto School of Veterinary Medicine and Animal Sciences, University of São Paulo, Butantã, Sao Paulo, Brazil

Mattis Bertlich The Department of Otorhinolaryngology, Head and Neck Surgery, University Hospital, Munich, Federal Republic of Germany

Alexander Birbrair Department of Radiology, Columbia University Medical Center, New York, NY, USA

Department of Pathology, Federal University of Minas Gerais, Pampulha, Belo Horizonte, Brazil

Daniela Bruckner Department of Ophthalmology, University Clinic of Ophthalmology and Optometry, Research Program for Experimental Ophthalmology and Glaucoma Research, Paracelsus Medical University/SALK, Salzburg, Austria

Martin Canis The Department of Otorhinolaryngology, Head and Neck Surgery, University Hospital, Munich, Federal Republic of Germany

Andressa Daronco Cereta School of Veterinary Medicine and Animal Sciences, University of São Paulo, Butantã, Sao Paulo, Brazil

Vishnu Chintalgattu Department of CardioMetabolic Disorders, Amgen Research and Discovery, Amgen Inc., South San Francisco, CA, USA

Leda M. C. Coimbra-Campos Department of Pathology, Federal University of Minas Gerais, Pampulha, Belo Horizonte, Brazil

Michail S. Davidoff University Medical Center Hamburg-Eppendorf, Hamburg Museum of Medical History, Hamburg, Germany

Mercedes Fernandez Angiogenesis in Liver Disease Research Group, IDIBAPS Biomedical Research Institute, Hospital Clinic, University of Barcelona, Barcelona, Spain

Biomedical Research Networking Center on Hepatic and Digestive Disease (CIBEREHD), Spanish National Institute of Health, Barcelona, Spain

Jyoti Gautam Department of Pharmaceutical and Biomedical Sciences, University of Georgia, Athens, GA, USA

Andrée Gauthier-Fisher CReATe Fertility Centre, University of Toronto, Toronto, ON, Canada

Chi F. Hung Division of Pulmonary, Critical Care and Sleep Medicine, University of Washington, Seattle, WA, USA

Motohiro Komaki Department of Highly Advanced Stomatology, Graduate School of Dentistry, Kanagawa Dental University, Yokohama City, Kanagawa, Japan

Enis Kostallari Division of Gastroenterology and Hepatology, Mayo Clinic, Rochester, MN, USA

Yuya Kunisaki Kyushu University Hospital, Center for Cellular and Molecular Medicine, Fukuoka, Japan

Limor Landsman Department of Cell and Developmental Biology, Sackler Faculty of Medicine, Tel Aviv University, Tel Aviv, Israel

Linda L. Lee Department of CardioMetabolic Disorders, Amgen Research and Discovery, Amgen Inc., South San Francisco, CA, USA

Clifford L. Librach CReATe Fertility Centre, University of Toronto, Toronto, ON, Canada

Department of Obstetrics and Gynecology, University of Toronto, Toronto, ON, Canada

Department of Physiology, University of Toronto, Toronto, ON, Canada

Institute of Medical Sciences, University of Toronto, Toronto, ON, Canada

Department of Obstetrics and Gynecology, Women's College Hospital, Toronto, ON, Canada

Marc Mejias Angiogenesis in Liver Disease Research Group, IDIBAPS Biomedical Research Institute, Hospital Clinic, University of Barcelona, Barcelona, Spain

Biomedical Research Networking Center on Hepatic and Digestive Disease (CIBEREHD), Spanish National Institute of Health, Barcelona, Spain

Maria Angelica Miglino School of Veterinary Medicine and Animal Sciences, University of São Paulo, Butantã, Sao Paulo, Brazil

Nuria Pell Angiogenesis in Liver Disease Research Group, IDIBAPS Biomedical Research Institute, Hospital Clinic, University of Barcelona, Barcelona, Spain

Marta Ramirez Angiogenesis in Liver Disease Research Group, IDIBAPS Biomedical Research Institute, Hospital Clinic, University of Barcelona, Barcelona, Spain

Herbert A. Reitsamer Department of Ophthalmology, University Clinic of Ophthalmology and Optometry, Research Program for Experimental Ophthalmology and Glaucoma Research, Paracelsus Medical University/SALK, Salzburg, Austria

Francisco J. Rivera Institute of Mol. Regenerative Medicine, Spinal Cord Injury and Tissue Regeneration Center Salzburg (SCI-TReCS), Paracelsus Medical University Salzburg, Salzburg, Austria

Laboratory of Stem Cells and Neuroregeneration, Institute of Anatomy, Histology and Pathology, Faculty of Medicine and Center for Interdisciplinary Studies on the Nervous System (CISNe), Universidad Austral de Chile, Valdivia, Chile

Patricia Romagnolli School of Veterinary Medicine and Animal Sciences, University of São Paulo, Butantã, Sao Paulo, Brazil

Lynn M. Schnapp Division of Pulmonary, Critical Care, Allergy and Sleep Medicine, Medical University of South Carolina, Charleston, SC, USA

Vijay H. Shah Division of Gastroenterology and Hepatology, Mayo Clinic, Rochester, MN, USA

Peter Szaraz CReATe Fertility Centre, University of Toronto, Toronto, ON, Canada

Andrea Trost Department of Ophthalmology, University Clinic of Ophthalmology and Optometry, Research Program for Experimental Ophthalmology and Glaucoma Research, Paracelsus Medical University/SALK, Salzburg, Austria

Carole L. Wilson Division of Pulmonary, Critical Care, Allergy and Sleep Medicine, Medical University of South Carolina, Charleston, SC, USA

Yao Yao Department of Pharmaceutical and Biomedical Sciences, University of Georgia, Athens, GA, USA

Chapter 1
Pericytes in the Retina

Andrea Trost, Daniela Bruckner, Francisco J. Rivera, and Herbert A. Reitsamer

Abstract Pericytes (PCs) are specialized cells located abluminal of endothelial cells (ECs) on capillaries, embedded within the same basement membrane. They are essential regulators of vascular development, remodeling, and blood-retina-barrier (BRB) tightness and are therefore important components to maintain tissue homeostasis. The perivascular localization and expression of contractile proteins suggest that PCs participate in capillary blood flow regulation and neurovascular coupling. Due to their ability to differentiate into various cell types in vitro, they are regarded as potential cells for tissue repair and therapeutic approaches in regenerative medicine. Altered function or loss of PCs is associated with a multitude of CNS diseases, including diabetic retinopathy (DR). In this chapter, we will provide a short overview of retinal vascular development, the origin of PCs, and focus on PCs in retinopathy of prematurity (ROP) and in the diabetic retina. Further, animal models to study the fate of PCs and the potential role of (retinal) PCs in regeneration and wound healing will be discussed.

Keywords Pericyte · Retina · Origin · Pericyte marker · PDGFRb · NG2 · tbx18 · Diabetic retinopathy (DR) · Retinopathy of prematurity (ROP) · Wound healing · Regeneration

A. Trost (✉) · D. Bruckner · H. A. Reitsamer
Department of Ophthalmology, University Clinic of Ophthalmology and Optometry, Research Program for Experimental Ophthalmology and Glaucoma Research, Paracelsus Medical University/SALK, Salzburg, Austria
e-mail: a.zurl@salk.at

F. J. Rivera
Institute of Mol. Regenerative Medicine, Spinal Cord Injury and Tissue Regeneration Center Salzburg (SCI-TReCS), Paracelsus Medical University Salzburg, Salzburg, Austria

Laboratory of Stem Cells and Neuroregeneration, Institute of Anatomy, Histology and Pathology, Faculty of Medicine and Center for Interdisciplinary Studies on the Nervous System (CISNe), Universidad Austral de Chile, Valdivia, Chile

A. Birbrair (ed.), *Pericyte Biology in Different Organs*,
Advances in Experimental Medicine and Biology 1122,
https://doi.org/10.1007/978-3-030-11093-2_1

Abbreviations

Angs	Angiopoietins
BBB	Blood-brain barrier
BM	Bone marrow
BRB	Blood-retina barrier
CNS	Central nervous system
DME	Diabetic macular edema
DR	Diabetic retinopathy
ECs	Endothelial cells
INL	Inner nuclear layer
IPL and OPL	Inner and outer plexiform layer
MSCs	Mesenchymal stem cells
NG2	Neuron-glial antigen 2
NVU	Neurovascular unit
ON	Optic nerve
ONL	Outer nuclear layer
P0	Postnatal day 0
PCs	Pericytes
PDGFRb	PDGF-receptor beta
RPE	Retinal pigment epithelial cells
tbx 18	T-box family transcription factor 18
TGF-b	Transforming growth factor beta
VEGF	Vascular endothelial growth factor
vSMCs	Vascular smooth muscle cells

1.1 Retinal Structure and Function

The human eye is composed of three different layers. The outermost layer is formed by the cornea and sclera. The middle layer is divided into an anterior part (iris and ciliary body) and a posterior part (choroid). The light-sensitive organ, the retina, forms the innermost layer and lines the inner surface of the eye, extending from the papilla to the ora serrata. In the center of the retina, axons of the ganglion cells are bundled within the optic nerve (ON) running to the visual cortex in the brain. In addition to cells of the oligodendroglial lineage, the ON contains incoming blood vessels, that vascularize the inner retina. Temporal of the ON, the blood vessel free fovea, the sharpest point of vision and most essential part of the retina for human vision, is located (Fig. 1.1a). The retina can be divided into the neurosensory retina and the retinal pigment epithelium. The neurosensory retina is composed of three layers of nerve cell bodies and two layers of synapses. The outer nuclear layer (ONL) contains cell bodies of the rods and cones. Bipolar, horizontal, and amacrine

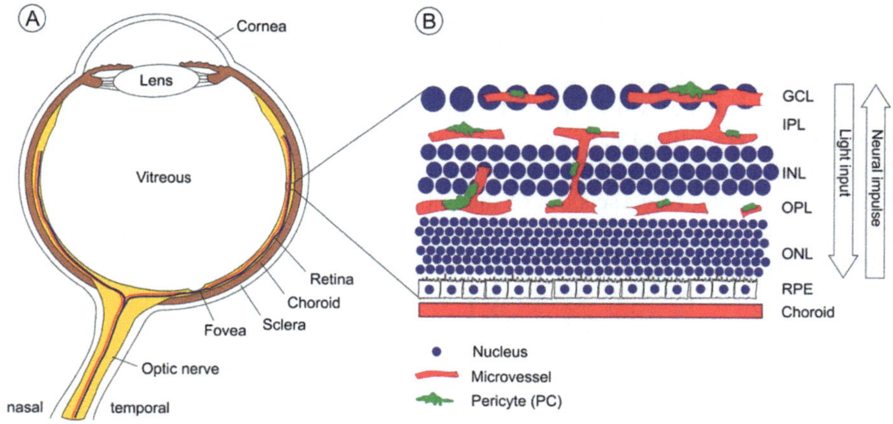

Fig. 1.1 (**a**) Schematic drawing of an eye cross section. (**b**) Schematic drawing of a retinal cross section: the neurosensory retina is composed of three layers of nerve cell bodies and two layers of synapses. The outer nuclear layer (ONL) contains cell bodies of the rods and cones. Bipolar, horizontal, and amacrine cells are located in the inner nuclear layer (INL). The innermost layer, the ganglion cell layer, contains cell bodies of ganglion cells and displaced amacrine cells. Within the inner and outer plexiform layer (IPL and OPL), located between the nerve cell layers, synaptic contacts occur. Nutrition to retina is provided by the choroid and the inner retinal vasculature: the three retinal vascular plexi are located in the GCL (superficial), in the IPL (intermediate), and in the OPL (deep)

cells are located in the inner nuclear layer (INL). The innermost layer, the ganglion cell layer contains cell bodies of ganglion cells and displaced amacrine cells. Within the inner and outer plexiform layer (IPL and OPL), located between the nerve cell layers, synaptic contacts occur (Fig. 1.1b) (Kolb 1995). One of the main functions of the retina is the conversion of light into an electric impulse, the first stage of image processing. The light passes through the entire retina, to reach the pigment molecules in the photoreceptors. The light signal is converted into an electrical impulse and transmitted to the bipolar cells and the ganglion cells where the signal is finally sent through the ON to the visual cortex. Photoreceptor cell metabolism and functioning of the visual cycle is maintained by a monolayer of retinal pigment epithelial cells (RPE), located in direct contact to the ONL.

1.2 Retinal Vascular Development

Nutrition of the metabolically highly active neural retina is provided by two vascular beds: the outer retina, including photoreceptors and RPE cells, is supplied by diffusion from the choriocapillaries. The inner retina is nourished by retinal blood vessels (Fig. 1.1b). During embryonal development, the inner retina is metabolically supported by the hyaloid vasculature, an arterial network in the vitreous. In humans, the hyaloid vasculature is replaced by retinal vasculature around 15 weeks of gestation and by the formation of the primary plexus. This remodels into three parallel connected vascular networks located in the nerve fiber layer and the inner and outer plexiform layer, until retinal vascularization is completed by 38–40 weeks of gestation (Lutty and McLeod 2017). In mouse, however, retinal vascular development starts by sprouting of vessels out of the optic nerve head at birth (P0). At this time point, a cellular network of astrocytes, already developed, provides a template for blood vessel sprouting and for the establishment of the primary vascular network. Within the first 3 postnatal weeks, the three vascular plexi are developed (Dorrell et al. 2002; Fruttiger 2002, 2007; Selvam et al. 2017): the vessels spread along the inner retinal surface to the ora serrata until the inner vascular layer is completed at postnatal day 7 to 10 (P7–P10). They subsequently spread into the retina and form the deep capillary layer. Finally, the intermediate capillary is formed from P14–P21.

The mammalian retina is dedicated to the central nervous system (CNS) since it derives from the neural tube and is formed through evagination from the diencephalon. Like in other CNS tissues, paracellular and transendothelial transport from the vasculature to the surrounding retinal tissue is highly regulated by the BRB, ensuring an optimal chemical composition of the neuronal microenvironment. The BRB is composed of the inner BRB (retinal capillary endothelial cells) and the outer BRB (retinal pigment epithelial cells). Although BRB tightness is mainly mediated by tight and adherent junctions between ECs, PCs have been proven to be an essential constituent of the BRB and blood-brain barrier (BBB). The contribution of PCs to the BRB/BBB is discussed in the subheading entitled: "Pericytes and their impact on the blood-retina barrier (BRB)."

1.3 Identification of (Retinal) Pericytes (PCs)

Currently, there is an urgent need to study and determine the role of retinal PCs in health and disease. To achieve this goal, the proper identification of this barely explored cell type is essential. However, following the gene and protein expression pattern, an identification of a specific PC molecular signature has been quite challenging. Up to now, no unique PC marker has been identified. Currently, the established marker panel for PC characterization comprises PDGF-receptor beta (PDGFRβ), neuron-glial antigen 2 (NG2, Cspg4, Fig. 1.2a), CD13 (brain PCs), Desmin, Vimentin, and RGS5 (Armulik et al. 2011). More recently, additional markers like Gli1 (Kramann et al. 2015) and Tbx18 (Guimaraes-Camboa et al. 2017) were proposed to identify PCs. A recent comparative single-cell transcriptomal study identified genes specifically expressed in (brain) PCs: in addition to the known markers Pdgfrb, Cspg4, Rgs5, and Anpep, Kcnj8, Cd248, Abcc9, Vtn, and S1pr3 were identified (Vanlandewijck et al. 2018). Of note, many markers used to identify PCs (e.g., NG2/Cspg4, PDGFRb) are also positive in vascular smooth

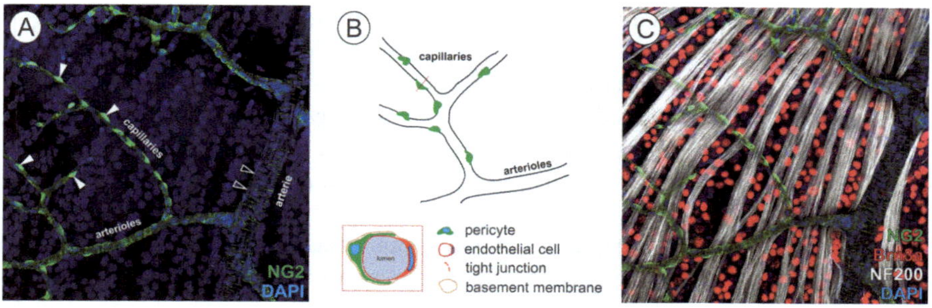

Fig. 1.2 (**a**) NG2-specific labeling of pericytes (PCs, green, arrowheads) on capillaries and vSMCs (green, open arrowheads) on arterioles and arteries in the ganglion cell layer (GCL) of a retinal whole mount preparation. (**b**) Schematic drawing of PCs (green) on capillaries, showing also a capillary cross section comprising PCs and endothelial cells. (**c**) Retinal whole mount, showing retinal ganglion cells (RGCs, labeled with Brn3a in red), axons of RGCs (labeled with NF200 in white), and PCs/vSMCs (green) in the GCL

muscle cells (vSMCs); therefore the localization of PCs on microvessels is an important identification criteria. Further, alpha-SMA allows for discrimination of PCs and vSMCs, being negative in PCs on capillaries, but positive in vSMCs on arterioles/arteries in vivo. Therefore, a combination of NG2 and PDGFRb with a-SMA and their capillary localization can be recommended for the identification of retinal PCs (Trost et al. 2013). Depending on their differentiation stage and tissue localization, PCs may exhibit a heterogeneous morphology. However, in general, PCs located on microvessels show a spherical-shaped soma with a prominent nucleus and processes that extend longitudinal on and around the vessel wall, covering several ECs (Armulik et al. 2011; Shepro and Morel 1993). As a part of the CNS, the retina exhibits a high PC density with a PC to EC ratio of about 1:1 to 1:3 (Armulik et al. 2011; Frank et al. 1990), which is constant in healthy conditions.

1.4 Origin of (Retinal) PCs

PCs are generated during embryonic and postnatal life (Winkler et al. 2011). During developmental stages two different primary sources have been described, a neuroectodermal and a mesodermal origin. Quail chick transplantations and a multitude of reporter mouse models have been used to investigate and discriminate neuroectodermal and mesodermal origin of PCs in different organs. Neuroectodermal or neural crest origin of PCs and vSMCs has been demonstrated by quail chick transplantations of brain anlagen and mesoderm in cerebral blood vessels (Korn et al. 2002) or in brain surface vessels by using Foxs1$^{+/ß\text{-}Gal}$ mice (Heglind et al. 2005). Using a Wnt-1-Cre mouse model, neural crest origin was proven for PCs of embryonic hyaloid blood vessels (Gage et al. 2005), however without providing PC characterization by additional markers. Along this line, a neural crest origin was demonstrated for the majority of retinal and choroidal PCs and vSMCs using a Sox10-Cre mouse model (Trost et al. 2013) and of cortical grey matter PCs using an inducible Sox10-CreER^{T2} mouse (E7.5 induction) (Simon et al. 2012). These findings were recently confirmed in the brain, eye, and thymus PCs and vSMCs using Cre-driven Sox10 and Wnt1 mouse models (Wang et al. 2017). To keep in mind, Wnt1 is a strict marker for neural crest cells and is not detected after migration of neural crest cells; however, Sox10 is expressed in postmigratory neural crest cells, as well as in glial cells of the developing nervous system and during adulthood. To distinguish in the adult tissue, cells with an embryonic neural crest origin vs. cells that derived from a postnatal Sox10-expressing tissue source, tamoxifen induction around E8.5–E12.5 in an inducible Sox10-CreER^{T2} mouse model (Hong and Saint-Jeannet 2005), would be desirable.

In addition to a neural crest origin of PCs and vSMCs (forebrain) (Etchevers et al. 2001), also a mesodermal origin has been described for PCs and vSMCs in the midbrain, brainstem, spinal cord, and peripheral organs (Korn et al. 2002). These experiments highlight that PCs and vSMC of both, neural crest and mesodermal origin, coexist within the same organ. Using an XlacZ4 reporter under the control of an adipose tissue-specific promoter, mesodermal origin was demonstrated in PCs and vSMCs throughout the vascular bed (Tidhar et al. 2001). Recent studies suggest a heterogeneous mesodermal origin of PCs: Yamamoto et al. described CD31^{+}F4/80^{+} macrophages as potential source for a subset of cerebrovascular NG2^{+} PCs (Yamamoto et al. 2017), however lacking verification with a second PC marker. In line with this, tissue myeloid progenitor cells are suggested to differentiate into a subset of dermal PCs in embryonic skin vasculature through transforming growth factor-ß signaling (Yamazaki et al. 2017): using a Vav-iCre mouse model to label hematopoietic cells, one third of NG2^{+} and PDGFRb^{+} PCs revealed reporter expression in the embryonic skin (E15.5). Further investigations on the subtype of hematopoietic cells revealed that tissue localized myeloid cells contributes to PC development in the skin vasculature (CD11b-Cre mouse model). The depletion of the respective cells [myeloid lineage (PU.1$^{-/-}$, Ncx1$^{-/-}$) and macrophages (Csf1$^{op/op}$)] resulted in a defective PC development in skin and brain. In addition, the authors concluded a limited contribution of neural crest cells to PC development in the skin (E15.5) using a Wnt-1-Cre reporter mouse and further suggested a minimal contribution of ECs (Tie2-Cre mouse model) (Yamazaki et al. 2017). In contrast, endocardial ECs have been demonstrated to give rise to a subset of cardiac PCs in the murine embryonic heart through endothelial-mesenchymal transition (Chen et al. 2016): using EC-specific reporter mice (cdh5-CreER^{T2}, Tie2-Cre, Nfatc1-CreER^{T2}), reporter-specific expression was detected in PDGFRb^{+} NG2^{+} cardiac PCs.

Postnatally, PCs can be recruited by proliferative expansion of preexisting PCs or from the bone marrow (BM) during vasculature remodeling, e.g., under ischemic conditions (Kokovay et al. 2006) or in tumors (Song et al. 2005). In line with this, BM-derived neovascular PCs were also reported following bFGF-induced corneal neovascularization (Ozerdem et al. 2005).

A multitude of studies investigated the origin of PCs and vSMCs, which is of great interest in developmental biology and further for the diagnosis and treatment of diseases associated with PC dysfunction. Up to now, these cells revealed heterogeneous origins, even within the same tissue. As recent data suggest that the differentiation potential of PCs is limited depending on their origin and tissue localization (Herrmann et al. 2016), knowledge about the origin may be crucial for the use of distinct PCs in regenerative approaches. In line with this, several PC subtypes have been identified, participating in tissue repair and regeneration. Since both PCs and vSMCs express the respective reporter, a common ancestor as well as a morphological and biochemical continuum from vSMC to PCs can be assumed. Indeed, an evolutionary conserved gradual phenotype change along the arteriovenous axis was proposed between PCs and (venous) SMCs as well as SMCs form arteries and arterioles applying single-cell transcriptomal analysis (Vanlandewijck et al. 2018). However, the authors detected no evidence for the existence of brain PC subtypes, but they clearly demonstrated an organotypicity, between PCs isolated from the brain and the lung (Vanlandewijck et al. 2018).

1.5 Animal Models to Study PCs

The contribution of PCs to angiogenesis, vessel stabilization, and BBB has been studied in mouse models targeting signaling pathways essential for PC-EC interaction and communication. Several molecules, like platelet-derived growth factor beta (PDGF-B), transforming growth factor beta (TGF-b), vascular endothelial growth factor (VEGF), and angiopoietins (Angs) (Armulik et al. 2011; Ribatti et al. 2011), are involved in the modulation and control of PC-EC interaction. ECs and PCs are interdependent; defects or impaired signaling in one cell type affect the other cell type.

1.5.1 PDGFRb/PDGFB

PDGFRb is expressed on PCs and vascular SMCs, and PDGFRb positive PCs/vSMC are recruited via endothelial bound PDGF-B to angiogenic sprouts (Betsholtz 2004; Gerhardt and Betsholtz 2003). The PDGF-B/PDGFRb pathway is crucial for PC proliferation, migration, and recruitment to angiogenic active sites. Disruption of PDGF-B/PDGFRb signaling components lead to impaired PC recruitment, reduced PC coverage, dilated capillaries, microaneurysm, and BBB breakdown (Bell et al. 2010; Lindahl et al. 1997; Tallquist et al. 2003; Winkler et al. 2010). Finally, impaired PDGFRb signaling and subsequent PC loss results in pathologically altered blood vessels and impaired tissue homeostasis in these mice. PDGFRb or PDGF-B knockout mice are not viable and die as embryos, displaying a loss of PC coverage on brain capillaries and showing evidence of lethal hemorrhages (Leveen et al. 1994; Lindahl et al. 1997; Soriano 1994).

Cre recombinase under the control of the PDGFRb promoter fragment (−4.7/+0.1 kb) crossed with a reporter strain was used to label PCs and vSMCs in vivo (Cuttler et al. 2011), showing a close correlation between the endogenous PDGFRb staining and the transgenic reporter (Cuttler et al. 2011). However, during embryonic development, PDGFRb is expressed broadly throughout the embryo and therefore not suitable to trace selectively the fate of PCs (Guimaraes-Camboa et al. 2017). However, in adult animals, PDGFRb expression is confined to PCs and vSMCs. To trace the fate of PCs, an inducible PDGFRb-CreER^{T2} mouse model has been generated (Sheikh et al. 2015), revealing distinct PC labeling in the retina after tamoxifen induction at P5, P6, and P7 (Park et al. 2017). A second inducible PDGFRb-P2A-CreER^{T2} mouse model was established, labeling 84.17 ± 3.48% of NG2^{+}/PDGFRb^{+} PCs after P1/P2/P3 tamoxifen induction via the lactating mother if combined with the reporter line Rosa-tdtomato (Cuervo et al. 2017). Interestingly, a significantly reduced labeling of NG2^{+} PCs (41.94 ± 18.67%) was detected by crossing PDGFRb-P2A-CreER^{T2} with the reporter line Rosa-mT/mG (Cuervo et al. 2017). These discrepancies between the recombination efficiency of different mouse reporter lines were reported previously (Liu et al. 2013) and should be taken into consideration if using reporter lines.

1.5.2 NG2 Proteoglycan

In 2001, NG2 proteoglycan (neural glial antigen 2, CSPG4) was first described as a PC-specific marker (Ozerdem et al. 2001), showing its impact on angiogenesis by reduced proliferation of both PCs and ECs in the retina of NG2 null mice (Ozerdem and Stallcup 2004). Until now, NG2-based mouse models have been used predominantly in the oligodendroglial lineage and myelin research fields; nevertheless several reports exist using these particular animal models to study PCs. For instance, Schallek et al. imaged retinal PCs noninvasively in the living eye using the NG2-dsRed mouse model (Schallek et al. 2013). This model was also used to investigate cerebral blood flow (Hall et al. 2014). Furthermore, inducible and non-inducible NG2-Cre reporter mouse models were used to study regional blood flow in brain (Hill et al. 2015). Similarly, the EYFP-NG2 mouse model was used to perform live cell imaging of PCs (Zehendner et al. 2013). Tamoxifen induction at different retinal vascular developmental stages in a NG2-CreER^{T2} reporter mouse revealed specific PC and vSMC labeling in all retinal vascular layers, although the recombination efficiency (15%) was rather low (Bruckner et al. 2018).

1.5.3 *Tbx18*

The T-box family transcription factor tbx 18 has been shown to be essential in kidney vascular development, being expressed in vSMCs and PCs (Xu et al. 2014). Recently, the specific expression of tbx18 in PCs and vSMCs was identified in the adult mouse and an inducible reporter mouse model established (tbx18-CreER^{T2}) to trace the fate of PCs and vSMCs. The authors verified a specific reporter expression in 90–95% of PDGFRb^{+}/NG2^{+}/CD146^{+} PCs and vSMCs in different tissues, including the retina using the Rosa26^{tdtomato} reporter mouse line (Guimaraes-Camboa et al. 2017), following 3x i.p. tamoxifen injections to adult animals (8-week-old).

Currently, three powerful fate mapping models, the PDGFRb-CreER^{T2}, PDGFRb-P2A-CreER^{T2}, and the tbx18-CreER^{T2}, labeling specifically the majority of PCs and vSMC following tamoxifen induction in the adulthood, are available. Combining the NG2-CreER^{T2} mouse with a more efficient reporter mouse may also increase the applicability of this fourth model. In general, these models allow to investigate the in vivo behavior of PCs under healthy conditions as well as their response and cell fate upon injury or under other pathological conditions. Of note, all these models label both, PCs and vSMCs, therefore being not able to distinguish between these two cell types based on the reporter expression. Nevertheless, taking the vascular diameter into consideration, PCs can be only identified on microvessels, the capillaries.

1.6 PCs and Their Impact on the Blood-Retina Barrier (BRB)

CNS homeostasis is maintained by a highly coordinated neurovascular unit (NVU), composed of ECs, PCs (capillaries), vSMC (arteries), glial cells, and neurons. The interrelation of these cells provides the formation and maintenance of the BRB/BBB, ensuring a tightly regulated barrier function and tissue/CNS homeostasis. In the last decades, PCs have been proven to be a substantial component of the BRB and BBB. Loss or dysfunction of PCs and disrupted BRB/BBB is associated with a variety of neuropathological diseases, like DR (Eshaq et al. 2017), Alzheimer's disease (Sweeney et al. 2018), multiple sclerosis, or CADASIL (Ghosh et al. 2015). The impact of PC on BBB function has been studied in mouse models with modified PC-EC cell signaling, drawing emphasis on the PDGFB/PDGFRb signaling pathway. The use of PDGFRb signaling-deficient mice demonstrated that reduced PC coverage in brain results in BBB breakdown and an accumulation of plasma-derived proteins, resulting in secondary neuronal degenerative alterations (Bell et al. 2010). In line with this, PC deficiency was associated with increased BBB permeability in further studies in adult (Armulik et al. 2010) and embryonic tissue (Daneman et al. 2010). PCs' contribution to the formation of the BRB in the developing retinal vasculature was demonstrated by the increased expression of the tight junction protein ZO-1 after direct contact of ECs and PCs and further astrocytes (Kim et al. 2009), however with the limitation that the authors used aSMA to identify PCs. A recent study emphasized that PC loss during embryonic development differs in its impact on BBB stability from adult PC loss: a reduction of PC coverage during vascular maturation, by PDGF-B depletion at P5, resulted in severe hemorrhage, disrupted vascular plexus formation, and macrophage infiltration (Park et al. 2017). PC dropout in capillaries of adult PDGFRb-Cre-ERT2 mice via diphtheria toxin resulted in a pronounced ablation of PDGFRb+ and NG2+ PCs and SMA+ vSMCs in retinal vessels, however without apparent changes in vascular remodeling or leakage in the retina and brain (Park et al. 2017). These results indicate that PC loss from stabilized (retinal) vessels is not sufficient to induce alterations in BRB and retinal vessel integrity, although adequate PC coverage is essential for BRB formation and maturation during vascular development. Nevertheless, in many age-related diseases, the breakdown of BBB and/or BRB is discussed as one of the main disease contributing factors.

1.7 PCs in Retinal Diseases

Pathological retinal neovascularization is a main factor of sight-threatening diseases, including diabetic retinopathy (DR), retinopathy of prematurity (ROP), and age-related macular degeneration (AMD). Therefore, understanding angiogenic mechanisms in the physiological as well as pathological retina is essential to identify involved cells, being potential targets for therapeutic interventions. The potential role of PC in the course of ROP and DR will be discussed.

1.7.1 PCs in Retinopathy of Prematurity (ROP)

The impact of PCs in neovascularization in ROP as well as the interaction of EC and PC in pathologic neovascularization is poorly understood. Contradictory reports, regarding the presence of PC on neovascular vessels/tufts in ROP, are published and will be discussed. ROP affects prematurely born babies, receiving intensive neonatal care, including oxygen therapy. Excess oxygen and subsequent hypoxia mainly contributed to ROP development, however, in case of controlled oxygen supplementation other risk factors get more prominent (gestational age, lack of growth factor, low birth weight, pre−/postnatal growth). The normal growth of blood vessels is directed by low oxygen in non-vascularized retinal areas and is completed by 38–40 weeks of gestation (see retinal vascular development). In utero, the partial pressure of oxygen is rather low (50 mmHg), which is increased at ambient room air (160 mmHg) and even further in case of oxygen supply during neonatal care. Subsequently, vascular development is reduced or even ceased resulting in avascular, thus hypoxic retina. This in turn leads to excess production of oxygen-sensitive growth factors, e.g., VEGF, resulting in neovascularization, including pathological vessels, also growing into the a-vascular vitreous [reviewed in Liegl et al. (2016)]. Although these vessels may regress during retinal development, fibrous scar tissue may remain and cause traction to the retina, potentially resulting in retinal detachment and blindness. Furthermore, due to impaired BRB, these pathological vessels are often leaky. In case of retinal neovascularization, laser photocoagulation and anti-VEGF therapies are used to treat infants, whereas the latter is discussed controversially regarding dosage and long-term safety and efficacy (Sankar et al. 2018). The aim to find a balance between low oxygen tension to minimize ROP and high oxygen tension to prevent brain hypoxia and death remains challenging as shown by a meta-analysis study (Askie et al. 2011). Another important factor will be the supplementation of IGF-1 in preterm infants to increase very low serum levels, enabling normal postnatal growth. First animal studies supported the role of supplemental IGF-1 in preventing ROP (Vanhaesebrouck et al. 2009). Pharmacokinetic and dosing studies of IGF-1/IGFBP-3 were performed in preterm infants (Lofqvist et al. 2009), and its impact on ROP investigated in a phase II study (NTC01096784), showing no effect on ROP, but reduction in bronchopulmonary dysplasia and intraventricular hemorrhage (Hansen-Pupp et al. 2017).

As retinal vasculature in rodents develops after birth, animal models are used to study pathologies in retinal vascular development postnatal. To gain insight into mechanisms of ROP, neovascularization following oxygen-induced retinopathy (OIR), mimicking the neovascular response in ROP, is analyzed in a multitude of studies. An established OIR model involves hyperoxia (75% oxygen) from postnatal day (P)7 to P12, inducing capillary regression in central retina. Following return to room air at P12, hypoxia triggers neovascularization which peaks around P17 (Smith et al. 1994). Lee et al. analyzed the impact of CCN1 expression on angiogenic processes under OIR. CCN1, an extracellular matrix protein active during angiogenesis, is expressed in ECs during vascular development. However, under ischemic conditions, CCN1 expression is detected on PCs. Following OIR-induced retinal neovascularization, retinal tufts, growing toward the vitreous, are formed and covered by PCs. PC-specific deletion of CCN1 under ischemic conditions results in reduced angiogenic signals and reduced neovascular outgrowth (Lee et al. 2017), that was associated with reduced expression of Wnt5a. The authors emphasized a change in EC-PC communication in angiogenesis under ischemic conditions. In line with the study of Lee et al., a recent study reported regular PC coverage in the three-layered vascular plexus as well as in neovascular tufts at P17 following OIR (Choi et al. 2018). Further, N-cadherin, essential for the maintenance of the BRB and EC-PC interaction, was detected in neovascular tufts, indicating an intact EC-PC interaction under ischemic conditions (Choi et al. 2018). However, whether this EC-PC interaction is intact and sufficient to mediate BRB tightness remains to be elucidated. In contrast, previous studies suggested a loss of NG2+ PCs following OIR in rats; however, PC specificity is hard to assess with the presented pictures (Saito et al. 2007). In line with them, Hughes et al. reported impaired mural cell differentiation and loss in the neovasculature of an obliterative as well as the surviving vasculature of a nonobliterative OIR model, both resulting in reduced mural cell coverage and enlarged vessels (Hughes et al. 2007). The authors suggested impaired perfusion and EC cell death as possible causes for mural cell changes. Reduction of mural cells in the hypoxic phase, through the application of a PDGFR inhibitor (STI571), increased neovascularization in ROP rats and was further correlated with an increase in VEGF and VEGFR-2 mRNA (Wilkinson-Berka et al. 2004). Concordantly, PDGF-B+/− mice, showing a 30% reduction in PC coverage at P7, developed twice as many new blood vessels as wild-type littermates under ROP (Hammes et al. 2002). This indicates that PC deficiency results in reduced inhibition of EC proliferation, especially under high VEGF levels in hypoxic conditions, resulting in the formation of new, potential pathologic blood vessels. In contrast, a more recent study suggested that the development of the deep capillary network rather than the recruitment of PCs to the vessel sprouts is important to mediate resistance to vascular regression processes induced by hyperoxia (Hoffmann et al. 2005). Regarding the tightness of neovascular vessels, increased BRB permeability was demonstrated in newly formed vessels by HRP leakage in the hypoxic phase of a cat ROP model; however, these newly developed vessels matured and became tight (Chan-Ling et al. 1992).

Although contradictory results are published regarding the impact of PC coverage on newly formed vessels in hypoxic retina, the presence of PCs seems to be vital for the maintenance of normal angiogenesis by controlling EC proliferation. While more recent studies report an intact PC-EC interaction and PC coverage (Choi et al. 2018; Lee et al. 2017), the physical contact may not be sufficient to prevent neovascularization in ROP/OIR. Further, the tightness of these neovascular tufts has not been tested in these studies and remains to be elucidated and compared to previous studies reporting BRB leakage in neovasculature after OIR. All together, these findings suggest an important role of PCs in neovascularization in OIR, rendering PCs as potential targets in therapeutic approaches to treat ROP.

1.7.1.1 Cell-Based Therapy Approaches in ROP

Cell-based therapies to ameliorate and repair retinopathic insults were studied in models of OIR. First attempts to ameliorate ischemia-induced neovascularization were performed using adult bone marrow-derived myeloid progenitor cells. Upon intravitreal injection during the hyperoxia or normoxia phase (P9–P12), these HIF-1a positive cells migrated to avascular regions of the OIR damaged retina and accelerated vascular repair and normalization. An even better effect was obtained injecting BM cells prior to exposure to hyperoxia (P2–P7) (Ritter et al. 2006). Detailed characterization of transplanted cells did not reveal NG2 or CD31 expression, and the authors excluded a differentiation into PCs or ECs. However, transplanted cells were positive for microglial/macrophagic marker and their differentiation into microglial cells concluded, improving and stabilizing the response of retinal vessels to hypoxia (Ritter et al. 2006). In contrast, EC-specific IGFBP-3 expression increased the differentiation of hematopoietic stem cells into

PCs and astrocytes and resulted in increased PC coverage and reduced PC apoptosis in OIR mice, followed by normalization in vessel morphology (Kielczewski et al. 2011). Further, the authors suggested a beneficial effect of reducing the number of activated microglial cells by IGFBP-3 expression, which is opposed to findings of Ritter et al. suggesting a benefit of differentiated microglial cells. The IGFBP-3 mediated reduction in preretinal neovascularization is already under investigation in clinical ROP studies as discussed above. Besides endothelial progenitor cells, adipose-derived stem cells, or bone marrow-derived cells, the transplantation of mesenchymal stem cells (MSCs) is considered beneficial for tissue repair and therapeutic angiogenesis (Sieveking and Ng 2009). In line with this, a TGF-ß1-mediated suppression of excessive neovascularization in an OIR mouse model was demonstrated after i.p. injection of human placental amniotic membrane-derived MSCs (AMSCs). The cells migrated to the retina; however, the lack of integration into the vascular network suggested absence of differentiation of AMSCs into ECs or PCs and therefore a regulation of angiogenesis through paracrine mechanisms (Kim et al. 2016b). Adipose-derived stem cells (ASCs) are able to differentiate into PCs and were tested for their ability to stabilize retinal vessels in models of retinal vasculopathy. Following OIR, intravitreally injected, differentiated ASCs were able to integrate into retinal vasculature and accelerated recovery from OIR through enhanced vessel regrowth. Further, injection of these cells before OIR vessel destabilization resulted in reduced retinal capillary dropout. Additionally, injection of ASCs prevented capillary dropout in a DR model (Akimba mouse) (Mendel et al. 2013). Additional approaches using cell-based therapies for therapeutic revascularization of ischemic injury is reviewed in Trinh et al. (2016). Although many studies report a beneficial effect following transplantation of diverse cells on OIR pathology, the precise mechanisms of neovascularization need additional investigations. Further, administration routes, mobilization/homing, time points, and appropriate cell numbers or survival and differentiation of transplanted cells need to be evaluated preclinical to achieve optimized therapeutic use in neovascularization.

1.7.2 PCs in Diabetic Retinopathy (DR)

The second pathology including retinal neovascularization and involvement of PCs is diabetic retinopathy (DR). DR is a major complication of diabetes leading to visual impairment and ultimately blindness (Yau et al. 2012). DR is characterized by increased vascular permeability and microaneurysms, which is most likely caused by PC loss (Frank 2004; Hammes et al. 2002). This PC loss and formation of microaneurysms are hallmarks of an early disease state, leading to decreased retinal microcirculation and subsequent hypoxia. Progressive retinal hypoxia, loss of retinal capillaries, and oxidative stress activate the release of inflammatory cytokines and hypoxia-induced expression of angiogenic growth factors (e.g., vascular endothelial growth factor, VEGF), resulting in retinal neovascularization: new blood vessels develop, which now extend into the normally a-vascular vitreous and anterior chamber angle, causing loss of vision and neovascular glaucoma, respectively. Neovascularization can be considered as a central element in the progression of DR. In a recent review, the effects of reactive metabolites, transient and chronic hyperglycemia in correlation to neurodegeneration, edema, and neovascularization are summarized (Hammes 2018).

DR has long been considered a vascular disease based on vascular pathologies detected in diabetic patients during ophthalmologic examinations. Increasing reports demonstrating dysfunction and degeneration of nonvascular cells [reviewed in Kern and Barber (2008)] changed the classification of DR to a neurodegenerative disease. The exact pathomechanisms for the development and progression of DR are still unclear, and the mechanism of PC loss in DR remains controversial. PC loss through apoptosis and destructive pathways under hyperglycemic conditions has been suggested (Behl et al. 2008). Altered glutamate excitation, reduced trophic factor signaling, oxidative stress, deposition of hyperglycemia-induced advanced glycation end products, alterations in intracellular pathways, dysregulation of endogenous neuroprotective factors, and neuroinflammation are among many potential causes for an increased apoptosis [reviewed in Barber et al. (2011), Eshaq et al. (2017) and Hernandez et al. (2016)]. Others suggest active PC depletion by non-apoptotic mechanisms via growth factor systems like angiopoietin-2/Tie-2 (Pfister et al. 2008). Further, the impact, interdependence, and chronology of neuronal and vascular degeneration in DR are still unclear and discussed controversially.

1.7.2.1 BRB Breakdown in DR

The breakdown of BRB in DR has been reviewed in an excellent article (Klaassen et al. 2013). Although the molecular mechanisms of PC loss and vascular leakage have been studied in different DR mouse models [e.g., Enge et al. (2002) and Huang et al. (2011)], PC loss and subsequent BRB breakdown are not fully understood. Disruption of endothelial-specific PDGF-B and subsequent reduced PC coverage resulted in a loss of neuronal layers and folding of the photoreceptor layer in the retina, resembling signs of DR (Enge et al. 2002; Lindblom et al. 2003). In line with this, a mutation in PDGFRb resulted in a CNS restricted reduced PC coverage, impaired vascular permeability, and aberrant BRB development, associated with retinal ganglion cell (RGC) apoptosis, mimicking features of non-proliferative DR (Jadeja et al. 2013). Consistently, transient inhibition of PC recruitment by an i.p. injection of anti-PDGFRb antibody (APB5) to developing retinal vessels (P1) resulted in BRB breakdown in adult mouse retina (Ogura et al. 2017), resembling DR pathology with sustained vascular abnormalities and autonomous disease progression. Furthermore, a clear correlation of APB5 concentration and retinal collapse could be demonstrated. However, APB5 injection in adult mice failed to induce PC loss from mature vessels, indicating that PDGFRb signaling is not essential for the maintenance of EC-PC associations. On the other hand, retinal overexpression of Ang-2 induces PC migration and vascular pathology, suggesting that Ang-2 plays an important role in diabetic vasoregression via PC destabilization (Pfister et al. 2010). The mouse model Ins2^{Aktia}VEGF$^{+/-}$, mimicking signs of advanced clinical DR (diabetic macular edema, proliferative DR), was established to study molecular mechanisms at later stages of DR (Wisniewska-Kruk et al. 2014). As the mechanisms of PC loss and BRB breakdown are not fully resolved until now and (PC-) specific treatment strategies are missing, current therapeutic interventions aim to reduce neovascularization in proliferative DR.

1.7.2.2 Current Treatments in DR and Potential Therapeutic Strategies

As poor glycemic control is the most relevant risk factor for DR development, tight blood glucose and blood pressure control are inevitable. In case of proliferative DR, surgical interventions like laser photocoagulation (Hammes 2005), vitrectomy (Newman 2010), and intravitreal anti-VEGF injections (Krebs et al. 2013) are current approaches to slow down the progression of DR. Although anti-VEGF therapies mediate the desired anti-angiogenic effect and significantly lower the incidence of vision loss, they also block the neuroprotective and neurotrophic VEGF effects, which may result in a decreased RGCs survival as demonstrated in rat ischemia-reperfusion experiments (Nishijima et al. 2007). Nevertheless, the preventive impact of an intravitreal ranibizumab (anti-VEGF(−A)) injection was tested in early diabetic rats, showing an amelioration in PC loss and increased RGC survival (Xiao et al. 2017). However, as anti-VEGF treatments in many patients do not inhibit disease progression (Ip et al. 2015), blocking of different angiogenic factors (PDGFRb, bFGF, HGF, CTGF) is necessary to target VEGF-independent pathways when aiming for a more effective therapy.

To date, a multitude of substances has been tested in animal models for their neuro- and vascular protective properties in DR progression, e.g., growth factors like insulin-like growth factor, pigment epithelium-derived factor, brain-derived neurotrophic factor, and nerve growth factor (Hernandez et al. 2016). Focusing on anti-inflammatory therapy approaches, in addition to corticosteroids, several agents are in preclinical testing and clinical trials: intravitreal injection of infliximab, a TNF-a inhibitor, showing positive effects on visual acuity in human patients with a diabetic macular edema (DME) (Sfikakis et al. 2010), or topical application of nepafenac, a potent cyclooxygenase inhibitor, revealing a reduction of DME development after cataract surgery in patients with DR (Pollack et al. 2016).

1.7.2.3 Cell-Based Therapy Approaches in DR

Endothelial progenitor cells (EPCs) are discussed for their use in therapeutic revascularization and potential vascular repair in diabetic patients [reviewed in Shaw et al. (2011)]. The beneficial effect of the intravitreal injection of outgrowth endothelial cells (OECs) in ischemic retina (OIR model) was demonstrated by decreased avascular areas and reduced pathologic neovascularization, due to the incorporation of OECs in the resident vasculature (Medina et al. 2010). In line with this, CD34+ cells are regarded as important therapeutic option for revascularization of ischemic vascular areas and have been tested successfully in clinical trials (Mackie and Losordo 2011). The use of CD34+ cells isolated from diabetic patients revealed impaired ability to attach and assimilate into the retinal vasculature after transplantation, highlighting that the diabetic environment alter EPC phenotype and function (Caballero et al. 2007). The transplantation of diabetic, transient TGF-ß-inhibited peripheral CD34+ cells resulted in retinal vascular repair in a retinal ischemia reperfusion injury mouse model, suggesting the potential use of (dysfunctional) cells of diabetic patients for autologous transplantation following ex vivo treatment (Bhatwadekar et al. 2010). A different cell-based therapeutic approach in DR was performed by intravitreal injection of TGF-ß1-preconditioned mouse adipose-derived stem cells (mASCs) into a model of DR (Akimba mouse). Injection of DiI-labeled mASCs at 5 weeks of age revealed retinal microvasculature integration of these cells 4 weeks later, suggesting their differentiation into PCs. Additionally, retinal stabilization was further suggested by DiI-labeled mASCs without direct incorporation on retinal vessels, by conditioning the retinal microenvironment (Mendel et al. 2013). The ability of mASCs to prevent diabetic retinal microvascular dropout was tested by their injection at P9 and resulted in reduced capillary loss 2 months later (Mendel et al. 2013). A similar approach was performed by intravitreal injection of adipose-derived mesenchymal stem cells (MSCs) into diabetic mice. Although an increase in intraocular neurotrophic factors and reduced retinal oxidative damage was detected, the cells remained in the vitreous and did not differentiate into neural- or perivascular-like cells (Ezquer et al. 2016). Therefore, the benefit of MSC administration in this model was mediated by the generation of a neuroprotective, anti-apoptotic microenvironment, thereby reducing RGC loss. Recently, human embryonic stem cells were differentiated into cells revealing similar phenotypic and functional characteristics of PCs. These cells (hESC-PVPCs) were intravitreally injected into one eye of diabetic rats and an amelioration in vascular leakage detected in comparison to the contralateral vehicle-injected eyes (Kim et al. 2016a). Whether these cells compensate for early PC loss or additionally stabilize the vasculature remains to be elucidated. Further, whether an equal amelioration can be achieved by injecting (retinal) PCs instead of hESC-PVPCs will be of high interest. An important parameter for the survival and integration of transplanted cells is the microenvironment in the recipient tissue, being rather inhospitable under diabetic conditions. In line with this, Ingram et al. demonstrated diminished vasculogenesis of diabetic versus nondiabetic cord blood-derived progenitor cells upon implantation (Ingram et al. 2008). In general, the combination of two or more cell populations for transplantation may result in superior neovascularization and regeneration, replacing different cell types of a vessel. However, absence of standard cell isolation protocols and the use of heterogeneous cell populations without detailed characterization impair the possibility to compare beneficial effects in animal and clinical studies, even within the same cell population.

Although some of these substances and cells revealed promising therapeutic potential, counteracting neuronal and vascular degeneration, ameliorating DR progression in experimental and clinical studies, their safe and efficient future therapeutic application in the clinic is uncertain. Even though great efforts and improvements in understanding DR pathogenesis and mechanisms have been achieved, DR development and progression seem to be a multifactorial, complicated process, and thus, the development of new therapeutic approaches remains challenging.

1.8 Potential Role of PCs in Tissue Regeneration and Wound Healing

The capacity of PCs to differentiate into mesenchymal cell types (e.g., osteoblasts, chondrocytes, adipocytes) has been confirmed in a multitude of studies, and the contribution of in vitro differentiated PCs to tissue repair and regeneration after transplantation has been demonstrated in several injury models. The osteogenic potential of retinal PCs has been proven in vitro and in a diffusion chamber assay in vivo, also demonstrating formation of cartilage, adipose, and fibrous tissue (Doherty et al. 1998; Farrington-Rock et al. 2004). Apart from retinal PCs, the differentiation potential of PCs isolated from the microvasculature of human muscle into skeletal muscle was demonstrated in vitro and also in vivo after transplantation in dystrophic mice (Dellavalle et al. 2007). The contribution of a PC subpopulation to fiber development was further confirmed in an alkaline phosphatase-based lineage-tracing mouse model, during postnatal growth and during acute and chronic tissue regeneration (Dellavalle et al. 2011). In line with this, PCs purified from different tissues exhibited myogenic potential and regenerated myotubes in cardiotoxin-injured skeletal muscle after transplantation (Crisan et al. 2008). The cells further exhibited osteogenic, chondrogenic, and adipogenic potentials and expressed MSC markers. Comparing $CD34^-/CD146^+$ PCs isolated from human adipose tissue and bone marrow, differences in the trilineage differentiation potential were detected, suggesting a tissue-dependent regenerative potential of PCs in vitro (Herrmann et al. 2016). Tissue dependency was also demonstrated for myocardial PCs, which were not able to differentiate into the myogenic lineage, as proven for skeletal muscle PCs (Chen et al. 2015). Furthermore, the ability of CNS PCs to differentiate into non-mesenchymal cells was proven by their differentiation into cells of the neural and glial lineage (Dore-Duffy et al. 2006; Paul et al. 2012). In addition, the multilineage potential of PCs isolated from ischemic brain was demonstrated, suggesting their potential repair capacity at the site of brain injuries (Nakagomi et al. 2015). However, beside tissue-dependent differences, the lack of a commonly used isolation and cultivation protocol (culture conditions like media, serum, cultivation/differentiation time period) impedes comparison of experimental findings. Nevertheless, this in vitro multilineage potential raised expectations that endogenous PCs may contribute to wound healing and regeneration by replacing specific cells types lost due to injury or disease. As the unequivocal identification of PCs is hampered by its differing morphological phenotype and protein expression according to the tissue located in and the differentiation state, a multitude of PC subtypes has been described and characterized until now. A subpopulation of PCs was shown to give rise to scar-forming stromal cells in an injured spinal cord (SCI) model using a Glast-CreER transgenic mouse (Goritz et al. 2011). In line with this, the participation of PCs in scar formation was described for Nestin$^-$/NG2$^+$ PCs (Birbrair et al. 2014) and PDGFRb$^+$ PCs (Matsushita et al. 2015) after SCI. Further, PCs have been described to differentiate into myofibroblasts in two kidney fibrosis models, using an inducible FoxD1-CreER^{T2} mouse model (Humphreys et al. 2010). Identifying PCs as potential precursors of myofibroblast and collagen producing cells highlighted them as target for anti-fibrotic therapies. Recently, PDGFRb$^+$ brain PCs have been proposed as potential source of perivascular neural stem cells following transient brain ischemia/reperfusion injury (Nakata et al. 2017). To confirm these findings, the use of a fate-mapping mouse models, like the PDGFRb-P2A-CreER^{T2} (Cuervo et al. 2017), will be desirable.

Whether PC heterogeneity is present within the same tissue is under debate, the distinct responsiveness of subpopulations of PCs to injury/stimuli, however, argues in favor of this hypothesis (Santos et al. 2017). In contrast, a recent work suggested the lack of brain PC subtypes but clearly demonstrated a difference between PCs isolated from the brain versus the lung (Vanlandewijck et al. 2018). The existence of retinal PC subtypes, however, remains unknown. Recent studies using fate-mapping mouse models (PDGFRb, tbx18) reported a > 85% efficiency in targeting retinal PC, verified by co-labeling with additional PC markers (Cuervo et al. 2017; Guimaraes-Camboa et al. 2017; Park et al. 2017), suggesting a quite homogeneous PC population in the retina regarding marker expression. Whether labeling of just a subpopulation of PCs in other reporter models [e.g., NG2 Hill et al. (2015)] is due to a specific lower recombination efficiency or due to the presence of distinct subtypes remains to be elucidated. The expression of different markers as well as the presence of additional cell types with perivascular location may hamper a confident identification of proper PCs, emerging the concept of PC subtypes.

Although a multitude of publications describe a potential contribution of PC (subtypes) to tissue repair and regeneration, in vivo (trans-) differentiation using fate mapping mouse models has not been proven until now. The recently established tbx18-CreER^{T2} mouse model, labeling ~90% of PDGFRb$^+$/NG2$^+$/CD146$^+$ PCs and vSMCs, was used to investigate the in vivo transdifferentiation potential of PCs (Guimaraes-Camboa et al. 2017). The authors tested five conditions, revealing no significant contribution of tbx18$^+$ cells to other cell lineages but retaining PDGFRb expression. Comparing the brain, heart, skeletal muscle, and brown and white adipose depots of 8- and 87-week-old reporter animals, absence of transdifferentiation was detected. In line with this, tbx18$^+$-PCs did not transdifferentiate following trans-aortic constriction (fibrosis model), high-fat diet (adipogenic progenitors), myofiber degeneration (myogenic potential), or cortical stab wound (neuronal transdifferentiation or scar tissue formation) (Guimaraes-Camboa et al. 2017). Although FACS sorted tbx18$^+$ PCs displayed lineage plasticity in vitro, they did not reveal adipogenic, fibrogenic, myogenic, or neuronal progenitor potential in vivo in the injury models tested. Hence, plasticity observed in vitro or following transplantation in vivo might be a result of the artificial cell culture environment. Nevertheless, whether the lack of in vivo transdifferentiation potential of PCs holds true in different injury models has to be tested. Importantly, this study highlights the use of inducible reporter mouse models to perform lineage-tracing studies. And, although PCs have been excluded as endogenous progenitors for regeneration in these tissues, the use and beneficial effect of in vitro cultivated PCs in regeneration persists.

1.9 A Commentary on Likely Future Trends or Directions

In addition to their central role in vascular development and maintenance, the multipotency of PCs emphasizes their role in tissue repair and regeneration. The potential of PC-associated applications seems to be continuously increasing, ranging from anti-fibrotic therapies, anti-angiogenic therapies to transplanted PCs for regenerative approaches. The identification of PCs remains challenging; however, the use of inducible fate-mapping models promises facilitation to gain insight into PC function and mechanism under physiological and pathological conditions. Whether PCs can be stimulated endogenously to participate in tissue regeneration and replacement of damaged cells remains elusive. In case of effective transdifferentiation of PCs into damaged cells in vitro or in vivo, their integration into the existing (neuronal) cell network will be essential for successful tissue regeneration, repair, and function. In fact, we need a deeper understanding of (retinal) PC function, interaction, and differentiation capacity in order to translate findings obtained from animal models to humans.

References

Armulik A, Genove G, Betsholtz C (2011) Pericytes: developmental, physiological, and pathological perspectives, problems, and promises. Dev Cell 21:193–215

Armulik A, Genove G, Mae M, Nisancioglu MH, Wallgard E, Niaudet C, He L, Norlin J, Lindblom P, Strittmatter K, Johansson BR, Betsholtz C (2010) Pericytes regulate the blood-brain barrier. Nature 468:557–561

Askie LM, Brocklehurst P, Darlow BA, Finer N, Schmidt B, Tarnow-Mordi W, Ne OCG (2011) NeOProM: neonatal oxygenation prospective Meta-analysis collaboration study protocol. BMC Pediatr 11:6

Barber AJ, Gardner TW, Abcouwer SF (2011) The significance of vascular and neural apoptosis to the pathology of diabetic retinopathy. Invest Ophthalmol Vis Sci 52:1156–1163

Behl Y, Krothapalli P, Desta T, DiPiazza A, Roy S, Graves DT (2008) Diabetes-enhanced tumor necrosis factor-alpha production promotes apoptosis and the loss of retinal microvascular cells in type 1 and type 2 models of diabetic retinopathy. Am J Pathol 172:1411–1418

Bell RD, Winkler EA, Sagare AP, Singh I, LaRue B, Deane R, Zlokovic BV (2010) Pericytes control key neurovascular functions and neuronal phenotype in the adult brain and during brain aging. Neuron 68:409–427

Betsholtz C (2004) Insight into the physiological functions of PDGF through genetic studies in mice. Cytokine Growth Factor Rev 15:215–228

Bhatwadekar AD, Guerin EP, Jarajapu YP, Caballero S, Sheridan C, Kent D, Kennedy L, Lansang MC, Ruscetti FW, Pepine CJ, Higgins PJ, Bartelmez SH, Grant MB (2010) Transient inhibition of transforming growth factor-beta1 in human diabetic CD34+ cells enhances vascular reparative functions. Diabetes 59:2010–2019

Birbrair A, Zhang T, Files DC, Mannava S, Smith T, Wang ZM, Messi ML, Mintz A, Delbono O (2014) Type-1 pericytes accumulate after tissue injury and produce collagen in an organ-dependent manner. Stem Cell Res Ther 5:122

Bruckner D, Kaser-Eichberger A, Bogner B, Runge C, Schrödl F, Strohmaier C, Silva ME, Zaunmair P, Couillard-Despres S, Aigner L, Rivera FJ, Reitsamer HA, Trost A (2018) Retinal pericytes: characterization of vascular development-dependent induction time points in an inducible NG2 reporter mouse model. Curr Eye Res 43(10):1274–1285. https://doi.org/10.1080/02713683.2018.1493130. Epub 2018 Jul 18. PMID:29939774.

Caballero S, Sengupta N, Afzal A, Chang KH, Li Calzi S, Guberski DL, Kern TS, Grant MB (2007) Ischemic vascular damage can be repaired by healthy, but not diabetic, endothelial progenitor cells. Diabetes 56:960–967

Chan-Ling T, Tout S, Hollander H, Stone J (1992) Vascular changes and their mechanisms in the feline model of retinopathy of prematurity. Invest Ophthalmol Vis Sci 33:2128–2147

Chen Q, Zhang H, Liu Y, Adams S, Eilken H, Stehling M, Corada M, Dejana E, Zhou B, Adams RH (2016) Endothelial cells are progenitors of cardiac pericytes and vascular smooth muscle cells. Nat Commun 7:12422

Chen WC, Baily JE, Corselli M, Diaz ME, Sun B, Xiang G, Gray GA, Huard J, Peault B (2015) Human myocardial pericytes: multipotent mesodermal precursors exhibiting cardiac specificity. Stem Cells 33:557–573

Choi SH, Chung M, Park SW, Jeon NL, Kim JH, Yu YS (2018) Relationship between Pericytes and endothelial cells in retinal neovascularization: a histological and immunofluorescent Study of retinal angiogenesis. Korean J Ophthalmol 32:70–76

Crisan M, Yap S, Casteilla L, Chen CW, Corselli M, Park TS, Andriolo G, Sun B, Zheng B, Zhang L, Norotte C, Teng PN, Traas J, Schugar R, Deasy BM, Badylak S, Buhring HJ, Giacobino JP, Lazzari L, Huard J, Peault B (2008) A perivascular origin for mesenchymal stem cells in multiple human organs. Cell Stem Cell 3:301–313

Cuervo H, Pereira B, Nadeem T, Lin M, Lee F, Kitajewski J, Lin CS (2017) PDGFRbeta-P2A-CreER(T2) mice: a genetic tool to target pericytes in angiogenesis. Angiogenesis 20:655–662

Cuttler AS, LeClair RJ, Stohn JP, Wang Q, Sorenson CM, Liaw L, Lindner V (2011) Characterization of Pdgfrb-Cre transgenic mice reveals reduction of ROSA26 reporter activity in remodeling arteries. Genesis 49:673–680

Daneman R, Zhou L, Kebede AA, Barres BA (2010) Pericytes are required for blood-brain barrier integrity during embryogenesis. Nature 468:562–566

Dellavalle A, Maroli G, Covarello D, Azzoni E, Innocenzi A, Perani L, Antonini S, Sambasivan R, Brunelli S, Tajbakhsh S, Cossu G (2011) Pericytes resident in postnatal skeletal muscle differentiate into muscle fibres and generate satellite cells. Nat Commun 2:499

Dellavalle A, Sampaolesi M, Tonlorenzi R, Tagliafico E, Sacchetti B, Perani L, Innocenzi A, Galvez BG, Messina G, Morosetti R, Li S, Belicchi M, Peretti G, Chamberlain JS, Wright WE, Torrente Y, Ferrari S, Bianco P, Cossu G (2007) Pericytes of human skeletal muscle are myogenic precursors distinct from satellite cells. Nat Cell Biol 9:255–267

Doherty MJ, Ashton BA, Walsh S, Beresford JN, Grant ME, Canfield AE (1998) Vascular pericytes express osteogenic potential in vitro and in vivo. J Bone Miner Res 13:828–838

Dore-Duffy P, Katychev A, Wang X, Van Buren E (2006) CNS microvascular pericytes exhibit multipotential stem cell activity. J Cereb Blood Flow Metab 26:613–624

Dorrell MI, Aguilar E, Friedlander M (2002) Retinal vascular development is mediated by endothelial filopodia, a preexisting astrocytic template and specific R-cadherin adhesion. Invest Ophthalmol Vis Sci 43:3500–3510

Enge M, Bjarnegard M, Gerhardt H, Gustafsson E, Kalen M, Asker N, Hammes HP, Shani M, Fassler R, Betsholtz C (2002) Endothelium-specific platelet-derived growth factor-B ablation mimics diabetic retinopathy. EMBO J 21:4307–4316

Eshaq RS, Aldalati AMZ, Alexander JS, Harris NR (2017) Diabetic retinopathy: breaking the barrier. Pathophysiology 24:229–241

Etchevers HC, Vincent C, Le Douarin NM, Couly GF (2001) The cephalic neural crest provides pericytes and smooth muscle cells to all blood vessels of the face and forebrain. Development 128:1059–1068

Ezquer M, Urzua CA, Montecino S, Leal K, Conget P, Ezquer F (2016) Intravitreal administration of multipotent mesenchymal stromal cells triggers a cytoprotective microenvironment in the retina of diabetic mice. Stem Cell Research and Therapy 7:42

Farrington-Rock C, Crofts NJ, Doherty MJ, Ashton BA, Griffin-Jones C, Canfield AE (2004) Chondrogenic and adipogenic potential of microvascular pericytes. Circulation 110:2226–2232

Frank RN (2004) Diabetic retinopathy. N Engl J Med 350:48–58

Frank RN, Turczyn TJ, Das A (1990) Pericyte coverage of retinal and cerebral capillaries. Invest Ophthalmol Vis Sci 31:999–1007

Fruttiger M (2002) Development of the mouse retinal vasculature: angiogenesis versus vasculogenesis. Invest Ophthalmol Vis Sci 43:522–527

Fruttiger M (2007) Development of the retinal vasculature. Angiogenesis 10:77–88

Gage PJ, Rhoades W, Prucka SK, Hjalt T (2005) Fate maps of neural crest and mesoderm in the mammalian eye. Invest Ophthalmol Vis Sci 46:4200–4208

Gerhardt H, Betsholtz C (2003) Endothelial-pericyte interactions in angiogenesis. Cell Tissue Res 314:15–23

Ghosh M, Balbi M, Hellal F, Dichgans M, Lindauer U, Plesnila N (2015) Pericytes are involved in the pathogenesis of cerebral autosomal dominant arteriopathy with subcortical infarcts and leukoencephalopathy. Ann Neurol 78:887–900

Goritz C, Dias DO, Tomilin N, Barbacid M, Shupliakov O, Frisen J (2011) A pericyte origin of spinal cord scar tissue. Science 333:238–242

Guimaraes-Camboa N, Cattaneo P, Sun Y, Moore-Morris T, Gu Y, Dalton ND, Rockenstein E, Masliah E, Peterson KL, Stallcup WB, Chen J, Evans SM (2017) Pericytes of multiple organs do not behave as mesenchymal stem cells in vivo. Cell Stem Cell 20:345–359.e5

Hall CN, Reynell C, Gesslein B, Hamilton NB, Mishra A, Sutherland BA, O'Farrell FM, Buchan AM, Lauritzen M, Attwell D (2014) Capillary pericytes regulate cerebral blood flow in health and disease. Nature 508:55–60

Hammes HP (2005) Pericytes and the pathogenesis of diabetic retinopathy. Horm Metab Res 37(Suppl 1):39–43

Hammes HP (2018) Diabetic retinopathy: hyperglycaemia, oxidative stress and beyond. Diabetologia 61:29–38

Hammes HP, Lin J, Renner O, Shani M, Lundqvist A, Betsholtz C, Brownlee M, Deutsch U (2002) Pericytes and the pathogenesis of diabetic retinopathy. Diabetes 51:3107–3112

Hansen-Pupp I, Hellstrom A, Hamdani M, Tocoian A, Kreher NC, Ley D, Hallberg B (2017) Continuous longitudinal infusion of rhIGF-1/rhIGFBP-3 in extremely preterm infants: evaluation of feasibility in a phase II study. Growth Hormon IGF Res 36:44–51

Heglind M, Cederberg A, Aquino J, Lucas G, Ernfors P, Enerback S (2005) Lack of the central nervous system- and neural crest-expressed forkhead gene Foxs1 affects motor function and body weight. Mol Cell Biol 25:5616–5625

Hernandez C, Dal Monte M, Simo R, Casini G (2016) Neuroprotection as a therapeutic target for diabetic retinopathy. J Diabetes Res 2016:9508541

Herrmann M, Bara JJ, Sprecher CM, Menzel U, Jalowiec JM, Osinga R, Scherberich A, Alini M, Verrier S (2016) Pericyte plasticity—comparative investigation of the angiogenic and multilineage potential of pericytes from different human tissues. Eur Cell Mater 31:236–249

Hill RA, Tong L, Yuan P, Murikinati S, Gupta S, Grutzendler J (2015) Regional blood flow in the Normal and ischemic brain is controlled by arteriolar smooth muscle cell contractility and not by capillary Pericytes. Neuron 87:95–110

Hoffmann J, Feng Y, vom Hagen F, Hillenbrand A, Lin J, Erber R, Vajkoczy P, Gourzoulidou E, Waldmann H, Giannis A, Wolburg H, Shani M, Jaeger V, Weich HA, Preissner KT, Hoffmann S, Deutsch U, Hammes HP (2005) Endothelial survival factors and spatial completion, but not pericyte coverage of retinal capillaries determine vessel plasticity. FASEB J 19:2035–2036

Hong CS, Saint-Jeannet JP (2005) Sox proteins and neural crest development. Semin Cell Dev Biol 16:694–703

Huang H, Gandhi JK, Zhong X, Wei Y, Gong J, Duh EJ, Vinores SA (2011) TNFalpha is required for late BRB breakdown in diabetic retinopathy, and its inhibition prevents leukostasis and protects vessels and neurons from apoptosis. Invest Ophthalmol Vis Sci 52:1336–1344

Hughes S, Gardiner T, Baxter L, Chan-Ling T (2007) Changes in pericytes and smooth muscle cells in the kitten model of retinopathy of prematurity: implications for plus disease. Invest Ophthalmol Vis Sci 48:1368–1379

Humphreys BD, Lin SL, Kobayashi A, Hudson TE, Nowlin BT, Bonventre JV, Valerius MT, McMahon AP, Duffield JS (2010) Fate tracing reveals the pericyte and not epithelial origin of myofibroblasts in kidney fibrosis. Am J Pathol 176:85–97

Ingram DA, Lien IZ, Mead LE, Estes M, Prater DN, Derr-Yellin E, DiMeglio LA, Haneline LS (2008) In vitro hyperglycemia or a diabetic intrauterine environment reduces neonatal endothelial colony-forming cell numbers and function. Diabetes 57:724–731

Ip MS, Domalpally A, Sun JK, Ehrlich JS (2015) Long-term effects of therapy with ranibizumab on diabetic retinopathy severity and baseline risk factors for worsening retinopathy. Ophthalmology 122:367–374

Jadeja S, Mort RL, Keighren M, Hart AW, Joynson R, Wells S, Potter PK, Jackson IJ (2013) A CNS-specific hypomorphic Pdgfr-beta mutant model of diabetic retinopathy. Invest Ophthalmol Vis Sci 54:3569–3578

Kern TS, Barber AJ (2008) Retinal ganglion cells in diabetes. J Physiol 586:4401–4408

Kielczewski JL, Hu P, Shaw LC, Li Calzi S, Mames RN, Gardiner TA, McFarland E, Chan-Ling T, Grant MB (2011) Novel protective properties of IGFBP-3 result in enhanced pericyte ensheathment, reduced microglial activation, increased microglial apoptosis, and neuronal protection after ischemic retinal injury. Am J Pathol 178:1517–1528

Kim JH, Kim JH, Yu YS, Kim DH, Kim KW (2009) Recruitment of pericytes and astrocytes is closely related to the formation of tight junction in developing retinal vessels. J Neurosci Res 87:653–659

Kim JM, Hong KS, Song WK, Bae D, Hwang IK, Kim JS, Chung HM (2016a) Perivascular progenitor cells derived from human embryonic stem cells exhibit functional characteristics of Pericytes and improve the retinal vasculature in a rodent model of diabetic retinopathy. Stem Cells Transl Med 5:1268–1276

Kim KS, Park JM, Kong T, Kim C, Bae SH, Kim HW, Moon J (2016b) Retinal angiogenesis effects of TGF-beta1 and paracrine factors secreted from human placental stem cells in response to a pathological environment. Cell Transplant 25:1145–1157

Klaassen I, Van Noorden CJ, Schlingemann RO (2013) Molecular basis of the inner blood-retinal barrier and its breakdown in diabetic macular edema and other pathological conditions. Prog Retin Eye Res 34:19–48

Kokovay E, Li L, Cunningham LA (2006) Angiogenic recruitment of pericytes from bone marrow after stroke. J Cereb Blood Flow Metab 26:545–555

Kolb H (1995) Simple anatomy of the retina. In: Kolb H, Fernandez E, Nelson R (eds) Webvision: the organization of the retina and visual system. University of Utah Health Sciences Center, Salt Lake City, UT

Korn J, Christ B, Kurz H (2002) Neuroectodermal origin of brain pericytes and vascular smooth muscle cells. J Comp Neurol 442:78–88

Kramann R, Schneider RK, DiRocco DP, Machado F, Fleig S, Bondzie PA, Henderson JM, Ebert BL, Humphreys BD (2015) Perivascular Gli1+ progenitors are key contributors to injury-induced organ fibrosis. Cell Stem Cell 16:51–66

Krebs I, Schmetterer L, Boltz A, Told R, Vecsei-Marlovits V, Egger S, Schonherr U, Haas A, Ansari-Shahrezaei S, Binder S (2013) A randomised double-masked trial comparing the visual outcome after treatment with ranibizumab or bevacizumab in patients with neovascular age-related macular degeneration. Br J Ophthalmol 97:266–271

Lee S, Elaskandrany M, Lau LF, Lazzaro D, Grant MB, Chaqour B (2017) Interplay between CCN1 and Wnt5a in endothelial cells and pericytes determines the angiogenic outcome in a model of ischemic retinopathy. Sci Rep 7:1405

Leveen P, Pekny M, Gebre-Medhin S, Swolin B, Larsson E, Betsholtz C (1994) Mice deficient for PDGF B show renal, cardiovascular, and hematological abnormalities. Genes Dev 8:1875–1887

Liegl R, Hellstrom A, Smith LE (2016) Retinopathy of prematurity: the need for prevention. Eye Brain 8:91–102

Lindahl P, Johansson BR, Leveen P, Betsholtz C (1997) Pericyte loss and microaneurysm formation in PDGF-B-deficient mice. Science 277:242–245

Lindblom P, Gerhardt H, Liebner S, Abramsson A, Enge M, Hellstrom M, Backstrom G, Fredriksson S, Landegren U, Nystrom HC, Bergstrom G, Dejana E, Ostman A, Lindahl P, Betsholtz C (2003) Endothelial PDGF-B retention is required for proper investment of pericytes in the microvessel wall. Genes Dev 17:1835–1840

Liu J, Willet SG, Bankaitis ED, Xu Y, Wright CV, Gu G (2013) Non-parallel recombination limits Cre-LoxP-based reporters as precise indicators of conditional genetic manipulation. Genesis 51:436–442

Lofqvist C, Niklasson A, Engstrom E, Friberg LE, Camacho-Hubner C, Ley D, Borg J, Smith LE, Hellstrom A (2009) A pharmacokinetic and dosing study of intravenous insulin-like growth factor-I and IGF-binding protein-3 complex to preterm infants. Pediatr Res 65:574–579

Lutty GA, McLeod DS (2017) Development of the hyaloid, choroidal and retinal vasculatures in the fetal human eye. Prog Retin Eye Res 62:58–76

Mackie AR, Losordo DW (2011) CD34-positive stem cells: in the treatment of heart and vascular disease in human beings. Tex Heart Inst J 38:474–485

Matsushita T, Lankford KL, Arroyo EJ, Sasaki M, Neyazi M, Radtke C, Kocsis JD (2015) Diffuse and persistent blood-spinal cord barrier disruption after contusive spinal cord injury rapidly recovers following intravenous infusion of bone marrow mesenchymal stem cells. Exp Neurol 267:152–164

Medina RJ, O'Neill CL, Humphreys MW, Gardiner TA, Stitt AW (2010) Outgrowth endothelial cells: characterization and their potential for reversing ischemic retinopathy. Invest Ophthalmol Vis Sci 51:5906–5913

Mendel TA, Clabough EB, Kao DS, Demidova-Rice TN, Durham JT, Zotter BC, Seaman SA, Cronk SM, Rakoczy EP, Katz AJ, Herman IM, Peirce SM, Yates PA (2013) Pericytes derived from adipose-derived stem cells protect against retinal vasculopathy. PLoS One 8:e65691

Nakagomi T, Kubo S, Nakano-Doi A, Sakuma R, Lu S, Narita A, Kawahara M, Taguchi A, Matsuyama T (2015) Brain vascular pericytes following ischemia have multipotential stem cell activity to differentiate into neural and vascular lineage cells. Stem Cells 33:1962–1974

Nakata M, Nakagomi T, Maeda M, Nakano-Doi A, Momota Y, Matsuyama T (2017) Induction of perivascular neural stem cells and possible contribution to neurogenesis following transient brain ischemia/reperfusion injury. Transl Stroke Res 8:131–143

Newman DK (2010) Surgical management of the late complications of proliferative diabetic retinopathy. Eye (Lond) 24:441–449

Nishijima K, Ng YS, Zhong L, Bradley J, Schubert W, Jo N, Akita J, Samuelsson SJ, Robinson GS, Adamis AP, Shima DT (2007) Vascular endothelial growth factor-a is a survival factor for retinal neurons and a critical neuroprotectant during the adaptive response to ischemic injury. Am J Pathol 171:53–67

Ogura S, Kurata K, Hattori Y, Takase H, Ishiguro-Oonuma T, Hwang Y, Ahn S, Park I, Ikeda W, Kusuhara S, Fukushima Y, Nara H, Sakai H, Fujiwara T, Matsushita J, Ema M, Hirashima M, Minami T, Shibuya M, Takakura N, Kim P, Miyata T, Ogura Y, Uemura A (2017) Sustained inflammation after pericyte depletion induces irreversible blood-retina barrier breakdown. JCI Insight 2:e90905

Ozerdem U, Alitalo K, Salven P, Li A (2005) Contribution of bone marrow-derived pericyte precursor cells to corneal vasculogenesis. Invest Ophthalmol Vis Sci 46:3502–3506

Ozerdem U, Grako KA, Dahlin-Huppe K, Monosov E, Stallcup WB (2001) NG2 proteoglycan is expressed exclusively by mural cells during vascular morphogenesis. Dev Dyn 222:218–227

Ozerdem U, Stallcup WB (2004) Pathological angiogenesis is reduced by targeting pericytes via the NG2 proteoglycan. Angiogenesis 7:269–276

Park DY, Lee J, Kim J, Kim K, Hong S, Han S, Kubota Y, Augustin HG, Ding L, Kim JW, Kim H, He Y, Adams RH, Koh GY (2017) Plastic roles of pericytes in the blood-retinal barrier. Nat Commun 8:15296

Paul G, Ozen I, Christophersen NS, Reinbothe T, Bengzon J, Visse E, Jansson K, Dannaeus K, Henriques-Oliveira C, Roybon L, Anisimov SV, Renstrom E, Svensson M, Haegerstrand A, Brundin P (2012) The adult human brain harbors multipotent perivascular mesenchymal stem cells. PLoS One 7:e35577

Pfister F, Feng Y, vom Hagen F, Hoffmann S, Molema G, Hillebrands JL, Shani M, Deutsch U, Hammes HP (2008) Pericyte migration: a novel mechanism of pericyte loss in experimental diabetic retinopathy. Diabetes 57:2495–2502

Pfister F, Wang Y, Schreiter K, vom Hagen F, Altvater K, Hoffmann S, Deutsch U, Hammes HP, Feng Y (2010) Retinal overexpression of angiopoietin-2 mimics diabetic retinopathy and enhances vascular damages in hyperglycemia. Acta Diabetol 47:59–64

Pollack A, Staurenghi G, Sager D, Mukesh B, Reiser H, Singh RP (2016) Prospective randomised clinical trial to evaluate the safety and efficacy of nepafenac 0.1% treatment for the prevention of macular oedema associated with cataract surgery in patients with diabetic retinopathy. Br J Ophthalmol 101:423–427

Ribatti D, Nico B, Crivellato E (2011) The role of pericytes in angiogenesis. Int J Dev Biol 55:261–268

Ritter MR, Banin E, Moreno SK, Aguilar E, Dorrell MI, Friedlander M (2006) Myeloid progenitors differentiate into microglia and promote vascular repair in a model of ischemic retinopathy. J Clin Invest 116:3266–3276

Saito Y, Geisen P, Uppal A, Hartnett ME (2007) Inhibition of NAD(P)H oxidase reduces apoptosis and avascular retina in an animal model of retinopathy of prematurity. Mol Vis 13:840–853

Sankar MJ, Sankar J, Chandra P (2018) Anti-vascular endothelial growth factor (VEGF) drugs for treatment of retinopathy of prematurity. Cochrane Database Syst Rev 1:CD009734

Santos GSP, Prazeres P, Mintz A, Birbrair A (2017) Role of pericytes in the retina. Eye (Lond) 32:483–486

Schallek J, Geng Y, Nguyen H, Williams DR (2013) Morphology and topography of retinal pericytes in the living mouse retina using in vivo adaptive optics imaging and ex vivo characterization. Invest Ophthalmol Vis Sci 54:8237–8250

Selvam S, Kumar T, Fruttiger M (2017) Retinal vasculature development in health and disease. Prog Retin Eye Res 63:1–19

Sfikakis PP, Grigoropoulos V, Emfietzoglou I, Theodossiadis G, Tentolouris N, Delicha E, Katsiari C, Alexiadou K, Hatziagelaki E, Theodossiadis PG (2010) Infliximab for diabetic macular edema refractory to laser photocoagulation: a randomized, double-blind, placebo-controlled, crossover, 32-week study. Diabetes Care 33:1523–1528

Shaw LC, Neu MB, Grant MB (2011) Cell-based therapies for diabetic retinopathy. Curr Diab Rep 11:265–274

Sheikh AQ, Misra A, Rosas IO, Adams RH, Greif DM (2015) Smooth muscle cell progenitors are primed to muscularize in pulmonary hypertension. Sci Transl Med 7:308ra159

Shepro D, Morel NM (1993) Pericyte physiology. FASEB J 7:1031–1038

Sieveking DP, Ng MK (2009) Cell therapies for therapeutic angiogenesis: back to the bench. Vasc Med 14:153–166
Simon C, Lickert H, Gotz M, Dimou L (2012) Sox10-iCreERT2 : a mouse line to inducibly trace the neural crest and oligodendrocyte lineage. Genesis 50:506–515
Smith LE, Wesolowski E, McLellan A, Kostyk SK, D'Amato R, Sullivan R, D'Amore PA (1994) Oxygen-induced retinopathy in the mouse. Invest Ophthalmol Vis Sci 35:101–111
Song S, Ewald AJ, Stallcup W, Werb Z, Bergers G (2005) PDGFRbeta+ perivascular progenitor cells in tumours regulate pericyte differentiation and vascular survival. Nat Cell Biol 7:870–879
Soriano P (1994) Abnormal kidney development and hematological disorders in PDGF beta-receptor mutant mice. Genes Dev 8:1888–1896
Sweeney MD, Sagare AP, Zlokovic BV (2018) Blood-brain barrier breakdown in Alzheimer disease and other neurodegenerative disorders. Nat Rev Neurol 14:133–150
Tallquist MD, French WJ, Soriano P (2003) Additive effects of PDGF receptor beta signaling pathways in vascular smooth muscle cell development. PLoS Biol 1:E52
Tidhar A, Reichenstein M, Cohen D, Faerman A, Copeland NG, Gilbert DJ, Jenkins NA, Shani M (2001) A novel transgenic marker for migrating limb muscle precursors and for vascular smooth muscle cells. Dev Dyn 220:60–73
Trinh TLP, Li Calzi S, Shaw LC, Yoder MC, Grant MB (2016) Promoting vascular repair in the retina: can stem/progenitor cells help? Eye Brain 8:113–122
Trost A, Schroedl F, Lange S, Rivera FJ, Tempfer H, Korntner S, Stolt CC, Wegner M, Bogner B, Kaser-Eichberger A, Krefft K, Runge C, Aigner L, Reitsamer HA (2013) Neural crest origin of retinal and choroidal pericytes. Invest Ophthalmol Vis Sci 54:7910–7921
Vanhaesebrouck S, Daniels H, Moons L, Vanhole C, Carmeliet P, De Zegher F (2009) Oxygen-induced retinopathy in mice: amplification by neonatal IGF-I deficit and attenuation by IGF-I administration. Pediatr Res 65:307–310
Vanlandewijck M, He L, Mae MA, Andrae J, Ando K, Del Gaudio F, Nahar K, Lebouvier T, Lavina B, Gouveia L, Sun Y, Raschperger E, Rasanen M, Zarb Y, Mochizuki N, Keller A, Lendahl U, Betsholtz C (2018) A molecular atlas of cell types and zonation in the brain vasculature. Nature 554:475–480
Wang D, Wu F, Yuan H, Wang A, Kang GJ, Truong T, Chen L, McCallion AS, Gong X, Li S (2017) Sox10(+) cells contribute to vascular development in multiple organs-brief report. Arterioscler Thromb Vasc Biol 37:1727–1731
Wilkinson-Berka JL, Babic S, De Gooyer T, Stitt AW, Jaworski K, Ong LG, Kelly DJ, Gilbert RE (2004) Inhibition of platelet-derived growth factor promotes pericyte loss and angiogenesis in ischemic retinopathy. Am J Pathol 164:1263–1273
Winkler EA, Bell RD, Zlokovic BV (2010) Pericyte-specific expression of PDGF beta receptor in mouse models with normal and deficient PDGF beta receptor signaling. Mol Neurodegener 5:32
Winkler EA, Bell RD, Zlokovic BV (2011) Central nervous system pericytes in health and disease. Nat Neurosci 14:1398–1405
Wisniewska-Kruk J, Klaassen I, Vogels IM, Magno AL, Lai CM, Van Noorden CJ, Schlingemann RO, Rakoczy EP (2014) Molecular analysis of blood-retinal barrier loss in the Akimba mouse, a model of advanced diabetic retinopathy. Exp Eye Res 122:123–131
Xiao A, Zhou Q, Shao Y, Zhong HF (2017) Effect of intravitreal injection of ranibizumab on retinal ganglion cells and microvessels in the early stage of diabetic retinopathy in rats with streptozotocin-induced diabetes. Exp Ther Med 13:3360–3368
Xu J, Nie X, Cai X, Cai CL, Xu PX (2014) Tbx18 is essential for normal development of vasculature network and glomerular mesangium in the mammalian kidney. Dev Biol 391:17–31
Yamamoto S, Muramatsu M, Azuma E, Ikutani M, Nagai Y, Sagara H, Koo BN, Kita S, O'Donnell E, Osawa T, Takahashi H, Takano KI, Dohmoto M, Sugimori M, Usui I, Watanabe Y, Hatakeyama N, Iwamoto T, Komuro I, Takatsu K, Tobe K, Niida S, Matsuda N, Shibuya M, Sasahara M (2017) A subset of cerebrovascular pericytes originates from mature macrophages in the very early phase of vascular development in CNS. Sci Rep 7:3855

Yamazaki T, Nalbandian A, Uchida Y, Li W, Arnold TD, Kubota Y, Yamamoto S, Ema M, Mukouyama YS (2017) Tissue myeloid progenitors differentiate into Pericytes through TGF-beta Signaling in developing skin vasculature. Cell Rep 18:2991–3004
Yau JW, Rogers SL, Kawasaki R, Lamoureux EL, Kowalski JW, Bek T, Chen SJ, Dekker JM, Fletcher A, Grauslund J, Haffner S, Hamman RF, Ikram MK, Kayama T, Klein BE, Klein R, Krishnaiah S, Mayurasakorn K, O'Hare JP, Orchard TJ, Porta M, Rema M, Roy MS, Sharma T, Shaw J, Taylor H, Tielsch JM, Varma R, Wang JJ, Wang N, West S, Xu L, Yasuda M, Zhang X, Mitchell P, Wong TY, Meta-Analysis for Eye Disease Study Group (2012) Global prevalence and major risk factors of diabetic retinopathy. Diabetes Care 35:556–564
Zehendner CM, Wedler HE, Luhmann HJ (2013) A novel in vitro model to study pericytes in the neurovascular unit of the developing cortex. PLoS One 8:e81637

Chapter 2
Pancreatic Pericytes in Glucose Homeostasis and Diabetes

Limor Landsman

Abstract Glucose homeostasis relies on tightly regulated insulin secretion from pancreatic beta-cells, and its loss in diabetes is associated with the dysfunction of these cells. Beta-cells reside in the islets of Langerhans, which are highly vascularized by a dense capillary network comprised of endothelial cells and pericytes. While the requirement of the endothelium for the proper pancreatic function is well established, the role of pancreatic pericytes has only recently begun to unveil. Recent studies described multiple roles for pancreatic pericytes in glucose homeostasis, highlighting their function as both regulators of islet blood flow and as a source of critical signals that support proper beta-cell function and mass. Furthermore, recent findings point to the contribution of pericytic abnormalities to beta-cell dysfunction in type 2 diabetes, implicating the involvement of pancreatic pericytes in both the initiation and the progression of this disease. This newly gained data implicate pancreatic pericytes as critical components of the cellular network required for glucose regulation.

Keywords Pancreatic pericytes · Islets of Langerhans · Islet vasculature · Islet pericytes · Islet capillaries · Beta-cells · Beta-cell function · Beta-cell dysfunction · Beta-cell mass · Glucose homeostasis · Glucose-stimulated insulin secretion · Type 2 diabetes

---

L. Landsman (✉)
Department of Cell and Developmental Biology, Sackler Faculty of Medicine,
Tel Aviv University, Tel Aviv, Israel
e-mail: limorl@tauex.tau.ac.il

A. Birbrair (ed.), *Pericyte Biology in Different Organs*,
Advances in Experimental Medicine and Biology 1122,
https://doi.org/10.1007/978-3-030-11093-2_2

2.1 Introduction

2.1.1 *The Pancreas and Its Role in Glucose Regulation and Diabetes*

The pancreas is required for both food digestion and blood glucose regulation. The pancreas is frequently described as "two organs in one," as it comprises of two different cellular compartments, the exocrine and the endocrine, each is responsible for a distinct pancreatic function. Pancreatic exocrine cells, which produce and transport digestive enzymes to the gut, consist of the majority of the adult pancreatic tissue (Longnecker et al. 2018). Pancreatic endocrine cells, which produce and secrete hormones to regulate blood glucose levels, are organized in islets of Langerhans, structures that are embedded in the exocrine pancreas (Mastracci and Sussel 2012). Cells of both the exocrine and the endocrine compartment are epithelial cells of endoderm origin (Gittes 2009).

The islets of Langerhans comprise multiple endocrine cell populations, primarily insulin-producing β-cells, glucagon-producing α-cells, and somatostatin-producing δ-cells (Mastracci and Sussel 2012). Both α- and β-cells are equipped with a sophisticated mechanism allowing them to sense minute fluctuation in blood glucose levels and respond by accurately secreting hormones to the bloodstream. An increase in blood glucose levels prompts insulin secretion from β-cells, in a process termed "glucose-stimulated insulin secretion." In response to insulin, glucose is being uptaken by the liver, muscles, adipose tissues, and other tissues, lowering its levels in the blood. In contrast, glucagon acts to break glucose storages, to ensure sufficient glucose levels in the blood for proper brain function. Thus, pancreatic α- and β-cells cooperate to maintain tightly regulated blood glucose levels (Mastracci and Sussel 2012).

Diabetes is a chronic disease, now reaching pandemic proportions (DeFronzo et al. 2015). It is characterized by hyperglycemia due to insufficient β-cell function and mass, either alone or accompanied by insulin resistance. Type 2 diabetes, the most predominant form of this disease, was regarded for many years as a disease of insulin resistance. However, there is now a widespread recognition that β-cell failure is a critical factor in this disease (Ashcroft and Rorsman 2012). β-Cell failure could result from their insufficient mass, impaired function, or combination of the two (Costes et al. 2013; Dor and Glaser 2013). Loss of the β-cell mature phenotype, in a process named "β-cell dedifferentiation," was recently shown to contribute to β-cell dysfunction during type 2 diabetes (Talchai et al. 2012; Cinti et al. 2016). However, the underlying cause(s) of β-cell failure in diabetes is yet to be uncovered (Cerasi 2011).

2.1.2 *The Islet Vasculature*

The pancreas vasculature integrates this organ's endocrine and exocrine compartments, allowing both intra-islets and the islet-acinar interactions (Bonner-Weir 1993). While the islets comprise only 1–2% of the pancreatic volume, they receive approximately 10% of its blood flow. Islet capillaries are highly permeable, and their endothelial cells are highly fenestrated to facilitate a quick pancreatic response to fluctuations in blood glucose levels (Bonner-Weir 1993; Richards et al. 2010). Each β-cell is in contact with a capillary to allow proper sensing of blood glucose and the subsequent hormone secretion to the bloodstream (Bonner-Weir 1993; Granot et al. 2009). In addition to transporting hormones and nutrient to and from the peripheral circulation, the islet vasculature is required for proper pancreas development and function (Lammert et al. 2001; Nikolova et al. 2006; Reinert et al. 2013; Brissova et al. 2014; Kragl et al. 2016; Hogan and Hull 2017). Similar to capillaries of other organs, the islet vasculature constitutes of endothelial cells and pericytes. While the role of endothelial cells in pancreas function has been extensively studied (Richards et al. 2010; Hogan and Hull 2017), the role of pericytes has only recently begun to unveiled.

2.1.3 Pancreatic Pericytes

Together with endothelial cells, pericytes make the islets dense capillary network (Armulik et al. 2011; Tang et al. 2013; Sasson et al. 2016) (Fig. 2.1). Similar to pericytes of other organs, pancreatic pericytes are contractile cells that extend primary cytoplasmic processes along the endothelial tube (Armulik et al. 2011; Tang et al. 2013; Sasson et al. 2016) (Fig. 2.1). Pericytes constitute less than 3% of the islet cell population and cover around 40% of their capillary area (Almaça et al. 2018). These cells express the pericytic markers PDGFRβ (platelet-derived growth factor receptor β), NG2 (neural/glial antigen 2), αSMA (α-smooth muscle actin), and desmin (Armulik et al. 2011; Tang et al. 2013; Sasson et al. 2016). Pericytes of the adult pancreas originate from the embryonic pancreatic mesenchyme, a multicellular layer surrounding the developing pancreatic epithelium (Harari et al., 2019).

In addition to the islets, pericytes are found around capillaries of the exocrine pancreas (Morikawa et al. 2002; Sakhneny et al. 2018). Although the vasculature of the endocrine and the exocrine display distinct characteristics (Bonner-Weir 1993), whether islet pericytes differ from these reside in the exocrine pancreas awaits investigation. Although pancreatic stellate cells, which reside in the exocrine pancreas and contribute to the pathology of pancreatitis and pancreatic cancer, have pericytic characteristics (Apte et al. 2015), the relation between these two cell populations is unclear. Together with vascular smooth muscle cells (vSMCs) that encircle large blood vessels, pericytes make up the pancreatic mural cell population (Armulik et al. 2011).

While the role of endothelial cells in supporting β-cell function and mass has been vastly investigated (Richards et al. 2010; Hogan and Hull 2017), the understanding of the role of pericytes in these process lags behind. In recent years, work by us and others pointed to pericytes as critical players in pancreatic function. *Here, I review current knowledge on the contribution of these cells to glucose regulation and its loss in diabetes, highlighting the role of pancreatic pericytes as both regulators of islet blood flow, and as a source of critical signals that support proper β-cell function and mass.*

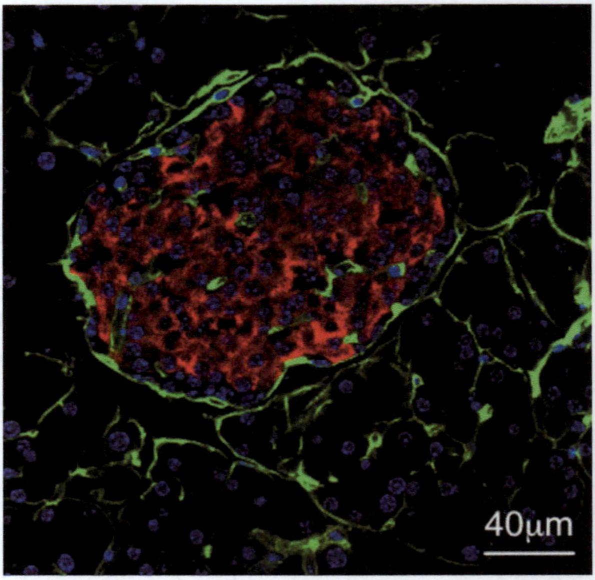

Fig. 2.1 Pancreatic pericytes are associated with islet of Langerhans. Mouse pancreatic section is stained with NG2 to label pericytes (green), insulin to label β-cells (red), and DAPI to label nuclei (blue)

2.2 Pancreatic Pericytes and Glucose Homeostasis

Proper β-cell function and mass have been shown to depend on cells of the islet microenvironment, including endothelial, neuronal, and immune cells (Nikolova et al. 2006; Eberhard and Lammert 2009; Tarussio et al. 2014; Brissova et al. 2014; Riley et al. 2015; Hogan and Hull 2017). Recent studies pointed to a role of pericytes in glucose homeostasis, implicating the direct and indirect roles of these cells in glucose-stimulated insulin secretion (Sasson et al. 2016; Houtz et al. 2016; Epshtein et al. 2017; Sakhneny et al. 2018; Almaça et al. 2018) (Illustrated in Fig. 2.2).

2.2.1 Pancreatic Pericytes and β-Cell Function

2.2.1.1 Pancreatic Pericytes and β-Cell Maturity

The ability of β-cells to properly secrete insulin depends on their mature phenotype. While mature β-cells are capable of precise glucose-stimulated insulin secretion, secretion of insulin from immature β-cells is not tightly linked to blood glucose levels (Bonner-Weir and Aguayo-Mazzucato 2016). The β-cell mature phenotype is maintained through the activity of an array of transcription factors (i.e., MafA, Pdx1, Unc3, and NeuroD1) (Bramswig and Kaestner 2014). These factors regulate the expression of functional genes required for glucose-stimulated insulin secretion, including insulin, the glucose transporter Glut2, the ATP-sensitive potassium

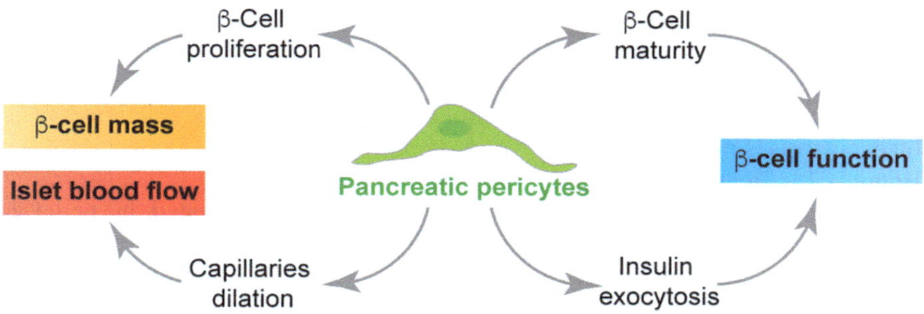

Fig. 2.2 Pancreatic pericytes regulate multiple aspects of glucose-stimulated insulin secretion. *Right,* pericytes regulate β-cell function through promoting both their mature phenotype and glucose-stimulated insulin exocytosis. *Upper left*, pericytes further regulate β-cell mass by stimulating their proliferation in the neonatal period. Lastly, pericytes modulate local blood flow in the islet through adjusting capillaries dilation and constriction (*lower left*)

channel Kir6.2, and the sulfonylurea receptor Sur1. In addition to cell-intrinsic factors, proper β-cell maturity and function rely on their appropriate interactions with cells in their microenvironment, including endocrine and non-endocrine cell populations (Eberhard and Lammert 2009). Loss of the β-cell mature phenotype, due to their dedifferentiation, contributes to the loss of glycemic control during type 2 diabetes (Dor and Glaser 2013).

We recently showed that pancreatic pericytes are required for proper glucose regulation (Sasson et al. 2016). To determine their role in vivo, we combined the use of transgenic mouse lines to selectively ablate pericytes in the pancreas (Sasson et al. 2016). Of note, other cell populations, such as pancreatic endothelial cells and pericytes outside of the pancreas, were not targeted by this manipulation. Shortly after ablation of pancreatic pericytes, mice became glucose intolerant due to diminished glucose-stimulated insulin secretion (Sasson et al. 2016). This data point to the requirement of pancreatic pericytes in supporting glucose homeostasis.

Depletion of pancreatic pericytes resulted in the loss of the mature β-cell phenotype, as evident by both their impaired function and the reduced levels of genes and proteins associated with β-cell maturity and function (Sasson et al. 2016). Islets lacking their pericytes displayed decreased levels of components of the glucose sensing and insulin secretion machinery, including insulin, Glut2, Kir6.2, and Sur1. Pericyte-depleted islets also exhibited diminished expression of transcription factors required for β-cell maturity, including MafA and Pdx1, pointing to β-cell dedifferentiation (Sasson et al. 2016). Loss of β-cell maturity was further observed when pericytes were depleted ex vivo in isolated islets, indicating that pericytes act within the islets to support β-cell function, independently of their role as regulators of blood flow (Sasson et al. 2016). These observations indicate that the mature β-cell phenotype, required for glucose-stimulated insulin secretion, depends on proper interactions of these cells with islet pericytes (Sasson et al. 2016).

Thus, pancreatic pericytes directly regulate β-cell maturity, in a blood flow independent manner, to control proper glucose regulation (Fig. 2.2).

2.2.1.2 Pancreatic Pericytes and Insulin Secretion

β-Cells store insulin in granules and release it in a tightly regulated manner in response to glucose. To ensure tight control of glucose levels, the regulation of insulin exocytosis is multilayered and involves complex signaling events (Rorsman and Braun 2013; Prentki et al. 2013).

Recently, the neurotrophin signaling pathway was implicated in pericyte-dependent β-cell function, through regulating glucose-stimulated insulin exocytosis. Houtz et al. showed that nerve growth factor (NGF) acutely regulates glucose-stimulated insulin secretion from β-cells (Houtz et al. 2016). The NGF receptor TrkA is expressed by β-cells, and its NGF-mediated activation promotes the release of insulin from its granules. Accordingly, selective loss of TrkA expression in β-cells leads to loss of glucose regulation (Houtz et al. 2016). Houtz et al. further showed that NGF is expressed by pancreatic pericytes and vSMCs when high glucose levels stimulate its secretion from these cells (Houtz et al. 2016). Transgenic mice lacking expression of NGF in their pericytes and vSMCs were glucose intolerant and displayed impaired glucose-stimulated insulin secretion (Houtz et al. 2016). *Thus, this study showed that through glucose-dependent NGF secretion, pericytes directly promote insulin secretion from β-cells* (Fig. 2.2).

2.2.2 *Pancreatic Pericytes and the Expansion of β-Cells*

After birth, the primary route of β-cell generation is their replication (Dor et al. 2004; Georgia and Bhushan 2004; Meier et al. 2008). β-Cell replication rates decline with age, and it is significantly higher during the neonatal period than during adulthood (Finegood et al. 1995; Gregg et al. 2012; Wang et al. 2015). While β-cell replication is primarily restricted to the neonatal period, the evidence is accumulating that adult β-cells maintain a proliferative capacity even as they age (Chen et al. 2009; Salpeter et al. 2010; Stolovich-Rain et al. 2012). For example, β-cells display compensatory proliferation during increased metabolic demand (Ouaamari et al. 2016).

β-Cell replication rate is regulated by both intrinsic and extrinsic cues, when cells of the islet microenvironment, including endothelial, neuronal, and immune cells, have been shown to promote this process (Nikolova et al. 2006; Eberhard and Lammert 2009; Tarussio et al. 2014; Brissova et al. 2014; Riley et al. 2015). Recently, we showed that pancreatic pericytes promote β-cell proliferation during the neonatal period (Epshtein et al. 2017). Depletion of pancreatic pericytes in vivo was sufficient to reduce the rate of β-cell replication in neonatal mice, pointing to the requirement of these cells for physiological expansion of β-cells (Epshtein et al. 2017) (Fig. 2.2).

Vascular basement membrane, a specialized extracellular matrix produced by endothelial cells and pericytes, is located within and around islets (Nikolova et al. 2006; Otonkoski et al. 2008; Kragl and Lammert 2010; Armulik et al. 2011; Stratman and Davis 2012). Components of the islet vascular basement membrane promote β-cell proliferation, through their binding to integrins expressed on these cells and the initiation of downstream signaling (Nikolova et al. 2006; Kragl and Lammert 2010; Diaferia et al. 2013). We recently showed that conditioned media of cultured neonatal pericytes stimulate β-cell proliferation (Epshtein et al. 2017). Furthermore, pericyte-stimulated β-cell replication was dependent on integrin activity (Epshtein et al. 2017). These findings suggest that pericytes produce basement membrane components to promote β-cell proliferation.

To conclude, pancreatic pericytes promote β-cell proliferation during the neonatal period to establish proper β-cell mass, likely through the production of vascular basement membrane components.

2.2.3 Pancreatic Pericytes in the Regulation of Blood Flow

In addition to their direct role in regulating β-cell function and mass, pancreatic pericytes may contribute to glucose homeostasis through regulating islet blood flow. The islet vascularization adapts to hormone secretion from endocrine cells, likely to facilitate the distribution of these hormones into the bloodstream (Schaeffer et al. 2011). In addition to this short-term response, islet vascularization adapts to prolonged metabolic demand, such as during insulin resistance, through capillaries dilation (Dai et al. 2013). However, whether the increase in islet blood flow promotes the lowering of blood glucose levels remains unclear (Hogan and Hull 2017).

The intra-islet density of pericytes is increased in response to insulin resistance and obesity, suggesting a role for these cells in the adaptation of the islet vasculature to high metabolic demand (Tang et al. 2013; Dai et al. 2013). Recently, pericyte-dependent dilation of islet capillaries was linked to a short-term increase in glucose levels, providing a potential mechanism through which increased demand for insulin facilitates its circulation (Almaça et al. 2018) (Fig. 2.2). As suggested by Almaça et al., glucose metabolism increases the levels of ATP and its metabolite adenosine, when the latter promotes the relaxation of islet pericytes, resulting in capillaries dilation (Almaça et al. 2018). Whereas this study implicated β-cells as the source of adenosine in the pancreas (Almaça et al. 2018), other cell types, including pericytes (Mandarino et al. 1994), are capable of metabolizing glucose and thus represent a potential source for this metabolite. Thus, glucose-dependent pericytes relaxation may represent a pericyte-autonomous or a nonautonomous mechanism through which these cells respond to the demand for insulin by increasing blood flow. In contrast to glucose, noradrenaline, produced by sympathetic innervation to the islets, promotes pericytes contraction and capillaries constriction to reduce blood flow (Almaça et al. 2018). These observations suggest that pericyte-dependent capillary dilation is a highly regulated process, which depends on proper cell-cell interactions within the islets of Langerhans. How pericyte-regulated local changes in islet blood flow influent glucose homeostasis remain to be elucidated.

To conclude, dilation of islet capillaries characterizes both prolonged and acute increase in insulin demands and is associated with changes in pericytes density and contractility.

2.3 Pericytes in the Loss of Pancreatic Function during Type 2 Diabetes

The clinical onset of type 2 diabetes occurs when β-cells fail to secrete sufficient amounts of insulin to maintain normoglycemia during insulin resistance (Costes et al. 2013; Dor and Glaser 2013; DeFronzo et al. 2015). The evidence is accumulating that abnormal pericyte function contributes to β-cell failure and disease development, whereas chronic hyperglycemia leads to loss of pancreatic pericytes (Hayden et al. 2007; Hayden et al. 2010; Sakhneny et al. 2018; Almaça et al. 2018) (Fig. 2.3). *Thus, pancreatic pericytes potentially play a role in both the initiation and progression of type 2 diabetes.*

2.3.1 Pancreatic Pericytes as Contributors to β-Cell Dysfunction and Type 2 Diabetes

While type 2 diabetes is influenced by several lifestyle factors, including age, pregnancy, and obesity, this disease has a strong genetic component (DeFronzo et al. 2015; Fuchsberger et al. 2016). Linkage and genome-wide association studies have identified more than 80 genes associated with an increased risk of type 2 diabetes, many of which are thought to be essential for β-cell function, development, or mass (Ashcroft and Rorsman 2012; Fuchsberger et al. 2016). In particular, polymorphism in *TCF7L2* (*TCF4*) is associated with an increased risk of diabetes (Grant et al. 2006). This gene encodes a member of T-cell factor/lymphoid enhancer factor (TCF/LEF) transcription factors family, which functions downstream of the canonical Wnt signaling pathway by recruiting β-catenin to target genes (Clevers and Nusse 2012). Polymorphism in *TCF7L2* has a strong correlation to type 2 diabetes in humans, when carriers of diabetes-associated single nucleotide polymorphism (SNP) display reduced basal and glucose-stimulated insulin secretion but maintain normal hepatic function (Grant et al. 2006; Helgason et al. 2007; Lyssenko et al. 2007).

Recently, we showed that Tcf7l2 is expressed at high levels by pancreatic pericytes (Sakhneny et al. 2018). Inactivation of this transcription factor specifically in these cells leads to glucose intolerance of transgenic mice, which was further augmented upon obesity. The impaired glucose regulation in Tcf7l2-deficient mice was due to compromised β-cell function and diminished glucose-stimulated insulin secretion (Sakhneny et al. 2018). Inactivation of pericytic Tcf7l2 was associated with impaired expression of genes required for β-cell function and maturity, includ-

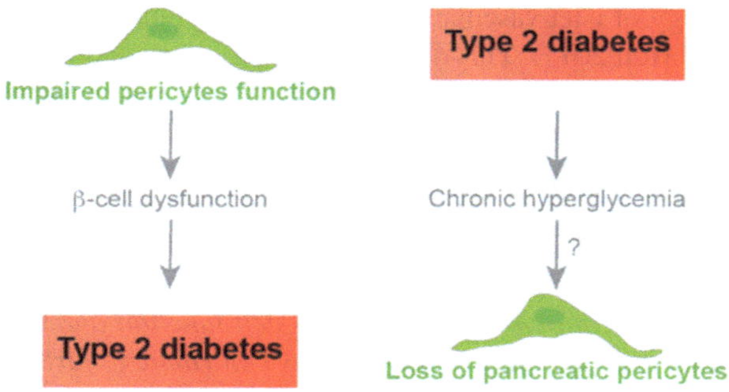

Fig. 2.3 Pancreatic pericytes can both contribute to and be affected by type 2 diabetes. *Left,* Impaired pericyte function leads to β-cell dysfunction and a blunted glucose-stimulated insulin secretion, thus contributing to diabetes progression. *Right*, type 2 diabetes leads to loss of pericyte coverage of islet capillaries, potentially due to the associated chronic hyperglycemia

ing transcription factors (MafA, Pdx1, and NeuroD1) and functional genes (Glut2, Kir6.2, and Sur1) (Sakhneny et al. 2018). *Thus, loss of pericytic Tcf7l2 activity is sufficient to cause β-cell dysfunction and impaired glucose regulation.*

Tcf7l2 is a transcription factor, and its inactivation in pancreatic pericytes alters their transcriptome (Clevers and Nusse 2012; Sakhneny et al. 2018). In particular, our analysis revealed the reduced expression of secreted factors in pericytes lacking Tcf7l2 (Sakhneny et al. 2018). Among these factors is the known Tcf7l2 target gene *BMP4* (bone morphogenetic protein 4) (van de Wetering et al. 2002). β-Cells were shown to depend on the activity of the BMP4 receptor, BMPR1A, for their proper function and gene expression in vivo (Goulley et al. 2007). Treating mice lacking pericytic Tcf7l2 activity with exogenous BMP4 was sufficient to rescue their impaired glucose-stimulated insulin secretion. Thus, impaired activation of the BMP4/BMPR1A pathway represents a potential mechanism through which abnormal pericytic Tcf7l2 activity leads to β-cell dysfunction and type 2 diabetes progression.

*To conclude, diabetes-associated genetic changes in pancreatic pericytes are sufficient to cause β-cell dysfunction and loss of glycemic control (*Fig. 2.3*), suggesting the abnormalities in pancreatic pericytes contributes to the development of type 2 diabetes.*

2.3.2 Hyperglycemia and Loss of Pancreatic Pericytes

One of the significant complications of diabetes is diabetic retinopathy, which is a leading cause of vision loss and visual disability (DeFronzo et al. 2015; Hammes 2017). Pericytes are considered the primary trigger of vascular damage in the diabetic retina, and diabetic retinopathy occurs when these cells are lost (Hammes 2017). Similar to pericytes in the retina, pancreatic pericytes were shown to be sensitive to chronic hyperglycemia that characterizes diabetes (Almaça et al. 2018).

The morphology of islet vasculature is abnormal in individuals with type 2 diabetes (Hogan and Hull 2017). This disease is further associated with morphological and phenotypical changes in islet-associated pericytes, resulting in widening of the islet exocrine interface (Hayden and Sowers 2007; Hayden et al. 2010). Islet capillaries of individuals with type 2 diabetes display significantly reduced coverage of pericytes (Almaça et al. 2018). More so, the decline in the coverage of islet capillaries by pericytes correlates with the duration of diabetes, pointing to a progressive loss of pancreatic pericytes in this disease (Almaça et al. 2018).

Considering their requirement for proper β-cell function and insulin secretion, the progressive loss of pancreatic pericytes may contribute to further deterioration in glucose regulation (Fig. 2.3).

2.4 Conclusions

The role of pericytes in proper pancreatic function has only recently begun to unveil. Recent studies by us and others showed that pancreatic pericytes play multiple roles in glucose regulation, including support of proper β-cell mass and function and adjustment of local blood flow in the islets. Evidence began to accumulate that type 2 diabetes is associated with functional changes in pancreatic pericytes. Furthermore, the progressive loss of these cells during diabetes may contribute to further deterioration in glycemic control. Additional study is required to elucidate the molecular basis of pericytes function, their interactions with other pancreatic cell types, and the contribution of pericyte dysfunction to diabetes progression. Nevertheless, the newly gained data implicate pancreatic pericytes as key components of the cellular network required for glucose regulation and potential targets for therapeutic intervention.

References

Almaça J, Weitz J, Rodriguez-Diaz R, Pereira E, Caicedo A (2018) The Pericyte of the pancreatic islet regulates capillary diameter and local blood flow. Cell Metab 27:630–644.e4. https://doi.org/10.1016/j.cmet.2018.02.016

Apte M, Pirola RC, Wilson JS (2015) Pancreatic stellate cell: physiologic role, role in fibrosis and cancer. Curr Opin Gastroenterol 31:416–423. https://doi.org/10.1097/MOG.0000000000000196

Armulik A, Genové G, Betsholtz C (2011) Pericytes: developmental, physiological, and pathological perspectives, problems, and promises. Dev Cell 21:193–215. https://doi.org/10.1016/j.devcel.2011.07.001

Ashcroft FM, Rorsman P (2012) Diabetes mellitus and the β cell: the last ten years. Cell 148:1160–1171. https://doi.org/10.1016/j.cell.2012.02.010

Bonner-Weir S (1993) The microvasculature of the pancreas, with emphasis on that of the islets of Langerhans. In: Go VLW et al (eds) The pancreas: biology, pathobiology, and disease. Raven Press, New York, pp 759–768

Bonner-Weir S, Aguayo-Mazzucato C (2016) Physiology: pancreatic [beta]-cell heterogeneity revisited. Nature 535:365–366. https://doi.org/10.1038/nature18907

Bramswig NC, Kaestner KH (2014) Transcriptional and epigenetic regulation in human islets. Diabetologia 57:451–454. https://doi.org/10.1007/s00125-013-3150-3

Brissova M, Aamodt K, Brahmachary P, Prasad N, Hong J-Y, Dai C, Mellati M, Shostak A, Poffenberger G, Aramandla R, Levy SE, Powers AC (2014) Islet microenvironment, modulated by vascular endothelial growth factor-a signaling, promotes β cell regeneration. Cell Metab 19:498–511. https://doi.org/10.1016/j.cmet.2014.02.001

Cerasi E (2011) β-Cell dysfunction vs insulin resistance in type 2 diabetes: the eternal "chicken and egg" question. Medicographia 33:35–41

Chen X, Zhang X, Chen F, Larson CS, Wang L-J, Kaufman DB (2009) Comparative study of regenerative potential of beta cells from young and aged donor mice using a novel islet transplantation model. Transplantation 88:496–503. https://doi.org/10.1097/TP.0b013e3181b0d2ee

Cinti F, Bouchi R, Kim-Muller JY, Ohmura Y, Sandoval PR, Masini M, Marselli L, Suleiman M, Ratner LE, Marchetti P, Accili D (2016) Evidence of β-cell dedifferentiation in human type 2 diabetes. J Clin Endocrinol Metab 101:1044–1054. https://doi.org/10.1210/jc.2015-2860

Clevers H, Nusse R (2012) Wnt/β-catenin signaling and disease. Cell 149:1192–1205. https://doi.org/10.1016/j.cell.2012.05.012

Costes S, Langen R, Gurlo T, Butler PC (2013) β-Cell failure in type 2 diabetes: a case of asking too much of too few? Diabetes 62:327–335. https://doi.org/10.2337/db12-1326

Dai C, Brissova M, Reinert RB, Nyman L, Liu EH, Thompson C, Shostak A, Shiota M, Takahashi T, Powers AC (2013) Pancreatic islet vasculature adapts to insulin resistance through dilation and not angiogenesis. Diabetes 62:4144–4153. https://doi.org/10.2337/db12-1657

DeFronzo RA, Ferrannini E, Groop L, Henry RR, Herman WH, Holst JJ, Hu FB, Kahn CR, Raz I, Shulman GI, Simonson DC, Testa MA, Weiss R (2015) Type 2 diabetes mellitus. Nat Rev Dis Primers 1:15019. https://doi.org/10.1038/nrdp.2015.19

Diaferia GR, Jimenez-Caliani AJ, Ranjitkar P, Yang W, Hardiman G, Rhodes CJ, Crisa L, Cirulli V (2013) β1 integrin is a crucial regulator of pancreatic β-cell expansion. Development 140:3360–3372. https://doi.org/10.1242/dev.098533

Dor Y, Glaser B (2013) β-Cell dedifferentiation and type 2 diabetes. N Engl J Med 368:572–573. https://doi.org/10.1056/NEJMcibr1214034

Dor Y, Brown J, Martinez O, Melton D (2004) Adult pancreatic beta-cells are formed by self-duplication rather than stem-cell differentiation. Nature 429:41–46. https://doi.org/10.1038/nature02520

Eberhard D, Lammert E (2009) The pancreatic beta-cell in the islet and organ community. Curr Opin Genet Dev 19:469–475. https://doi.org/10.1016/j.gde.2009.07.003

Epshtein A, Rachi E, Sakhneny L, Mizrachi S, Baer D, Landsman L (2017) Neonatal pancreatic pericytes support β-cell proliferation. Molecular Metabolism 6:1330–1338. https://doi.org/10.1016/j.molmet.2017.07.010

Finegood DT, Scaglia L, Bonner-Weir S (1995) Dynamics of beta-cell mass in the growing rat pancreas. Estimation with a simple mathematical model. Diabetes 44:249–256

Fuchsberger C, Flannick J, Teslovich TM et al (2016) The genetic architecture of type 2 diabetes. Nature 536:41–47. https://doi.org/10.1038/nature18642

Georgia S, Bhushan A (2004) Beta cell replication is the primary mechanism for maintaining postnatal beta cell mass. J Clin Investig 114:963–968. https://doi.org/10.1172/JCI200422098

Gittes GK (2009) Developmental biology of the pancreas: a comprehensive review. Dev Biol 326:4–35. https://doi.org/10.1016/j.ydbio.2008.10.024

Goulley J, Dahl U, Baeza N, Mishina Y, Edlund H (2007) BMP4-BMPR1A signaling in beta cells is required for and augments glucose-stimulated insulin secretion. Cell Metab 5:207–219. https://doi.org/10.1016/j.cmet.2007.01.009

Granot Z, Swisa A, Magenheim J, Stolovich-Rain M, Fujimoto W, Manduchi E, Miki T, Lennerz JK, Stoeckert CJ Jr, Meyuhas O, Seino S, Permutt MA, Piwnica-Worms H, Bardeesy N, Dor Y (2009) LKB1 regulates pancreatic β cell size, polarity, and function. Cell Metab 10:296–308. https://doi.org/10.1016/j.cmet.2009.08.010

Grant SFA, Thorleifsson G, Reynisdottir I, Benediktsson R, Manolescu A, Sainz J, Helgason A, Stefansson H, Emilsson V, Helgadottir A, Styrkarsdottir U, Magnusson KP, Walters GB, Palsdottir E, Jonsdottir T, Gudmundsdottir T, Gylfason A, Saemundsdottir J, Wilensky RL, Reilly MP, Rader DJ, Bagger Y, Christiansen C, Gudnason V, Sigurdsson G, Thorsteinsdottir U, Gulcher JR, Kong A, Stefansson K (2006) Variant of transcription factor 7-like 2 (TCF7L2) gene confers risk of type 2 diabetes. Nat Genet 38:320–323. https://doi.org/10.1038/ng1732

Gregg BE, Moore PC, Demozay D, Hall BA, Li M, Husain A, Wright AJ, Atkinson MA, Rhodes CJ (2012) Formation of a human β-cell population within pancreatic islets is set early in life. J Clin Endocrinol Metab 97:3197–3206. https://doi.org/10.1210/jc.2012-1206

Hammes H-P (2017) Diabetic retinopathy: hyperglycaemia, oxidative stress and beyond. Diabetologia 61:1–10. https://doi.org/10.1007/s00125-017-4435-8

Harari N, Sakhneny L, Khalifa-Malka L, Busch A, Hertel KJ, Hebrok M, Landsman L (2019) Pancreatic pericytes originate from the embryonic pancreatic mesenchyme. Developmental Biology In press

Hayden MR, Sowers JR (2007) Isletopathy in type 2 diabetes mellitus: implications of islet RAS, islet fibrosis, islet amyloid, remodeling, and oxidative stress. Antioxid Redox Signal 9:891–910. https://doi.org/10.1089/ars.2007.1610

Hayden MR, Karuparthi PR, Habibi J, Wasekar C, Lastra G, Manrique C, Stas S, Sowers JR (2007) Ultrastructural islet study of early fibrosis in the Ren2 rat model of hypertension. Emerging role of the islet pancreatic pericyte-stellate cell. JOP 8:725–738

Hayden MR, Yang Y, Habibi J, Bagree SV, Sowers JR (2010) Pericytopathy oxidative stress and impaired cellular longevity in the pancreas and skeletal muscle in metabolic syndrome and type 2 diabetes. Oxidative Med Cell Longev 3:290–303. https://doi.org/10.4161/oxim.3.5.13653

Helgason A, Pálsson S, Thorleifsson G, Grant SFA, Emilsson V, Gunnarsdottir S, Adeyemo A, Chen Y, Chen G, Reynisdottir I, Benediktsson R, Hinney A, Hansen T, Andersen G, Borch-Johnsen K, Jørgensen T, Schäfer H, Faruque M, Doumatey A, Zhou J, Wilensky RL, Reilly MP, Rader DJ, Bagger Y, Christiansen C, Sigurdsson G, Hebebrand J, Pedersen O, Thorsteinsdottir U, Gulcher JR, Kong A, Rotimi C, Stefansson K (2007) Refining the impact of TCF7L2 gene variants on type 2 diabetes and adaptive evolution. Nat Genet 39:218–225. https://doi.org/10.1038/ng1960

Hogan MF, Hull RL (2017) The islet endothelial cell: a novel contributor to beta cell secretory dysfunction in diabetes. Diabetologia 60:952–959

Houtz J, Borden P, Ceasrine A, Minichiello L, Kuruvilla R (2016) Neurotrophin signaling is required for glucose-induced insulin secretion. Dev Cell 39:329–345. https://doi.org/10.1016/j.devcel.2016.10.003

Kragl M, Lammert E (2010) Basement membrane in pancreatic islet function. Adv Exp Med Biol 654:217–234. https://doi.org/10.1007/978-90-481-3271-3_10

Kragl M, Schubert R, Karsjens H, Otter S, Bartosinska B, Jeruschke K, Weiss J, Chen C, Alsteens D, Kuss O, Speier S, Eberhard D, Müller DJ, Lammert E (2016) The biomechanical properties of an epithelial tissue determine the location of its vasculature. Nat Commun 7:13560. https://doi.org/10.1038/ncomms13560

Lammert E, Cleaver O, Melton D (2001) Induction of pancreatic differentiation by signals from blood vessels. Science 294:564–567. https://doi.org/10.1126/science.1064344

Longnecker DS, Gorelick F, Thompson ED (2018) Anatomy, histology, and fine structure of the pancreas, 2nd ed. Wiley-Blackwell, Chichester

Lyssenko V, Lupi R, Marchetti P, Del Guerra S, Orho-Melander M, Almgren P, Sjögren M, Ling C, Eriksson K-F, Lethagen A-L, Mancarella R, Berglund G, Tuomi T, Nilsson P, Del Prato S, Groop L (2007) Mechanisms by which common variants in the TCF7L2 gene increase risk of type 2 diabetes. J Clin Investig 117:2155–2163. https://doi.org/10.1172/JCI30706

Mastracci TL, Sussel L (2012) The endocrine pancreas: insights into development, differentiation, and diabetes. Wiley Interdiscip Rev Developmental Biology 1:609–628. https://doi.org/10.1002/wdev.44

Meier JJ, Butler AE, Saisho Y, Monchamp T, Galasso R, Bhushan A, Rizza RA, Butler PC (2008) Beta-cell replication is the primary mechanism subserving the postnatal expansion of beta-cell mass in humans. Diabetes 57:1584–1594. https://doi.org/10.2337/db07-1369

Mandarino LJ, Finlayson J, Hassell JR (1994) High glucose downregulates glucose transport activity in retinal capillary pericytes but not endothelial cells. Invest Ophthalmol Vis Sci 35:964–972.

Morikawa S, Baluk P, Kaidoh T, Haskell A, Jain RK, McDonald DM (2002) Abnormalities in pericytes on blood vessels and endothelial sprouts in tumors. Am J Pathol 160:985–1000. https://doi.org/10.1016/S0002-9440(10)64920-6

Nikolova G, Jabs N, Konstantinova I, Domogatskaya A, Tryggvason K, Sorokin L, Fässler R, Gu G, Gerber H-P, Ferrara N, Melton DA, Lammert E (2006) The vascular basement membrane: a niche for insulin gene expression and Beta cell proliferation. Dev Cell 10:397–405. https://doi.org/10.1016/j.devcel.2006.01.015

Otonkoski T, Banerjee M, Korsgren O, Thornell L-E, Virtanen I (2008) Unique basement membrane structure of human pancreatic islets: implications for beta-cell growth and differentiation. Diabetes Obes Metab 10(Suppl 4):119–127. https://doi.org/10.1111/j.1463-1326.2008.00955.x

Ouaamari El A, Dirice E, Gedeon N, Hu J, Zhou J-Y, Shirakawa J, Hou L, Goodman J, Karampelias C, Qiang G, Boucher J, Martinez R, Gritsenko MA, De Jesus DF, Kahraman S, Bhatt S, Smith RD, Beer H-D, Jungtrakoon P, Gong Y, Goldfine AB, Liew CW, Doria A, Andersson O, Qian W-J, Remold-O'Donnell E, Kulkarni RN (2016) SerpinB1 promotes pancreatic β cell proliferation. Cell Metab 23:194–205. https://doi.org/10.1016/j.cmet.2015.12.001

Prentki M, Matschinsky FM, Madiraju SRM (2013) Metabolic signaling in fuel-induced insulin secretion. Cell Metab 18:162–185. https://doi.org/10.1016/j.cmet.2013.05.018

Reinert RB, Brissova M, Shostak A, Pan FC, Poffenberger G, Cai Q, Hundemer GL, Kantz J, Thompson CS, Dai C, McGuinness OP, Powers AC (2013) Vascular endothelial growth factor-a and islet vascularization are necessary in developing, but not adult, pancreatic islets. Diabetes 62:4154–4164. https://doi.org/10.2337/db13-0071

Richards OC, Raines SM, Attie AD (2010) The role of blood vessels, endothelial cells, and vascular pericytes in insulin secretion and peripheral insulin action. Endocr Rev 31:343–363. https://doi.org/10.1210/er.2009-0035

Riley KG, Pasek RC, Maulis MF, Dunn JC, Bolus WR, Kendall PL, Hasty AH, Gannon M (2015) Macrophages are essential for CTGF-mediated adult β-cell proliferation after injury. Mol Metab 4:584–591. https://doi.org/10.1016/j.molmet.2015.05.002

Rorsman P, Braun M (2013) Regulation of insulin secretion in human pancreatic islets. Annu Rev Physiol 75:155–179. https://doi.org/10.1146/annurev-physiol-030212-183754

Sakhneny L, Rachi E, Epshtein A, Guez HC, Wald-Altman S, Lisnyansky M, Khalifa-Malka L, Hazan A, Baer D, Priel A, Weil M, Landsman L (2018) Pancreatic Pericytes support β-cell function in a Tcf7l2-dependent manner. Diabetes 67:437–447. https://doi.org/10.2337/db17-0697

Salpeter SJ, Klein AM, Huangfu D, Grimsby J, Dor Y (2010) Glucose and aging control the quiescence period that follows pancreatic beta cell replication. Development 137:3205–3213. https://doi.org/10.1242/dev.054304

Sasson A, Rachi E, Sakhneny L, Baer D, Lisnyansky M, Epshtein A, Landsman L (2016) Islet Pericytes are required for β-cell maturity. Diabetes 65:3008–3014. https://doi.org/10.2337/db16-0365

Schaeffer M, Hodson DJ, Lafont C, Mollard P (2011) Endocrine cells and blood vessels work in tandem to generate hormone pulses. J Mol Endocrinol 47:R59–R66. https://doi.org/10.1530/JME-11-0035

Stolovich-Rain M, Hija A, Grimsby J, Glaser B, Dor Y (2012) Pancreatic beta cells in very old mice retain capacity for compensatory proliferation. J Biol Chem 287:27407–27414. https://doi.org/10.1074/jbc.M112.350736

Stratman AN, Davis GE (2012) Endothelial cell-pericyte interactions stimulate basement membrane matrix assembly: influence on vascular tube remodeling, maturation, and stabilization. Microsc Microanal 18:68–80. https://doi.org/10.1017/S1431927611012402

Talchai C, Xuan S, Lin HV, Sussel L, Accili D (2012) Pancreatic β cell dedifferentiation as a mechanism of diabetic β cell failure. Cell 150:1223–1234. https://doi.org/10.1016/j.cell.2012.07.029

Tang S-C, Chiu Y-C, Hsu C-T, Peng S-J, Fu Y-Y (2013) Plasticity of Schwann cells and pericytes in response to islet injury in mice. Diabetologia 56:2424–2434. https://doi.org/10.1007/s00125-013-2977-y

Tarussio D, Metref S, Seyer P, Mounien L, Vallois D, Magnan C, Foretz M, Thorens B (2014) Nervous glucose sensing regulates postnatal β cell proliferation and glucose homeostasis. J Clin Investig 124:413–424. https://doi.org/10.1172/JCI69154

van de Wetering M, Sancho E, Verweij C, de Lau W, Oving I, Hurlstone A, van der Horn K, Batlle E, Coudreuse D, Haramis AP, Tjon-Pon-Fong M, Moerer P, Van Den Born M, Soete G, Pals S, Eilers M, Medema R, Clevers H (2002) The beta-catenin/TCF-4 complex imposes a crypt progenitor phenotype on colorectal cancer cells. Cell 111:241–250

Wang P, Fiaschi-Taesch NM, Vasavada RC, Scott DK, García-Ocaña A, Stewart AF (2015) Diabetes mellitus--advances and challenges in human β-cell proliferation. Nat Rev Endocrinol 11:201–212. https://doi.org/10.1038/nrendo.2015.9

Chapter 3
Pericytes in the Lung

Chi F. Hung, Carole L. Wilson, and Lynn M. Schnapp

Abstract The lung has numerous roles, including gas exchange, immune surveillance, and barrier function. Being a highly vascularized organ, the lung receives dual blood supply from both the pulmonary and bronchial circulation. Therefore, pericytes likely play a prominent role in lung physiology given their localization in the perivascular niche. New genetic approaches have increased our understanding of the origin and the diverse functions of lung pericytes. Lung pericytes are myofibroblast progenitors, contributing to development of fibrosis in mouse models. Lung pericytes are also capable of responding to danger signals and amplify the inflammatory response through elaboration of cytokines and adhesion molecules. In this chapter, we describe the molecular, anatomical, and phenotypical characterization of lung pericytes. We further highlight their potential roles in the pathogenesis of lung diseases including pulmonary fibrosis, asthma, and pulmonary hypertension. Finally, current gaps in knowledge and areas of ongoing investigation in lung pericyte biology are also discussed.

Keywords Lung pericytes · PDGFRβ · Lung development · Lung myofibroblasts · Pulmonary fibrosis · Lung injury · Lung inflammation

---

C. F. Hung
Division of Pulmonary, Critical Care and Sleep Medicine, University of Washington, Seattle, WA, USA

C. L. Wilson · L. M. Schnapp (✉)
Division of Pulmonary, Critical Care, Allergy and Sleep Medicine, Medical University of South Carolina, Charleston, SC, USA
e-mail: schnapp@musc.edu

A. Birbrair (ed.), *Pericyte Biology in Different Organs*, Advances in Experimental Medicine and Biology 1122,
https://doi.org/10.1007/978-3-030-11093-2_3

3.1 Introduction

A critical but vastly understudied cell in the lung is the pericyte. Pericytes are mesenchymal cells defined by their anatomical association with endothelial cells (Armulik et al. 2011; von Tell et al. 2006). Pericytes directly interact with microvascular endothelium, including capillaries, precapillary arterioles, postcapillary, and collecting venules, and are critical to blood vessel maturation and stability. However, it is increasingly clear that lung pericytes play important roles not only in vascular homeostasis but in lung injury and repair. In this chapter, we will review what is known about lung pericytes in normal lung homeostasis, including their developmental origins, discuss characterization of these cells in vitro and in vivo, and highlight current knowledge of lung pericytes in lung diseases, such as pulmonary fibrosis and pulmonary hypertension. Finally, we will speculate on the potential role of pericytes as a regenerative niche in the lung.

3.2 Definition of Lung Pericytes

Anatomic Characterization Pericytes are classically defined by ultrastructural localization demonstrating direct contact with microvascular endothelial cells through gaps in the shared basement membrane. Using transmission electron microscopy, Epling was the first to observe pericytes as "filament-containing cells" in bovine and porcine lungs (Epling 1966), and Weibel confirmed the presence of pericytes in capillaries of the lungs of humans and other mammals (Weibel 1974). Compared to organs such as the heart or pancreas, pericytes are more rare in the lung, and their frequency correlates with body size. For example, pericytes were readily detected in human and dog lungs but were sparser in rodent lungs and could not be found in lungs of the Etruscan shrew, the smallest mammal known (Weibel 1974). Pericytes are sensitive to oxygen levels and thus are less abundant in areas of gas exchange (Shepro and Morel 1993). The pericyte-to-endothelial ratio in the lung has been estimated to be about 1:10 (Shepro and Morel 1993), less than that in the retina and central nervous system (~1:1 to 1:3), where barrier function is paramount and pericytes play a key role in regulating permeability. By genetically tracing pericytes in mice, Kato et al. found a pericyte-to-endothelial ratio from 1:7 to 1:9 (Kato et al. 2018). Coverage of the abluminal surface of capillaries also varies by organ and ranges between 18 and 26% in bovine lungs, depending on developmental stage of the animal (Sims and Westfall 1983).

Perivascular stromal cells are found along all levels of the bronchovascular bundle. Classically, lung pericytes are defined at the level of the alveolar capillary bed (Armulik et al. 2011). Electron microscopy lungs show pericytes embedded within the capillary basement membrane of alveoli in humans, mice, and other mammals (Hung et al. 2013) (Epling 1966; Weibel 1974). In human lung samples, Weibel observed that while most of the pericyte cell body and its large cellular processes are separated from endothelial cells by the intervening basement membrane, finer processes penetrate the basement membrane and make contacts with the endothelial cells in capillaries (Weibel 1974). Some pericytes were even seen to bridge separate capillary segments (Weibel 1974). The nature of pericyte interactions with endothelial cells in larger microvessels (i.e., the arterioles and venules) of the lung has not yet been described at the ultrastructural level.

Developmental Origins Current understanding of lung pericytes indicates that they primarily originate from the mesoderm and mesothelium (Que et al. 2008). In the mouse, a multipotent cardiopulmonary progenitor positive for the transcription factors Gli1 and Isl1 and the signaling molecule Wnt2 gives rise to mesodermal lineages in the lung, including pericytes (Peng et al. 2013). We showed that the forkhead transcription factor FoxD1 is transiently expressed in pulmonary mesenchymal progenitors as they infiltrate the lung buds during mouse embryogenesis between days 11.5 and 12.5 (Hung et al. 2013). Expression of FoxD1 is then silenced by day 15.5 once the progenitors differentiate into mature mesenchymal cells, making this factor a useful marker for fate mapping these progenitors in the mouse lung (see section on Molecular Identification below). For most organs, including the lung, the PDGF-B-PDGFRβ signaling axis is critical in recruiting pericytes to their proper anatomical niche within the microvascular environment during embryogenesis (Armulik et al. 2011). Disruption of this axis by knockout of PDGFRβ or PDGF-B results in virtually total absence of pericytes in the mouse embryonic lung (Hellstrom et al. 1999). The developmental origin of human lung pericytes is not known, but we presume that they derive from a mesodermal precursor as well. As discussed below, PDGFRβ is a key, robust marker of lung pericytes.

Molecular Identification and Pericyte Subtypes Whereas identification of pericytes in early studies of lung tissue specimens used electron microscopy to define the cells at the ultrastructural level, this method is cumbersome and difficult to carry out in most laboratories. Advances in molecular characterization of pericytes and in fate-mapping transgenic models have enabled investigators to study lung pericytes with greater ease and flexibility than in the past. Nevertheless, lack of a truly pericyte-specific marker means that researchers need to use multiple criteria including location relative to endothelial cells in situ, morphology, and genetic markers to identify pericytes.

Only in the last decade or so have a limited number of studies been undertaken to isolate and characterize pericytes from "normal" human lung (typically from failed donor organs or tissue adjacent to carcinoma). The commonalities in these studies were that pericytes were selected from dissociated lung cells using magnetic bead-linked antibodies against one or two markers and then expanded in commercial medium especially designed for growth of pericytes. In culture, isolated lung pericytes have a stellate or elongated morphology with long extensive cellular processes (Yuan et al. 2015; Wilson et al. 2018) (Fig. 3.1). In addition, lung pericytes form primitive tubelike structures on the basement membrane preparation Matrigel (Wilson et al. 2018; Bagley et al. 2006) (Fig. 3.1).

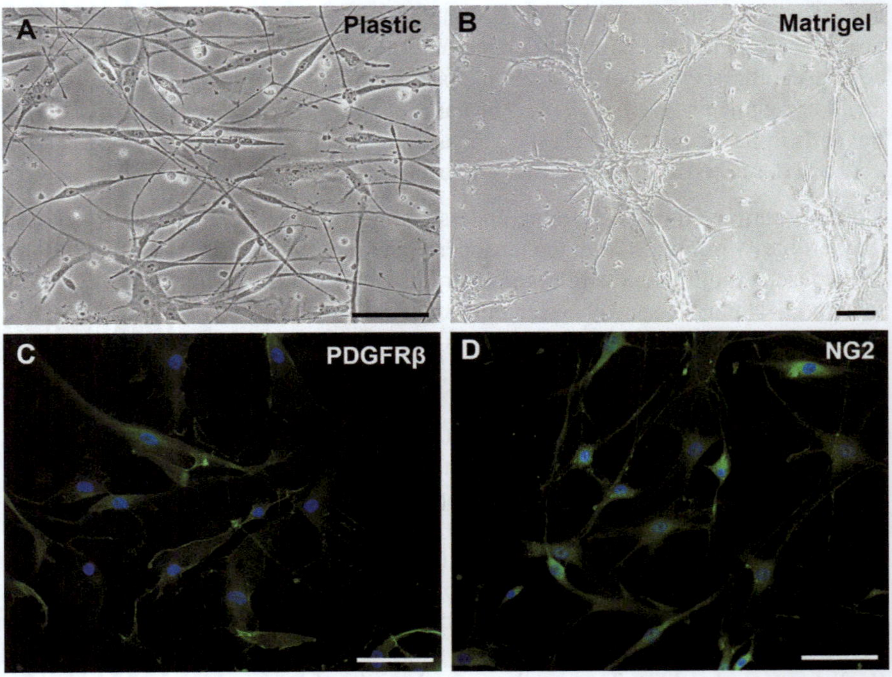

Fig. 3.1 Cell morphology and characteristics of isolated human lung pericytes. Representative phase-contrast light microscopic images of pericytes from normal lungs after plating on plastic (**a**) or Matrigel™ (**b**). Note the extensive number of elongated cellular processes emanating from the cells on plastic. On Matrigel™, pericytes assemble into primitive networks composed of nodes of cells from which cellular processes project and connect with other nodes (**b**). Human lung pericytes are positive for PDGFRβ (**c**) and NG2 (**d**) by immunofluorescence. Scale bar = 500 μm in (**a**), (**c**), and (**d**) and 100 μm in (**b**)

Table 3.1 summarizes the markers that were used for isolation of human lung pericytes and the other markers that were detected in these cells in different studies. Expression of the proteoglycan neural glial antigen-2 (NG2 or chondroitin sulfate proteoglycan 4) (Yuan et al. 2015; Wilson et al. 2018; Bagley et al. 2005; Bichsel et al. 2015; Ricard et al. 2014) was reported in all the studies (Fig. 3.2). NG2+ cells localized to perivascular regions in human lung tissue by immunofluorescence (Sava et al. 2017; Rock et al. 2011). Most studies showed that lung pericytes are PDGFRβ+, as expected (Fig. 3.2), and we used this marker in our own work for cell selection from both human and mouse lungs (Wilson et al. 2018; Hung et al. 2017a).

Other prominent mesenchymal markers found in isolated human lung pericytes include CD73 (ecto-5′-nucleotidase), CD90 (Thy1, also a fibroblast marker), and the hyaluronan receptor CD44, although not all studies examined the cells for these markers. Expression of endoglin (CD105) and PDGFRα is also characteristic of human lung pericytes. The monoclonal antibody 3G5-defined ganglioside antigen has emerged as a useful pericyte marker for the lung as well as the skin

Table 3.1 Gene markers of isolated human lung pericytes

Markers	Yuan et al. (2015) (3G5)	Bagley et al. (2006) (NG2)	Ricard et al. (2014) (3G5)	Bichsel et al. (2015) (CD73/ CD90)	Wilson et al. (2018) (CD140b)
CD44	ND	ND	ND	+	+
CD73	ND	ND	ND	+	+
CD90	+	+	ND	+	+
CD105 (endoglin)	ND	+	ND	+	ND
CD140a (PDGFRα)	ND	ND	ND	+	+
CD140b (PDGFRβ)	+	+	ND	ND	+
CD146	+	ND	ND	–	–
NG2 (CSPG4)	+	+	+	+	+
3G5	+	+	+	ND	ND
αSMA	+	+	–	–	–
Desmin	ND	+	ND	ND	ND

Marker(s) used for cell selection shown in parentheses. *ND* not done

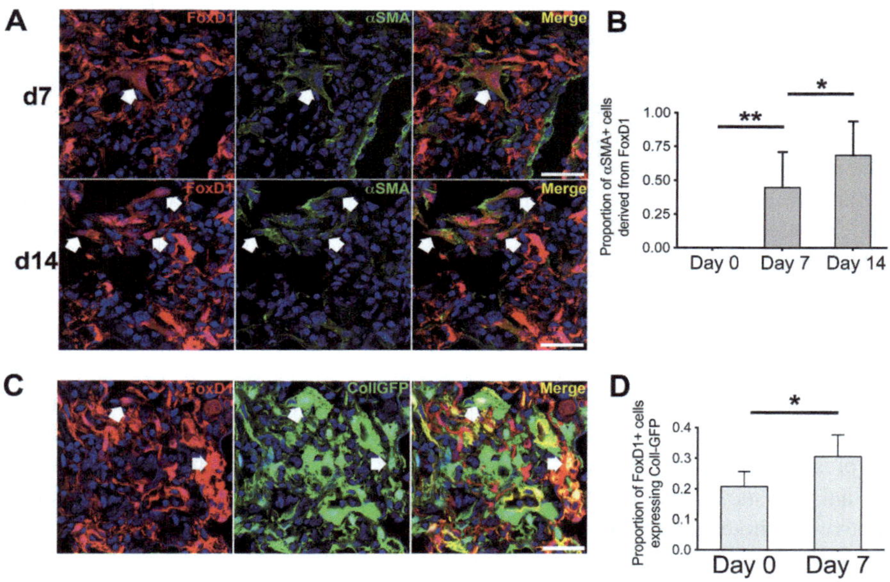

Fig. 3.2 Fate-mapping of Foxd1 progenitor-derived pericytes identifies them as a major source of myofibroblast precursors in lung injury. (**a**, **b**) Confocal images showing fibrotic foci on d7 and d14 after bleomycin lung injury in *Foxd1-Cre; Rs26-TdT-R* mice stained for the myofibroblast marker αSMA (green). Co-expression of myofibroblast marker αSMA and the tdTomato fate marker of Foxd1-derived pericytes indicated by (🡅). Graph indicating proportion of tdTomato+ cells co-expressing αSMA+ in fibroblastic foci at indicated time points after bleomycin injury (mean ± SEM, n = 3). (**c**, **d**) Confocal images showing fibrotic foci on d7 after bleomycin lung injury in triple transgenic mouse *Foxd1-Cre; Rs26-TdT-R; Coll-GFP*^{Tg} and graph summarizing proportion of tdTomato cells co-expressing Coll-GFP in fibrotic foci (mean ± SEM, n = 3) (Bar = 50 μm). Modified and reprinted with permission of the American Thoracic Society. Copyright © 2018 American Thoracic Society (Hung et al. 2013). The *American Journal of Respiratory and Critical Care Medicine* is an official journal of the American Thoracic Society

(Helmbold et al. 2001a, b) and was used by two investigator teams to isolate these cells (Yuan et al. 2015; Ricard et al. 2014). There are discrepant reports on CD146 (also known as MCAM or melanoma cell adhesion molecule) expression: one study found that the cells are positive for CD146 (Yuan et al. 2015), whereas two others did not (Wilson et al. 2018; Bichsel et al. 2015). The lack of significant CD146 expression is surprising as it is a typical marker of pericytes from other organs, such as the placenta and brain (Crisan et al. 2008), and was detected in mouse lung pericytes (Hung et al. 2013). Potential explanations for these conflicting data on CD146 positivity include the presence of subpopulations of lung pericytes, differences in underlying pathology of donor lung tissue, and/or differences in culture conditions. Also, while α-SMA has been described as a pericyte marker, in general, only mural cells associated with precapillary arterioles and postcapillary venules express this contractile protein, whereas quiescent capillary pericytes do not (Nehls and Drenckhahn 1991). Further complicating interpretations, while pericytes on arterioles and capillaries are NG2+, venular pericytes are NG2−, at least in rat mesenteric and subcutaneous tissue and skeletal muscle (Murfee et al. 2005). NG2 is induced in venular pericytes during angiogenesis and vascular remodeling (Murfee et al. 2006). Although this differential expression of NG2 has not yet been established in the microvasculature of the human lung, it raises the possibility that a cell selection strategy based on NG2 may not include the entire spectrum of pericytes. Because the markers are not pericyte specific [e.g., CD146 is expressed by endothelial cells (Bardin et al. 2001) and 3G5 by mast cells (Gushi et al. 2008)], it is equally important to exclude expression of other cell-specific markers such as CD31 and CD144 (endothelial cell markers PECAM and VE-cadherin, respectively), CD45 (pan leukocyte marker), and CD326 (EpCAM, an epithelial cell marker).

Animal Models Much of our knowledge on the origin and fate of lung pericytes comes from genetic approaches in mouse models, mostly using Cre-LoxP technology (Table 3.2). In these models, Cre recombinase expression is driven by a promoter selective for pericytes or pericyte progenitors, and the recombinase activates expression of a reporter allele by targeting LoxP sequences that flank a transcriptional stop site upstream of the reporter. This technique irreversibly labels cells with active promoter expression and their progeny. We showed that expression of the transcriptional factor FoxD1 marks progenitors destined to become pericytes in the lung (Hung et al. 2013). Using triple transgenic mice expressing Cre recombinase from the FoxD1 promoter and GFP from the collagen I promoter (Yata et al. 2003), in conjunction with a tdTomato reporter allele at the ROSA locus, we found that FoxD1-derived progenitors give rise to vascular smooth muscle cells and two populations of pericytes. The major population does not express collagen I or α-SMA and is positive for PDGFRβ; a second, minor population expresses collagen I and PDGFRα rather than PDGFRβ (Hung et al. 2013). In this study, all FoxD1-derived cells were negative for CD31, an important distinction given that a group has reported that FoxD1 positivity also marks endothelial progenitors in the mouse lung (Sims-Lucas et al. 2013).

Table 3.2 Transgenic mice for identifying and tracking lung pericytes

Promoter	Insert	Type	Major findings in uninjured lung	References
FoxD1	GFP-Cre	Knock-in	Cells are PDGFRβ+, CD146+, 60% NG2+, αSMA-; variable expression of collagen I and PDGFRα	Hung et al. (2013)
FoxD1	GFP-CreERT2	Knock-in, TAM inducible	Promoter active between days 11.5 and 15.5 in embryogenesis	Hung et al. (2013)
FoxJ1	CreER	Transgenic, TAM inducible	Expressed in NG2+ cells	Rock et al. (2011)
NG2	DsRedBAC	Transgenic	Cells are PDGFRβ+, CD146+, αSMA−	Ricard et al. (2014), Chow et al. (2013) and Akamatsu et al. (2013)
NG2	CreER™BAC	Transgenic, TAM inducible	Only 15% recombination efficiency; cells are PDGFRβ+, αSMA−	Rock et al. (2011)
PDGFRβ	Cre	Transgenic	PDGFRβ expressed in other lineages in addition to pericytes during embryogenesis	Guimaraes-Camboa et al. (2017)
PDGFRβ (BAC)	CreER	Transgenic, TAM inducible	Cells are NG2+	Kato et al. (2018)

TAM tamoxifen

One important limitation in lung pericyte research using lineage tracing animal models is the lack of a definitive marker for pericytes. Commonly recognized markers such as PDGFRβ, NG2, and CD146 are not uniformly expressed in all pericytes. Moreover, their expression may be spatially and temporally dynamic throughout development and after injury. For example, PDGFRβ promoter is active in multiple cell lineages during embryogenesis and is not specific to pericytes (Guimaraes-Camboa et al. 2017). Furthermore, activated myofibroblasts derived from non-pericyte populations following lung injury may upregulate expression of PDGFRβ (Henderson et al. 2013; Hung et al. 2018).

Pulse labeling of pericyte populations circumvents the dynamic expression of pericyte markers that are observed in other cell lineages during development and with physiologic stress. Modification of Cre recombinase to make its activity inducible allows temporal control in labeling pericytes. For example, a tamoxifen-inducible PDGFRβ-CreER transgenic animal enables postnatal marking of PDGFRβ+ cells; tamoxifen administration at days 1 through 3 after birth labeled cells tightly juxtaposed to endothelial cells in the lung (Kato et al. 2018). Isolated labeled cells were positive for expression of PDGFRβ and NG2 and were negative for markers of endothelial and epithelial cells (Kato et al. 2018).

There are important limitations to Cre-loxP models. As discussed previously, promoter activity driving Cre recombinase expression may not be restricted to pericytes, depending on the developmental stage and the presence of physiologic stress. Secondly, Cre recombinase activity is rarely 100% efficient. Promoter activity, accessibility of the target loxP sequences in the genome, and experimental conditions used to induce Cre activity in inducible models can all influence labeling efficiency. In one report, postnatal tamoxifen administration labeled only ~15% of the NG2+ lung cells in NG2-CreER mice (Rock et al. 2011), highlighting a potential drawback of these types of models. Furthermore, bioavailability of tamoxifen can influence labeling efficiency in tamoxifen-inducible Cre models. Some investigators have leveraged this property to study lineage-labeled single cells distributed throughout the lung by administering low-dose tamoxifen (Kato et al. 2018; Barkauskas et al. 2013). Finally, there is currently no single marker that identifies the entire lung pericyte population. The commonly used markers do not completely overlap, suggesting heterogeneity within the lung pericyte population. We showed that only a subset of PDGFRβ+ cells (~60%) express NG2 (Hung et al. 2013), suggesting that NG2 may not be an all-inclusive marker of mouse lung pericytes. Indeed, NG2 has been described as a marker of pericytes in angiogenesis and tissue remodeling rather than homeostasis (Murfee et al. 2006). Interestingly, Rock et al. found that a FoxJ1-CreER transgene also marks NG2+ pericytes in the lung (Rock et al. 2011), although there have been no further reports in the literature using this transgenic mouse for pericyte fate mapping. Our group investigated NG2 and CD146 expression in human pericytes. We found that PDGFRβ+ human lung pericytes have low CD146 expression unlike mouse lung pericytes (Hung et al. 2017a, b). Taken together, the studies suggest that the lung pericyte population is heterogeneous. Understanding functional differences in lung pericyte subpopulations is an area of ongoing investigation.

Cre-loxP technology can also induce specific gene deletions in pericyte-lineage lung cells. A cross of mice with loxP-flanked *Yap1* and *Wwtr1* (encoding TAZ protein) alleles to PDGFRβ-CreER mice generated pericyte-specific inducible deletions of these genes in postnatal lung (Kato et al. 2018). The disruptions in YAP1 and TAZ signaling in pericytes altered postnatal lung morphogenesis. Another novel application of Cre-loxP technology involves targeted ablation of pericytes within the lung. Mice are inherently insensitive to diphtheria toxin (DT) as they lack its receptor (DTR). We crossed ROSA26iDTR mice, in which the expression of the simian diphtheria toxin receptor (iDTR) is Cre inducible, to FoxD1-Cre mice to generate the double transgenic FoxD1-Cre;ROSA26iDTR model. In this mouse model, iDTR is expressed in FoxD1-derived cells, rendering them sensitive to ablation upon exposure to DT. Conventional delivery of DT by intraperitoneal injection resulted in neurologic side effects such as seizures within 72 hours of DT administration (unpublished data). This observation reflected the critical role of brain pericytes in the regulation of blood-brain barrier (Hung et al. 2017b). To circumvent this limitation, we administered low-dose DT by oropharyngeal aspiration. Using this method, we achieved approximately 40% ablation of PDGFRβ+ stromal cells in the lung at 7 days after DT delivery (Hung et al. 2017b). Long-term ablation, however, remains elusive as progenitor populations replenish ablated cells.

3.3 Pericytes During Lung Homeostasis and Repair

The exact roles of pericytes in lung homeostasis remain largely uncharacterized. Functional studies that disrupt the pericyte-endothelial signaling axis (e.g., PDGF-PDGFRß) result in early embryonic lethality due to abnormal vasculogenesis. Most models that induce pericyte loss focus on developing vessels rather than fully mature pericyte-endothelial units. In one study, pericytes were depleted by administering a PDGFRβ-blocking antibody in postnatal pups, which caused abnormal retinal capillary formation (Ogura et al. 2017) and demonstrated a critical role of pericytes in retinal angiogenesis. However, administration of the same antibody in adult mice did not result in pericyte coverage loss, suggesting the PDGF-PDGFRβ signaling axis is dispensable in the developed microvascular unit. Akt/Jagged1 signaling maintains perivascular stromal cell coverage in homeostasis in the heart and retina as disruption of endothelial Akt production led to pericyte apoptosis and breakdown of perivascular matrix (Kerr et al. 2016). Whether blockade of the Akt/Jagged1/Notch signaling axis results in similar findings in the lung remains unknown. When retinal pericytes were ablated with DT in adult mice, microangiopathy developed, suggesting that maintenance of vascular integrity is a critical function of retinal pericytes (Valdez et al. 2014). In our model of lung pericyte ablation with DT, we did not observe changes in lung permeability (Hung et al. 2017b), suggesting that lung pericytes may not be as critical to normal barrier function as in other organs. Thus, our understanding of pericyte biology in mature lungs under homeostatic conditions is limited.

The lung has limited ability to regenerate compared to skin and gut epithelium. Acute lung injury from a variety of insults leads to denudement of damaged alveolar epithelium and a concurrent local inflammatory response. This is normally followed by resolution of inflammation, transient scarring, and repopulation of the lost epithelial barrier. In some instances, a pathological response results in persistent scarring or fibrosis, leading to distortion of normal architecture and impairment of lung function. The precise cellular mechanisms that control regenerative versus pathological responses in the lung remain an area of intense interest.

Accumulating evidence suggests pericytes are important in regulating lung repair as progenitors of myofibroblasts and mesenchymal progenitor cells. The multidrug resistance transporter ATP-binding cassette G (ABCG2) was identified as a marker in lineage-tracing experiments that labeled mesenchymal progenitor cells (MPCs) that give rise to pericytes (Jun et al. 2011; Marriott et al. 2014; Chow et al. 2013; Gaskill et al. 2017). Differentiation of ABCG2-lineage cells to mature pericytes was important in attenuation of bleomycin-induced lung fibrosis and restoration of normal vascular function following injury (Gaskill et al. 2017). Overexpression of the Wnt/ß-catenin pathway in ABCG2-lineage cells led to increased pericyte specification without maturation. These mice developed worse lung fibrosis and increased microvascular dysfunction, suggesting MPC differentiation into mature, functional pericytes is integral to normal repair in the lung. There may be other pools of progenitors that give rise to renewal of pericytes in different physiologic contexts such as aging and apoptosis. Identification of the progenitor pool and understanding the molecular signals that direct regeneration will greatly advance the field's understanding of normal and abnormal repair in the lung interstitium, with relevance to many chronic pulmonary diseases such as COPD and interstitial lung diseases where effective medical therapy continues to be lacking.

One aspect of the regenerative response that has garnered significant attention is the identity of adult progenitor cells that are capable of self-renewal or even renewal of other cell lineages (multipotency). These elusive adult progenitor cells are thought to be of mesenchymal origin and are often referred as mesenchymal stem cells (MSCs). Pericytes have been postulated to be a source of MSCs in multiple organs (Crisan et al. 2008). MSCs are marked by expression of the transcription factor Gli1 and ABCG2 and comprise a small subset of the PDGFRβ+ cells in the lung (Marriott et al. 2014; Chow et al. 2013; Kramann et al. 2015). Although these MSCs have been called "pericytes," this term primarily reflects their anatomical location, as they do not express other markers associated with mature pericytes in the mouse, such as NG2 and CD146. Using a genetic tracing approach to isolate ABCG2+ cells, Chow et al. showed that lung MSCs can differentiate into NG2+ pericytes and endothelial cells in vitro (Chow et al. 2013). In contrast, a study using the transcription factor Tbx18 as a marker of pericytes found that Tbx18-lineage cells did not exhibit multipotency in vivo (Guimaraes-Camboa et al. 2017). Likewise, we did not detect expression of MSC markers on human lung pericytes (defined by NG2+ PDGFRß+) in vitro (Wilson et al. 2018). One possibility to account for the conflicting reports on pericyte multipotency is that cell culture environments often introduce non-physiologic cues that may induce phenotypic changes in pericytes, cues that are not representative of the native tissue environment where pericytes reside. To date, there is no evidence to suggest lung pericytes possess multipotent cell fate plasticity in vivo.

3.4 Pericytes in Lung Disease

Lung Inflammation and Injury Historically, focus on pericytes has been limited to their functions as structural mural cells that support vascular maturation and endothelial homeostasis. Accumulating evidence, however, suggests these specialized cells may have broader biological functions. One area that has captured the interest of investigators is the potential role of pericytes in immunity. Much of the evidence for pericytes in inflammation comes from studies in the central nervous system where pericytes elaborate chemoattractants, adhesion molecules, and paracrine signaling to amplify local inflammation (Jansson et al. 2014; Rustenhoven et al. 2017).

Similar to findings in other organs, our group has shown that mouse lung pericytes respond to pro-inflammatory signaling. Murine lung pericytes possess multiple functional TLRs and elaborate cytokines in response to specific TLR agonists in vitro (Table 3.3) (Hung et al. 2017a). Similarly, cultured human lung pericytes respond to inflammatory cues and upregulate cytokine expression. Compared to mouse lung pericytes, human lung pericytes exhibited a narrower range of TLR responses, with TLR 2/6 and TLR 4 signaling leading to the most robust pro-inflammatory responses (Table 3.3). We showed that depletion of pericytes by DT exposure attenuated the acute inflammatory response in a sterile lung injury model (Hung et al. 2017a, b). How lung pericyte ablation affects the inflammatory profile in other models of sterile lung injury and infectious models is yet to be defined. Based on our experience, we speculate that pericytes play an important supportive role in mediating local lung inflammation. Their perivascular position in the alveoli suggests they may play important roles in the activation of the endothelium through paracrine signaling and direct contact during inflammation. Furthermore, they reside in a prime location to sense pro-inflammatory alveolar components and may be one of the very first responders to alveolar damage. Beyond the production of inflammatory cytokines, pericytes may also direct trafficking of inflammatory cells that exit systemic circulation through upregulation of adhesion molecules as well as modification of the extracellular matrix in the perivascular space and the lung interstitium.

Pulmonary Fibrosis During pulmonary fibrosis, the main effector cell for matrix production is the lung myofibroblast. However, the origin of lung myofibroblasts remains controversial. In our transgenic mouse model work, we found that FoxD1-

Table 3.3 Inflammatory response of lung pericytes[a]

Species	Stimulus	Inflammatory cytokines and adhesion molecules
Mouse	IL-1r	CXCL1, CXCL2, CXCL10, CCL2, ICAM1
	TLR2/1	CXCL1, CXCL2, CXCL10, CCL2, TNFα, ICAM1
	TLR2/6	CXCL1, CXCL2, CXCL10, CCL2, TNFα, ICAM1
	TLR4	CXCL1, CXCL2, CXCL10, CCL2, TNFα, ICAM1
	TLR5	ICAM1
	TLR7	CXCL1, CXCL2, CXCL10, CCL2, TNFα, ICAM1
	TLR9	CXCL2, TNFα
Human	TLR2/6	CXCL1, CXCL2, CXCL8, CCL2, ICAM1
	TLR4	CXCL1, CXCL2, CXCL8, CCL2, ICAM1

[a]Pericytes were selected based on PDGFRβ positivity

derived cells express common markers used to identify pericytes such as PDGFRß, NG2, and CD146, while they lack expression of endothelial-, epithelial-, or hematopoietic-specific markers. In addition, FoxD1-derived cells are located adjacent to endothelial cells, suggesting they are indeed pericytes (Fig. 3.2) (Hung et al. 2013). Importantly, they contributed to >50% of α-smooth muscle actin (αSMA)-positive myofibroblasts in the bleomycin model of lung fibrosis (Hung et al. 2013) (Fig. 3.2). Our study and Rock et al. (Rock et al. 2011) found other cell types contribute significantly to the myofibroblast pool in the lung. In contrast, in both kidney and liver models of fibrosis, pericytes were the predominant source of myofibroblasts (Lin et al. 2008; Humphreys et al. 2010; Mederacke et al. 2013). These findings establish the novel concept that pericytes are key effectors and potential therapeutic targets in pulmonary fibrosis.

Pulmonary Arterial Hypertension Pulmonary Arterial Hypertension (PAH), defined by elevated mean arterial pressure ≥ 25 mmHg at rest, is characterized by progressive remodeling of the distal pulmonary arterials with resultant increase in pulmonary vascular resistance, leading to right heart failure and ultimately death. Pulmonary endothelial dysfunction is a key feature of PAH pathogenesis. The pulmonary artery wall includes a secondary layer of vascular smooth muscle cells (SMCs) or smooth muscle-like pericytes (Hemnes and Humbert 2017). Pericytes isolated from PAH lungs had impaired association with endothelial cells (Yuan et al. 2015). Another study showed increased number of pericytes (defined by 3G5 staining) surrounding pulmonary vessels of all sizes in explanted human PAH lungs and in rodent models of pulmonary hypertension (Ricard et al. 2014). Furthermore, the increased number of pericytes was detected before changes in arterial pressures were detected, suggesting the changes in pericyte were not secondary to hemodynamic derangements. Their data suggested that dysregulated FGF2 and IL-6 signaling from endothelial cells drive pericyte proliferation during PAH and contribute to pathogenesis.

Interestingly, there were two case reports of infants with Adams-Oliver syndrome, a rare genetic disorder characterized by congenital scalp defects and limb defects of unknown pathogenesis, who developed severe infantile pulmonary hypertension, intracranial bleeding, and cutis marmorata telangiectatica congenita (Patel et al. 2004). At autopsy, lack of pericyte coverage was found areas of vascular dilatation, while increased pericyte coverage and proliferation were found in association with vessel stenosis. They hypothesized that aberrant pericyte recruitment caused pulmonary hypertension in this syndrome (Patel et al. 2004). These studies support the role of pericyte recruitment and proliferation in the pathogenesis of pulmonary arterial hypertension.

Hereditary Hemorrhagic Telangiectasia Hereditary Hemorrhagic Telangiectasia (HHT) also known as Osler-Weber-Rendu syndrome is an autosomal dominant disorder characterized by arteriovenous malformations (AVM) affecting major organs, particularly the lung, liver, and brain. Lung AVMs are more commonly seen in HHT1 patients. Most HHT1 patients have mutations in *ENG* encoding endoglin

(McAllister et al. 1994), a receptor for transforming growth factor-β (TGFβ)/bone morphogenetic protein (BMP) expressed primarily in endothelial cells and pericytes. Impaired communication between endothelial cell and pericyte, including loss of normal TGFβ activation and signaling, is postulated to lead to impaired pericyte differentiation and vessel maturation in HHT (Thalgott and Lebrin 2015). One group reported that thalidomide increased pericyte number, α-SMA expression, and their recruitment to blood vessels in a mouse model of HHT, leading to vasculature stabilization (Thalgott and Lebrin 2015). A speculated mechanism was a thalidomide-induced increase of PDGF-B expression by endothelial cells.

Pericytes and Asthma Asthma is a common disease characterized by recurrent inflammatory episodes with airway and vascular remodeling. The airway remodeling includes myofibroblast accumulation, collagen deposition, and increased smooth muscle mass leading to subepithelial fibrosis and airway narrowing. Our understanding of the contributions of pericytes in asthma is limited. In a murine model of chronic house dust mite (HDM) exposure, investigators noted decreased capillary-associated pericyte coverage (defined by NG2 positivity) and increased subepithelial (peribronchial) pericytes. These findings were exaggerated when mice were treated with a PDFGRβ inhibitor and resulted in worsened airway hyperresponsiveness and airway smooth muscle thickening (Johnson et al. 2015). Given evidence supporting pericyte transition to myofibroblast in other conditions, it is interesting to speculate that pericytes likewise contribute to subepithelial fibrosis in asthma. However, more studies are needed to define a role of pericytes in asthma pathogenesis.

3.5 Conclusion

We are only beginning to appreciate the multiple roles pericytes play in lung disease pathogenesis. Studies in animal models suggest lung pericytes exhibit a degree of biological plasticity that was previously underappreciated. They participate in inflammation and fibrosis, and they also regulate microvascular permeability and stability (Fig. 3.3). Clinically, some disorders such as pulmonary hypertension and HHT are associated with changes in pericyte coverage. So are pericytes beneficial or harmful in disease? To date, there is not enough information on lung pericytes to answer this question. The timing of pericyte activation during disease, the interaction of pericytes with their cellular neighbors and their microenvironmental niche when homeostasis is disturbed, and the environmental cues that activate pericytes are some of the important knowledge gaps in lung pericyte biology that require further study. Refining methods to identify, isolate, and manipulate lung pericytes in the laboratory will be essential to advancing knowledge on these cells in pulmonary health and disease.

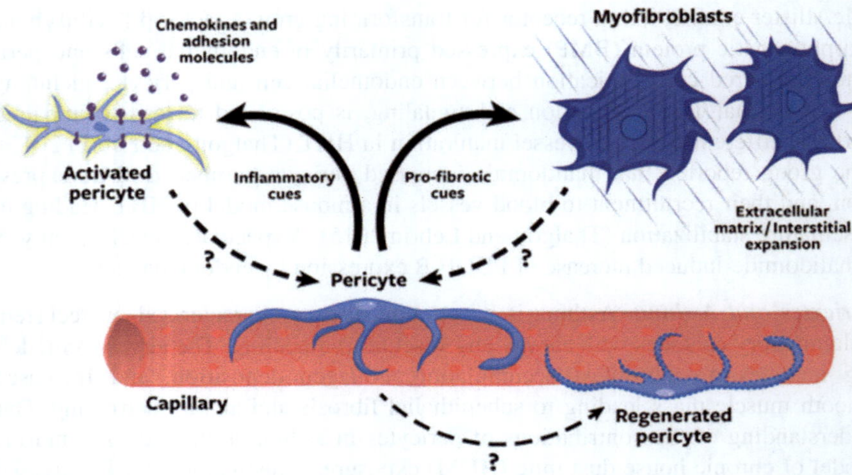

Fig. 3.3 Schematic of potential roles of pericytes in the lung

Acknowledgments The authors thank all the members of the Schnapp and Hung laboratories who have contributed to studies described in this chapter. We also thank Seth Bollenbecker for providing the artwork in Fig. 3.3.

References

Akamatsu T, Arai Y, Kosugi I, Kawasaki H, Meguro S, Sakao M, Shibata K, Suda T, Chida K, Iwashita T (2013) Direct isolation of myofibroblasts and fibroblasts from bleomycin-injured lungs reveals their functional similarities and differences. Fibrogenesis Tissue Repair 6(1):15. https://doi.org/10.1186/1755-1536-6-15. PubMed PMID: 23927729; PMCID: PMC3751789

Armulik A, Genove G, Betsholtz C (2011) Pericytes: developmental, physiological, and pathological perspectives, problems, and promises. Dev Cell 21(2):193–215. https://doi.org/10.1016/j.devcel.2011.07.001. PubMed PMID: 21839917

Bagley RG, Weber W, Rouleau C, Teicher BA (2005) Pericytes and endothelial precursor cells: cellular interactions and contributions to malignancy. Cancer Res 65(21):9741–9750. https://doi.org/10.1158/0008-5472.CAN-04-4337. PubMed PMID: 16266995

Bagley RG, Rouleau C, Morgenbesser SD, Weber W, Cook BP, Shankara S, Madden SL, Teicher BA (2006) Pericytes from human non-small cell lung carcinomas: an attractive target for anti-angiogenic therapy. Microvasc Res 71(3):163–174. https://doi.org/10.1016/j.mvr.2006.03.002. Epub 2006/04/19. PubMed PMID: 16624341

Bardin N, Anfosso F, Masse JM, Cramer E, Sabatier F, Le Bivic A, Sampol J, Dignat-George F (2001) Identification of CD146 as a component of the endothelial junction involved in the control of cell-cell cohesion. Blood 98(13):3677–3684. PubMed PMID: 11739172

Barkauskas CE, Cronce MJ, Rackley CR, Bowie EJ, Keene DR, Stripp BR, Randell SH, Noble PW, Hogan BL (2013) Type 2 alveolar cells are stem cells in adult lung. J Clin Invest 123(7):3025–3036. https://doi.org/10.1172/JCI68782. Epub 2013/08/08. PubMed PMID: 23921127; PMCID: PMC3696553

Bichsel CA, Hall SR, Schmid RA, Guenat OT, Geiser T (2015) Primary human lung pericytes support and stabilize in vitro perfusable microvessels. Tissue Eng Part A 21(15–16):2166–2176. https://doi.org/10.1089/ten.TEA.2014.0545. PubMed PMID: 25891384

Chow K, Fessel JP, Kaoriihida S, Schmidt EP, Gaskill C, Alvarez D, Graham B, Harrison DG, Wagner DH Jr, Nozik-Grayck E, West JD, Klemm DJ, Majka SM (2013) Dysfunctional resident lung mesenchymal stem cells contribute to pulmonary microvascular remodeling. Pulm Circ 3(1):31–49. https://doi.org/10.4103/2045-8932.109912. PubMed PMID: 23662173; PMCID: PMC3641738

Crisan M, Yap S, Casteilla L, Chen CW, Corselli M, Park TS, Andriolo G, Sun B, Zheng B, Zhang L, Norotte C, Teng PN, Traas J, Schugar R, Deasy BM, Badylak S, Buhring HJ, Giacobino JP, Lazzari L, Huard J, Peault B (2008) A perivascular origin for mesenchymal stem cells in multiple human organs. Cell Stem Cell 3(3):301–313. https://doi.org/10.1016/j.stem.2008.07.003. PubMed PMID: 18786417

Epling GP (1966) Electron microscopic observations of pericytes of small blood vessels in the lungs and hearts of normal cattle and swine. Anat Rec 155:513–530

Gaskill CF, Carrier EJ, Kropski JA, Bloodworth NC, Menon S, Foronjy RF, Taketo MM, Hong CC, Austin ED, West JD, Means AL, Loyd JE, Merryman WD, Hemnes AR, De Langhe S, Blackwell TS, Klemm DJ, Majka SM (2017) Disruption of lineage specification in adult pulmonary mesenchymal progenitor cells promotes microvascular dysfunction. J Clin Invest 127(6):2262–2276. https://doi.org/10.1172/JCI88629. Epub 2017/05/04. PubMed PMID: 28463231; PMCID: PMC5451236

Guimaraes-Camboa N, Cattaneo P, Sun Y, Moore-Morris T, Gu Y, Dalton ND, Rockenstein E, Masliah E, Peterson KL, Stallcup WB, Chen J, Evans SM (2017) Pericytes of multiple organs do not behave as mesenchymal stem cells in vivo. Cell Stem Cell 20(3):345–359.e5. https://doi.org/10.1016/j.stem.2016.12.006. PubMed PMID: 28111199; PMCID: PMC5337131

Gushi A, Tanaka M, Tsuyama S, Nagai T, Kanzaki T, Kanekura T, Matsuyama T (2008) The 3G5 antigen is expressed in dermal mast cells but not pericytes. J Cutan Pathol 35(3):278–284. https://doi.org/10.1111/j.1600-0560.2007.00809.x. PubMed PMID: 18251741

Hellstrom M, Kalen M, Lindahl P, Abramsson A, Betsholtz C (1999) Role of PDGF-B and PDGFR-beta in recruitment of vascular smooth muscle cells and pericytes during embryonic blood vessel formation in the mouse. Development 126(14):3047–3055. PubMed PMID: 10375497

Helmbold P, Nayak RC, Marsch WC, Herman IM (2001a) Isolation and in vitro characterization of human dermal microvascular pericytes. Microvasc Res 61(2):160–165. https://doi.org/10.1006/mvre.2000.2292. PubMed PMID: 11254395

Helmbold P, Wohlrab J, Marsch WC, Nayak RC (2001b) Human dermal pericytes express 3G5 ganglioside—a new approach for microvessel histology in the skin. J Cutan Pathol 28(4):206–210. PubMed PMID: 11426828

Hemnes AR, Humbert M (2017) Pathobiology of pulmonary arterial hypertension: understanding the roads less travelled. Eur Respir Rev 26(146). https://doi.org/10.1183/16000617.0093-2017

Henderson NC, Arnold TD, Katamura Y, Giacomini MM, Rodriguez JD, McCarty JH, Pellicoro A, Raschperger E, Betsholtz C, Ruminski PG, Griggs DW, Prinsen MJ, Maher JJ, Iredale JP, Lacy-Hulbert A, Adams RH, Sheppard D (2013) Targeting of alphav integrin identifies a core molecular pathway that regulates fibrosis in several organs. Nat Med 19(12):1617–1624. https://doi.org/10.1038/nm.3282. PubMed PMID: 24216753; PMCID: PMC3855865

Humphreys BD, Lin SL, Kobayashi A, Hudson TE, Nowlin BT, Bonventre JV, Valerius MT, McMahon AP, Duffield JS (2010) Fate tracing reveals the pericyte and not epithelial origin of myofibroblasts in kidney fibrosis. Am J Pathol 176(1):85–97. https://doi.org/10.2353/ajpath.2010.090517. PubMed PMID: 20008127; PMCID: PMC2797872

Hung C, Linn G, Chow YH, Kobayashi A, Mittelsteadt K, Altemeier WA, Gharib SA, Schnapp LM, Duffield JS (2013) Role of lung pericytes and resident fibroblasts in the pathogenesis of pulmonary fibrosis. Am J Respir Crit Care Med 188(7):820–830. https://doi.org/10.1164/rccm.201212-2297OC. PubMed PMID: 23924232; PMCID: 3826269

Hung CF, Mittelsteadt KL, Brauer R, McKinney BL, Hallstrand TS, Parks WC, Chen P, Schnapp LM, Liles WC, Duffield JS, Altemeier WA (2017a) Lung pericyte-like cells are functional interstitial immune sentinel cells. Am J Physiol Lung Cell Mol Physiol 312(4):L556–L567. https://doi.org/10.1152/ajplung.00349.2016. PubMed PMID: 28188224; PMCID: PMC5407093

Hung CF, Chow YH, Liles WC, Altemeier WA, Schnapp LM (2017b) Ablation of Pericyte-like cells in lungs by oropharyngeal aspiration of diphtheria toxin. Am J Respir Cell Mol Biol 56(2):160–167. https://doi.org/10.1165/rcmb.2016-0083MA. PubMed PMID: 27779900

Hung CF, Wilson CL, Chow YH, Schnapp LM (2018) Role of integrin alpha8 in murine model of lung fibrosis. PLoS One 13(5):e0197937. https://doi.org/10.1371/journal.pone.0197937. Epub 2018/05/29. PubMed PMID: 29813125

Jansson D, Rustenhoven J, Feng S, Hurley D, Oldfield RL, Bergin PS, Mee EW, Faull RL, Dragunow M (2014) A role for human brain pericytes in neuroinflammation. J Neuroinflammation 11:104. https://doi.org/10.1186/1742-2094-11-104. PubMed PMID: 24920309; PMCID: PMC4105169

Johnson JR, Folestad E, Rowley JE, Noll EM, Walker SA, Lloyd CM, Rankin SM, Pietras K, Eriksson U, Fuxe J (2015) Pericytes contribute to airway remodeling in a mouse model of chronic allergic asthma. Am J Physiol Lung Cell Mol Physiol 308(7):L658–L671. https://doi.org/10.1152/ajplung.00286.2014. Epub 2015/02/01. PubMed PMID: 25637607; PMCID: PMC4385988

Jun D, Garat C, West J, Thorn N, Chow K, Cleaver T, Sullivan T, Torchia EC, Childs C, Shade T, Tadjali M, Lara A, Nozik-Grayck E, Malkoski S, Sorrentino B, Meyrick B, Klemm D, Rojas M, Wagner DH, Jr., Majka SM. The pathology of bleomycin-induced fibrosis is associated with loss of resident lung mesenchymal stem cells that regulate effector T-cell proliferation. Stem Cells 2011;29(4):725–735. doi: https://doi.org/10.1002/stem.604. Epub 2011/02/12. PubMed PMID: 21312316; PMCID: PMC3322548

Kato K, Dieguez-Hurtado R, Park DY, Hong SP, Kato-Azuma S, Adams S, Stehling M, Trappmann B, Wrana JL, Koh GY, Adams RH (2018) Pulmonary pericytes regulate lung morphogenesis. Nat Commun 9(1):2448. https://doi.org/10.1038/s41467-018-04913-2. PubMed PMID: 29934496; PMCID: PMC6015030

Kerr BA, West XZ, Kim YW, Zhao Y, Tischenko M, Cull RM, Phares TW, Peng XD, Bernier-Latmani J, Petrova TV, Adams RH, Hay N, Naga Prasad SV, Byzova TV (2016) Stability and function of adult vasculature is sustained by Akt/Jagged1 signalling axis in endothelium. Nat Commun 7:10960. https://doi.org/10.1038/ncomms10960. Epub 2016/03/15. PubMed PMID: 26971877; PMCID: PMC4793084

Kramann R, Schneider RK, DiRocco DP, Machado F, Fleig S, Bondzie PA, Henderson JM, Ebert BL, Humphreys BD (2015) Perivascular Gli1+ progenitors are key contributors to injury-induced organ fibrosis. Cell Stem Cell 16(1):51–66. https://doi.org/10.1016/j.stem.2014.11.004. PubMed PMID: 25465115; PMCID: PMC4289444

Lin SL, Kisseleva T, Brenner DA, Duffield JS (2008) Pericytes and perivascular fibroblasts are the primary source of collagen-producing cells in obstructive fibrosis of the kidney. Am J Pathol 173(6):1617–1627. https://doi.org/10.2353/ajpath.2008.080433. ajpath.2008.080433 [pii]. Epub 2008/11/15. PubMed PMID: 19008372; PMCID: 2626374

Marriott S, Baskir RS, Gaskill C, Menon S, Carrier EJ, Williams J, Talati M, Helm K, Alford CE, Kropski JA, Loyd J, Wheeler L, Johnson J, Austin E, Nozik-Grayck E, Meyrick B, West JD, Klemm DJ, Majka SM (2014) ABCG2pos lung mesenchymal stem cells are a novel pericyte subpopulation that contributes to fibrotic remodeling. Am J Physiol Cell Physiol 307(8):C684–C698. https://doi.org/10.1152/ajpcell.00114.2014. PubMed PMID: 25122876; PMCID: PMC4200000

McAllister KA, Grogg KM, Johnson DW, Gallione CJ, Baldwin MA, Jackson CE, Helmbold EA, Markel DS, McKinnon WC, Murrell J et al (1994) Endoglin, a TGF-beta binding protein of endothelial cells, is the gene for hereditary haemorrhagic telangiectasia type 1. Nat Genet 8(4):345–351. https://doi.org/10.1038/ng1294-345. PubMed PMID: 7894484

Mederacke I, Hsu CC, Troeger JS, Huebener P, Mu X, Dapito DH, Pradere JP, Schwabe RF (2013) Fate tracing reveals hepatic stellate cells as dominant contributors to liver fibrosis independent of its aetiology. Nat Commun 4:2823. https://doi.org/10.1038/ncomms3823. Epub 2013/11/23. PubMed PMID: 24264436; PMCID: PMC4059406

Murfee WL, Skalak TC, Peirce SM (2005) Differential arterial/venous expression of NG2 proteoglycan in perivascular cells along microvessels: identifying a venule-specific phenotype. Microcirculation 12(2):151–160. https://doi.org/10.1080/10739680590904955. PubMed PMID: 15824037

Murfee WL, Rehorn MR, Peirce SM, Skalak TC (2006) Perivascular cells along venules upregulate NG2 expression during microvascular remodeling. Microcirculation 13(3):261–273. https://doi.org/10.1080/10739680600559153. PubMed PMID: 16627368

Nehls V, Drenckhahn D (1991) Heterogeneity of microvascular pericytes for smooth muscle type alpha-actin. J Cell Biol 113(1):147–154. PubMed PMID: 2007619; PMCID: PMC2288926

Ogura S, Kurata K, Hattori Y, Takase H, Ishiguro-Oonuma T, Hwang Y, Ahn S, Park I, Ikeda W, Kusuhara S, Fukushima Y, Nara H, Sakai H, Fujiwara T, Matsushita J, Ema M, Hirashima M, Minami T, Shibuya M, Takakura N, Kim P, Miyata T, Ogura Y, Uemura A (2017) Sustained inflammation after pericyte depletion induces irreversible blood-retina barrier breakdown. JCI Insight 2(3):e90905. https://doi.org/10.1172/jci.insight.90905. Epub 2017/02/15. PubMed PMID: 28194443; PMCID: PMC5291729 employees of Astellas Pharma Inc.

Patel MS, Taylor GP, Bharya S, Al-Sanna'a N, Adatia I, Chitayat D, Lewis MES, Human DG (2004) Abnormal pericyte recruitment as a cause for pulmonary hypertension in Adams–Oliver syndrome. Am J Med Genet A 129A(3):294–299. https://doi.org/10.1002/ajmg.a.30221

Peng T, Tian Y, Boogerd CJ, Lu MM, Kadzik RS, Stewart KM, Evans SM, Morrisey EE (2013) Coordination of heart and lung co-development by a multipotent cardiopulmonary progenitor. Nature 500(7464):589–592. https://doi.org/10.1038/nature12358. PubMed PMID: 23873040; PMCID: PMC3758448

Que J, Wilm B, Hasegawa H, Wang F, Bader D, Hogan BL (2008) Mesothelium contributes to vascular smooth muscle and mesenchyme during lung development. Proc Natl Acad Sci U S A 105(43):16626–16630. https://doi.org/10.1073/pnas.0808649105. PubMed PMID: 18922767; PMCID: PMC2567908

Ricard N, Tu L, Le Hiress M, Huertas A, Phan C, Thuillet R, Sattler C, Fadel E, Seferian A, Montani D, Dorfmuller P, Humbert M, Guignabert C (2014) Increased pericyte coverage mediated by endothelial-derived fibroblast growth factor-2 and interleukin-6 is a source of smooth muscle-like cells in pulmonary hypertension. Circulation 129(15):1586–1597. https://doi.org/10.1161/CIRCULATIONAHA.113.007469. PubMed PMID: 24481949

Rock JR, Barkauskas CE, Cronce MJ, Xue Y, Harris JR, Liang J, Noble PW, Hogan BL (2011) Multiple stromal populations contribute to pulmonary fibrosis without evidence for epithelial to mesenchymal transition. Proc Natl Acad Sci U S A 108(52):E1475–E1483. https://doi.org/10.1073/pnas.1117988108. PubMed PMID: 22123957; PMCID: PMC3248478

Rustenhoven J, Jansson D, Smyth LC, Dragunow M (2017) Brain pericytes as mediators of neuroinflammation. Trends Pharmacol Sci 38(3):291–304. https://doi.org/10.1016/j.tips.2016.12.001. Epub 2016/12/27. PubMed PMID: 28017362

Sava P, Ramanathan A, Dobronyi A, Peng X, Sun H, Ledesma-Mendoza A, Herzog EL, Gonzalez AL (2017) Human pericytes adopt myofibroblast properties in the microenvironment of the IPF lung. JCI Insight. 2(24). https://doi.org/10.1172/jci.insight.96352. PubMed PMID: 29263297; PMCID: PMC5752282

Shepro D, Morel NM (1993) Pericyte physiology. FASEB J 7(11):1031–1038. PubMed PMID: 8370472

Sims DE, Westfall JA (1983) Analysis of relationships between pericytes and gas exchange capillaries in neonatal and mature bovine lungs. Microvasc Res 25(3):333–342. PubMed PMID: 6855632

Sims-Lucas S, Schaefer C, Bushnell D, Ho J, Logar A, Prochownik E, Gittes G, Bates CM (2013) Endothelial progenitors exist within the kidney and lung mesenchyme. PLoS One 8(6):e65993. https://doi.org/10.1371/journal.pone.0065993. PubMed PMID: 23823180; PMCID: PMC3688860

Thalgott J, Lebrin F (2015) Pericytes as targets in hereditary hemorrhagic telangiectasia. Front Genet 6:37

Valdez CN, Arboleda-Velasquez JF, Amarnani DS, Kim LA, D'Amore PA (2014) Retinal microangiopathy in a mouse model of inducible mural cell loss. Am J Pathol 184(10):2618–2626. https://doi.org/10.1016/j.ajpath.2014.06.011. S0002-9440(14)00366-6 [pii]. Epub 2014/08/06. PubMed PMID: 25092275

von Tell D, Armulik A, Betsholtz C (2006) Pericytes and vascular stability. Exp Cell Res 312(5):623–629. https://doi.org/10.1016/j.yexcr.2005.10.019. PubMed PMID: 16303125

Weibel ER (1974) On pericytes, particularly their existence on lung capillaries. Microvasc Res 8(2):218–235. PubMed PMID: 4140459

Wilson CL, Stephenson SE, Higuero JP, Feghali-Bostwick C, Hung CF, Schnapp LM (2018) Characterization of human PDGFRβ-positive pericytes from IPF and non-IPF lungs. Am J Physiol Lung Cell Mol Physiol. https://doi.org/10.1152/ajplung.00289.2018

Yata Y, Scanga A, Gillan A, Yang L, Reif S, Breindl M, Brenner DA, Rippe RA (2003) DNase I-hypersensitive sites enhance alpha1(I) collagen gene expression in hepatic stellate cells. Hepatology 37(2):267–276. https://doi.org/10.1053/jhep.2003.50067. PubMed PMID: 12540776

Yuan K, Orcholski ME, Panaroni C, Shuffle EM, Huang NF, Jiang X, Tian W, Vladar EK, Wang L, Nicolls MR, Wu JY, de Jesus Perez VA (2015) Activation of the Wnt/planar cell polarity pathway is required for pericyte recruitment during pulmonary angiogenesis. Am J Pathol 185(1):69–84. https://doi.org/10.1016/j.ajpath.2014.09.013. PubMed PMID: 25447046; PMCID: PMC4278244

Chapter 4
Pericytes in Skeletal Muscle

Jyoti Gautam and Yao Yao

Abstract Skeletal muscle regeneration is a highly orchestrated process and involves the activation of many cellular and molecular pathways. Although satellite cells (SCs) are the major cell type responsible for muscle regeneration, pericytes show remarkable myogenic potential and various advantages as cell therapy in muscular disorders. This chapter first introduces the structure, marker expression, origin, and category of pericytes. Next, we discuss their functions in muscular dystrophy and/or muscle injuries, focusing on their myogenic, adipogenic, fibrogenic, chondrogenic, and osteogenic activities. Understanding this knowledge will promote the development of innovative cell therapies for muscle disorders, including muscular dystrophy.

Keywords Pericytes · Satellite cells · Mesoangioblasts · Myogenesis · Adipogenesis · Fibrosis · Chondrogenesis · Osteogenesis · Ossification · Muscle injury · Muscular dystrophy · Muscle regeneration · Differentiation

4.1 Introduction

Skeletal muscle, which constitutes around 40% of body weight in humans (Yin et al. 2013), is composed of a heterogeneous collection of muscle fibers. During development, myofibers are formed via fusion of mesoderm progenitors called myoblasts (Yablonka-Reuveni 2011). In neonatal/juvenile stages, the number of myofibers remains constant, but each myofiber grows in size through fusion with SCs and postnatal muscle stem/progenitor cells (Yin et al. 2013; Yablonka-Reuveni 2011).

J. Gautam • Y. Yao (✉)
Department of Pharmaceutical and Biomedical Sciences, University of Georgia, Athens, GA, USA
e-mail: yyao@uga.edu

A. Birbrair (ed.), *Pericyte Biology in Different Organs*, Advances in Experimental Medicine and Biology 1122,
https://doi.org/10.1007/978-3-030-11093-2_4

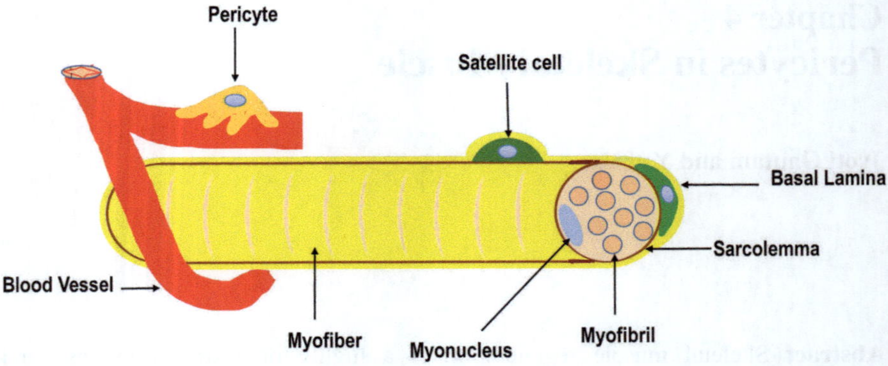

Fig. 4.1 Anatomical locations of pericytes and satellite cells in skeletal muscle. Pericytes are associated with capillaries in skeletal muscle, while satellite cells are located between the plasma membrane of myofiber (sarcolemma) and the basal lamina

SCs are located between the plasma membrane of myofiber (sarcolemma) and the basal lamina (Mauro 1961; Muir et al. 1965) (Fig. 4.1). These cells normally remain in the quiescent state but become activated in response to injury. Activated SCs proliferate and differentiate along the myogenic pathway to repair and/or replace damaged myofibers (Yin et al. 2013; Relaix and Zammit 2012; Dumont et al. 2015; Verdijk et al. 2014). As the primary myogenic cells in skeletal muscle, SCs have been tested for their therapeutic potential in a variety of muscular disorders, including muscular dystrophy. Although SCs are able to undergo myogenic differentiation and promote muscle regeneration (Collins et al. 2005; Montarras et al. 2005; Sacco et al. 2008; Boldrin et al. 2009), a few disadvantages limit their potential clinical use. First, cultured SC-derived myoblasts gradually lose their myogenic activity/regenerative property. Compared to freshly isolated SCs, cultured SCs induce regeneration with much lower efficiency after transplantation (Montarras et al. 2005; Gilbert et al. 2010; Hashimoto et al. 2004). Second, SCs have a low survival rate in vivo. It has been shown that less than 1% of SC-derived myoblasts survive the first day after transplantation (Beauchamp et al. 1999; Fan et al. 1996). Third, SCs are unable to cross the endothelial lining of blood vessels if given intravenously, which makes them only suitable for intramuscular delivery. Due to these drawbacks, alternative myogenic cells in skeletal muscle have been investigated.

In addition to SCs, many other cell populations in skeletal muscle have myogenic potential. These cells include pericytes, pericyte-like mesoangioblasts, muscle-derived stem cells, side population cells, and $CD133^+$ stem cells (Ten Broek et al. 2010; Asakura et al. 2002; Torrente et al. 2004; Galvez et al. 2006; Peault et al. 2007; Negroni et al. 2009; Dellavalle et al. 2007). Among them, we mainly focus on pericytes and pericyte-like mesoangioblasts owing to their unique features and promising therapeutic potential. Similar to SCs, pericytes demonstrate strong myogenic capability in vitro (Birbrair et al. 2013a, 2013b) and in vivo after muscle injury and/or during muscular dystrophy (Dellavalle et al. 2007, 2011; Birbrair et al. 2013b; Kostallari et al. 2015; Yao et al. 2016). Ablation of muscle pericytes has been shown to result in both myofiber hypotrophy and impaired establishment of stem cell quiescence (Kostallari et al. 2015). Unlike SCs, pericytes maintain high myogenic activity after in vitro amplification (Dellavalle et al. 2007; Costamagna et al. 2016; Quattrocelli et al. 2014; Bonfanti et al. 2015). In addition, pericytes can be delivered systemically. It has been shown that pericytes colonize host skeletal muscle and generate dystrophin-positive muscle fibers after intra-arterial injection into dystrophin-null mdx mice, a mouse model of Duchenne muscular dystrophy (DMD) (Dellavalle et al. 2007). Additionally, injected pericytes can also replenish the SC pool in dystrophic mice (Dellavalle et al. 2011). Furthermore, transplanted pericytes are incorporated into the vasculature and muscle and promote their regeneration in a limb ischemia model (Dar et al. 2012). These results suggest a great therapeutic potential of pericytes in muscular dystrophy and other muscle disorders. Here in this chapter, we first introduce pericyte physiology (structure, marker, origin, and category), followed by their functions (myogenesis, adipogenesis, fibrosis, chondrogenesis, osteogenesis, and others) in skeletal muscle.

4.2 Pericyte Physiology

4.2.1 Pericyte Structure and Density

Unlike SCs, pericytes are associated with capillaries in skeletal muscle (Cappellari and Cossu 2013; Attwell et al. 2016) (Fig. 4.1). At ultrastructural level, pericytes make numerous direct contacts with endothelial cells (Armulik et al. 2005, 2011; Sims 1986). It should be noted that the basement membrane is often absent at places where pericytes and endothelial cells are in close juxtaposition (Tilton et al. 1979).

Pericyte density varies greatly in different tissues/organs. The ratios of pericyte to endothelium have been estimated to be as high as 1:1 in the retina and CNS and as low as 1:100 in the capillary bed of skeletal muscle in the upper extremities (Armulik et al. 2005; Bodnar et al. 2016; Shepro and Morel 1993; Daneman and Prat 2015; von Tell et al. 2006). Quantitative analysis in rat and human skeletal muscle shows around 21–24% pericyte coverage/distribution around the capillaries (Sims 1986; Sims et al. 1994). The difference in pericyte coverage and density implies that pericytes may exert distinct functions in different tissues/organs. In skeletal muscle, pericytes have been shown to contribute to muscle development/homeostasis and muscle regeneration after injury.

4.2.2 Pericyte Markers

Many cellular markers have been used to identify pericytes. The most commonly used ones include neural-glial antigen 2 (NG2), platelet-derived growth factor receptor β (PDGFRβ), smooth muscle α-actin (α-SMA), desmin, CD13, regulator of G protein signaling 5 (RGS5), CD146, and nestin (Armulik et al. 2011; Birbrair et al. 2011; Kumar et al. 2017). In addition, alkaline phosphatase (AP) has also been reported as a pericyte marker in skeletal muscle (Dellavalle et al. 2007; Crisan et al. 2008). It should be noted, however, that none of these markers are pericyte-specific. They either also label other cells or only mark a subpopulation of pericytes. For example, NG2 is also expressed in macrophages and oligodendrocyte progenitor cells (Polito and Reynolds 2005; Moransard et al. 2011). PDGFRβ, α-SMA, and CD13 also label smooth muscle cells (Kumar et al. 2017; Hellstrom et al. 1999; Jin et al. 2008; Owens et al. 2004; Wang et al. 2015; Chen et al. 2016). Similarly, RGS5 expression is also found in smooth muscle cells and tumor-derived endothelial cells (Crisan et al. 2008; Silini et al. 2012), and CD146 is detected in endothelial cells (Chen et al. 2017). Nestin marks neural stem/progenitor cells and cancer stem cells, in addition to a subpopulation of pericytes (Birbrair et al. 2013a, 2011, 2014a; Neradil and Veselska 2015; Suzuki et al. 2010). Pericyte-specific markers are urgently needed and will significantly advance this field.

4.2.3 Pericyte Origins

Numerous lineage-tracing experiments show that pericytes have different embryonic origins in different organs (Armulik et al. 2011; Majesky 2007; Majesky et al. 2011). For example, pericytes in the CNS and thymus are derived from neural crest (Foster et al. 2008; Muller et al. 2008; Bergwerff et al. 1998; Etchevers et al. 2001; Korn et al. 2002), whereas pericytes in the gut, lung, and liver are derived from mesothelium (Wilm et al. 2005; Que et al. 2008; Asahina et al. 2011). Pericytes in the aorta have multiple developmental origins, including neural crest, somite, and secondary heart field (Armulik et al. 2011; Majesky 2007; Majesky et al. 2011). A recent study demonstrates that mesodermal mural cells originate from a clonal precursor mesenchymoangioblast (MB) (Kumar et al. 2017). These MBs differentiate into primitive $PDGFR\beta^{+}CD271^{+}CD73^{-}$ mesenchymal progenitors, which give rise to proliferative pericytes, smooth muscle cells, and mesenchymal stem/stromal cells (Kumar et al. 2017). MB-derived PCs can be further specified to $CD274^{+}$ capillary and $DLK1^{+}$ arteriolar PCs with a proinflammatory and contractile phenotype, respectively (Kumar et al. 2017). Furthermore, constitutive activation of Notch-1 has been shown to be critical for the conversion of mesoderm-derived somite cells and neural crest-derived frontonasal mesenchyme to a perivascular cell fate (Miller et al. 2017). The developmental origin of pericytes in skeletal muscle, however, remains not fully clear.

4.2.4 *Pericyte Subtypes*

As a heterogeneous population, various subtypes of pericytes have been identified. These subpopulations have been speculated to exert different functions. In earlier studies, pericytes have been categorized into three subtypes: precapillary, midcapillary, and postcapillary pericytes, based on their location and morphology (Krueger and Bechmann 2010; Nehls and Drenckhahn 1991). Precapillary pericytes have several circular branches, which tend to wrap around blood vessels. Midcapillary pericytes are spindle-shaped highly elongated cells, which extend mainly in the long axis of the vessels and have many short secondary processes. Postcapillary pericytes are shorter stellate-shaped cells that cover the abluminal surface of postcapillaries and postcapillary venules (Nehls and Drenckhahn 1991). These pericytes differ in their expression of α-SMA. Midcapillary pericytes show lack of α-SMA, while pre- and postcapillary pericytes are α-SMA$^+$ (Nehls and Drenckhahn 1991).

In addition, pericytes are also categorized into different subtypes according to their expression of cellular markers. For example, two subtypes of pericytes, named type 1 and type 2, have been described in skeletal muscle (Birbrair et al. 2013a; Majesky 2007). In the Nestin-GFP/NG2-DsRed double transgenic line, type-1 and type-2 pericytes are defined as Nestin-GFP$^-$/NG2-DsRed$^+$ and Nestin-GFP$^+$/NG2-DsRed$^+$ cells, respectively (Birbrair et al. 2013a, 2013b, 2014a, 2013c). It has been shown that type-1 pericytes generate α-SMA$^+$ classical pericytes, contributing to fat accumulation and fibrosis in skeletal muscle and other tissues (Birbrair et al. 2013a, 2013b, 2014a, 2013c). Type-2 pericytes, on the other hand, can give rise to either Tuj1$^+$ cells or α-SMA$^+$ classical pericytes and participate in muscle regeneration and normal angiogenesis (Birbrair et al. 2013a, 2013b, 2014b). Both type-1 and type-2 pericytes also express PDGFRβ and CD146 and are associated with capillaries (Birbrair et al. 2013a, 2013b).

4.3 Pericytes in Skeletal Muscle Injury and Muscular Dystrophy

4.3.1 *Myogenesis*

Skeletal muscle has the capability to regenerate itself in response to injury and/or degenerative diseases, such as muscular dystrophy. SCs, postnatal muscle progenitor cells, play an important role in this process. In addition, there is evidence showing that pericytes and pericyte-like cells also contribute to muscle regeneration/repair (Fig. 4.2). For example, like SCs, pericytes can express myogenic markers upon in vitro differentiation, although with different expression patterns and dynamics (Dellavalle et al. 2007). When transplanted into immunodeficient mdx mice, pericytes undergo myogenesis in vivo, increase dystrophin-positive fibers, and ameliorate muscle injury (Dellavalle et al. 2007). Next, it has been demonstrated that

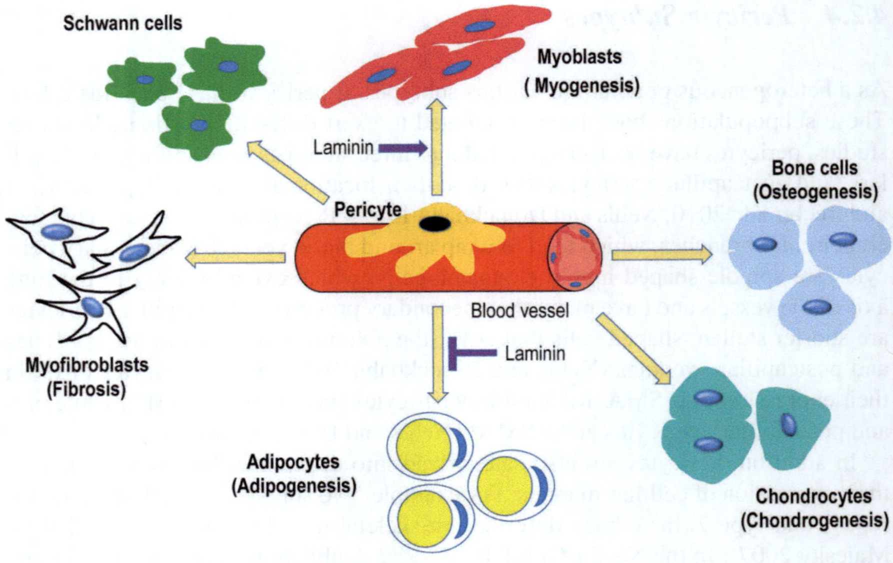

Fig. 4.2 Differentiation plasticity of skeletal muscle pericytes. In vitro and/or in vivo studies show that pericytes are able to differentiate into myoblasts, adipocytes, myofibroblasts, chondrocytes, osteoblasts, and possibly Schwann cells

AP^+ pericytes can generate SCs in early postnatal period and contribute to myogenesis in both cardiotoxin-induced acute muscle injury and chronic muscular dystrophy (Dellavalle et al. 2011). Consistent with these reports, we have demonstrated that skeletal muscle $PDGFR\beta^+$ pericytes can differentiate along either myogenic or adipogenic pathway depending on the availability of laminin signaling (Fig. 4.2), and that their myogenic differentiation contributes to muscle development and regeneration (Yao et al. 2016; Gautam et al. 2017). Further studies show that it is type-2 but not type-1 pericytes that have myogenic potential (Birbrair et al. 2013a, 2013b; Gautam et al. 2017). It has been reported that transplanted type-2 rather than type-1 pericytes undergo myogenesis, contributing to muscle regeneration in $BaCl_2$-induced muscle injury model (Birbrair et al. 2013b). In addition, intraperitoneal injection of human adipose tissue-derived pericytes significantly increased the life span of dystrophin/utrophin double knockout (mdx/utrn$^{-/-}$) mice (Valadares et al. 2014), another DMD model (Grady et al. 1997; Deconinck et al. 1997), suggesting a beneficial role of pericytes in DMD. It should be noted, however, that this beneficial effect is due to immune modulation rather than muscle regeneration.

Mesoangioblasts, multipotent cells that reside in the dorsal aorta of the developing embryo, are pericyte-like cells (Pierantozzi et al. 2016; Vezzani et al. 2016). Accumulating evidence shows that mesoangioblasts can improve muscle regeneration in various types of muscular dystrophy in mice, dogs, and humans (Galvez et al. 2006; Berry et al. 2007; Sampaolesi et al. 2003, 2006; Cossu et al. 2015). First, intra-arterial delivery of mesoangioblasts is able to reverse the dystrophic phenotype both morphologically and functionally in α-sarcoglycan-null mice, a limb-girdle dystrophy model (Galvez et al. 2006; Sampaolesi et al. 2003). Next, intra-arterial delivery of wild-type canine mesoangioblasts substantially enhances dystrophin expression and improves muscle morphology and function in dogs with golden retriever muscular dystrophy, which mimics the DMD in humans (Sampaolesi et al. 2006). Similar beneficial effects are observed in mdx/utrn$^{-/-}$ mice after administration of aorta-derived mesoangioblasts (Berry et al. 2007). Furthermore, the therapeutic effect of intra-arterial transplantation of mesoangioblasts in DMD has been tested in humans and found to be relatively safe and effective (Cossu et al. 2015).

4.3.2 Adipogenesis

Intramuscular adipose tissue (IMAT) deposition is the accumulation of adipocytes between muscle cells and beneath the muscle fascia (Pagano et al. 2015; Vettor et al. 2009). It is often seen after skeletal muscle injury or in muscular dystrophy (Lukjanenko et al. 2013) and has been shown to negatively affect skeletal muscle regeneration. Fibro/adipogenic progenitors (FAPs), muscle-resident mesenchymal stem cells that express platelet-derived growth factor receptor alpha (PDGFRα), are a major source of adipocytes in skeletal muscle (Uezumi et al. 2010; Motohashi et al. 2008). It has been reported that FAPs proliferate and differentiate into adipocytes in a DMD model, contributing to the accumulation of IMAT (Vettor et al. 2009; Takegahara et al. 2014), which correlates with a decrease in SC number (Heslop et al. 2000; Blau et al. 1983).

In addition to FAPs, pericytes also contribute to adipogenesis and fat accumulation in skeletal muscle (Fig. 4.2). For example, pericytes isolated from various tissues, including skeletal muscle, are able to undergo adipogenic differentiation in vitro (Crisan et al. 2008). Using lineage-tracing technique, we have demonstrated that PDGFRβ$^+$ pericytes are able to differentiate into perilipin$^+$ adipocytes in skeletal muscle in vivo in a congenital muscular dystrophy model (Yao et al. 2016). Further studies show that type-1 but not type-2 pericytes undergo adipogenesis, contributing to IMAT deposition and skeletal muscle fatty degeneration (Birbrair et al. 2013a, 2013b; Gautam et al. 2017).

4.3.3 Fibrosis

Fibrosis is a pathological condition characterized by excessive deposition of extracellular matrix and formation of fibrous connective tissue in an organ (Wynn 2008). It is observed in both pathological conditions and normal aging and generally referred to as "scarring" when induced in response to injury. Under chronic disease conditions, such as DMD, progressive fibrosis leads to skeletal muscle dysfunction

(Morales et al. 2013; Mann et al. 2011; Acuna et al. 2014). During normal aging, increased infiltration of fibrous tissue results in skeletal muscle stiffness, weakness, and atrophy (Ryall et al. 2008; Thompson 2009; Kragstrup et al. 2011; Walston 2012). Moreover, formation of fibrous tissue also affects the regeneration potential of skeletal muscle due to diminished innervation and contractile properties (Juhas and Bursac 2013; Birbrair et al. 2014c). The major cell type responsible for fibrosis is myofibroblasts, which are derived from a variety of cell types (Wynn 2008; Willis et al. 2006; Quan et al. 2006; Zeisberg et al. 2007; Duffield et al. 2013; Lin et al. 2008; Humphreys et al. 2010).

Recent studies suggest that pericytes may serve as a source of myofibroblasts during skeletal muscle fibrosis (Fig. 4.2). For example, ADAM12 (disintegrin and metalloproteinase domain-containing protein 12), which labels PDGFRβ$^+$/NG2$^+$ pericytes, has been shown to participate in the formation of fibrotic scar in skeletal muscle after acute injury (Dulauroy et al. 2012), suggesting that ADAM12 identifies a myofibroblast progenitor lineage with pericyte features. Additionally, type-1 but not type-2 pericytes are fibrogenic in vitro and in vivo and contribute to fibrous tissue deposition in skeletal muscle and many other organs (Birbrair et al. 2014a, 2013c). Furthermore, GLAST$^+$ pericytes have been found to participate in the formation of scar tissue in a spinal cord injury model, although GLAST does not label pericytes specifically (Goritz et al. 2011).

4.3.4 Chondrogenesis and Osteogenesis/Ossification

Heterotopic ossification is the formation of lamellar bone in non-osseous tissues, such as skeletal muscle, after traumatic injury or in pathological conditions. The general process of ossification involves an initial formation of ectopic cartilage and a subsequent formation of endochondral bone (Shimono et al. 2013). Skeletal muscle ossification has been observed following muscle injury due to aberrant activation of the chondrogenic and osteogenic pathways (Shimono et al. 2013; Bosse et al. 1994). There is evidence showing that pericytes have chondrogenic and osteogenic potential and contribute to skeletal muscle ossification (Fig. 4.2). First, osteoprogenitor cells isolated from human skeletal muscles express high levels of pericyte markers, including AP (Levy et al. 2001). Next, pericytes (CD146$^+$CD34$^-$CD45$^-$CD56$^-$AP$^+$NG2$^+$PDGFRβ$^+$ cells) isolated from human adult skeletal muscle show chondrogenic and osteogenic potential in vitro (Crisan et al. 2008). Like skeletal muscle pericytes, pericytes from retina also have chondrogenic potential in vitro (Farrington-Rock et al. 2004). Consistent with these reports, various studies have demonstrated that pericytes can undergo chondrogenic and osteogenic differentiation in vitro (Birbrair et al. 2013b, 2014c; Zhang et al. 2011; James et al. 2012). Whether pericytes contribute to skeletal muscle ossification in vivo, however, remains unknown.

Ossification of fibrous tissue, including skeletal muscle, is observed in fibrodysplasia ossificans progressiva (FOP) (Ramirez et al. 2014; Shore and Kaplan 2010), a genetic disorder caused by an autosomal dominant mutation of BMP receptor—activin-like kinase 2 (ALK2) (Shore et al. 2006). These findings suggest that ALK2 ligands (e.g., BMPs) may be primary inducers of heterotopic ossification. Consistent with this speculation, BMP-2, BMP-4, and BMP-9 are highly expressed in human lesions with heterotopic ossification (Gannon et al. 1997; Grenier et al. 2013) and promote osteogenesis when mixed with Matrigel and injected into skeletal muscle (Shimono et al. 2013; Chen et al. 2012; Nishimura et al. 2012). In addition, we have demonstrated that type-1 pericytes express both BMP-2 and BMP-4 in vitro (Gautam et al. 2017), again suggesting a possible role of BMP signaling in pericyte osteogenic differentiation.

4.3.5 Nerve Injury

Motor nerves innervate skeletal muscle (Carlson 1981). It has been shown that nerve innervation is essential for the mass and function of skeletal muscle (Delbono 2003, 2011). Upon denervation, which usually results from aging, trauma, or immobility, skeletal muscle undergoes reinnervation. Although the molecular mechanism underlying reinnervation remains largely unclear, Schwann cells, which cover nerve terminal branches with their processes, have been shown to play a critical role in this process. It has been shown that Schwann cells from denervated synaptic sites form a "bridge" to guide the growth of nerve terminal sprouts (Love and Thompson 1999; Sugiura and Lin 2011). There is evidence supporting that pericytes may contribute to skeletal muscle reinnervation via differentiating into myelinating Schwann cells (Fig. 4.2). For example, under optimized skeletal muscle culture conditions, type-2 pericytes are able to differentiate into oligodendrocyte progenitor cells (Birbrair et al. 2013a), which generate mature oligodendrocytes and Schwann cells (Zawadzka et al. 2010).

4.4 Conclusions

Pericytes are an alternative stem/progenitor population with myogenic potential in skeletal muscle. Recent studies show that pericytes actively affect muscle regeneration after injury or in muscular dystrophy through their myogenic and non-myogenic activities. It is hypothesized that enhancing pericyte myogenesis and inhibiting their non-myogenic differentiation have a therapeutic potential in muscle disorders. Future studies should focus on addressing the following key questions on pericytes. First, what determines the fate of pericytes in their differentiation (myogenesis, adipogenesis, fibrosis, osteogenesis, and others)? Second, what are the key signaling pathways that regulate pericyte differentiation and fate determination? Third, what is the difference between pericytes and other stem/progenitor cells in term of myogenic differentiation? Fourth, what are the molecular markers specific for pericytes and/or subtypes of pericytes? Fifth, do different subtypes of pericytes have distinct differentiation capability? Sixth, how are different pericyte subpopulations distributed in the vasculature in skeletal muscle? Answers to these questions will significantly enrich our knowledge in pericyte biology and promote the development of effective therapies for various muscle disorders, including muscular dystrophy.

References

Acuna MJ et al (2014) Restoration of muscle strength in dystrophic muscle by angiotensin-1-7 through inhibition of TGF-beta signalling. Hum Mol Genet 23(5):1237–1249

Armulik A, Abramsson A, Betsholtz C (2005) Endothelial/pericyte interactions. Circ Res 97(6):512–523

Armulik A, Genove G, Betsholtz C (2011) Pericytes: developmental, physiological, and pathological perspectives, problems, and promises. Dev Cell 21(2):193–215

Asahina K et al (2011) Septum transversum-derived mesothelium gives rise to hepatic stellate cells and perivascular mesenchymal cells in developing mouse liver. Hepatology 53(3):983–995

Asakura A et al (2002) Myogenic specification of side population cells in skeletal muscle. J Cell Biol 159(1):123–134

Attwell D et al (2016) What is a pericyte? J Cereb Blood Flow Metab 36(2):451–455

Beauchamp JR et al (1999) Dynamics of myoblast transplantation reveal a discrete minority of precursors with stem cell-like properties as the myogenic source. J Cell Biol 144(6):1113–1122

Bergwerff M et al (1998) Neural crest cell contribution to the developing circulatory system: implications for vascular morphology? Circ Res 82(2):221–231

Berry SE et al (2007) Multipotential mesoangioblast stem cell therapy in the mdx/utrn−/− mouse model for Duchenne muscular dystrophy. Regen Med 2(3):275–288

Birbrair A et al (2011) Nestin-GFP transgene reveals neural precursor cells in adult skeletal muscle. PLoS One 6(2):e16816

Birbrair A et al (2013a) Skeletal muscle pericyte subtypes differ in their differentiation potential. Stem Cell Res 10(1):67–84

Birbrair A et al (2013b) Role of pericytes in skeletal muscle regeneration and fat accumulation. Stem Cells Dev 22(16):2298–2314

Birbrair A et al (2013c) Type-1 pericytes participate in fibrous tissue deposition in aged skeletal muscle. Am J Physiol Cell Physiol 305(11):C1098–C1113

Birbrair A et al (2014a) Type-1 pericytes accumulate after tissue injury and produce collagen in an organ-dependent manner. Stem Cell Res Ther 5(6):122

Birbrair A et al (2014b) Type-2 pericytes participate in normal and tumoral angiogenesis. Am J Physiol Cell Physiol 307(1):C25–C38

Birbrair A et al (2014c) Pericytes: multitasking cells in the regeneration of injured, diseased, and aged skeletal muscle. Front Aging Neurosci 6:245

Blau HM, Webster C, Pavlath GK (1983) Defective myoblasts identified in Duchenne muscular dystrophy. Proc Natl Acad Sci U S A 80(15):4856–4860

Bodnar RJ et al (2016) Pericytes: a newly recognized player in wound healing. Wound Repair Regen 24(2):204–214

Boldrin L et al (2009) Mature adult dystrophic mouse muscle environment does not impede efficient engrafted satellite cell regeneration and self-renewal. Stem Cells 27(10):2478–2487

Bonfanti C et al (2015) PW1/Peg3 expression regulates key properties that determine mesoangioblast stem cell competence. Nat Commun 6:6364

Bosse A et al (1994) Collagens and growth factors in heterotopic ossification. Pathologe 15(4):216–225

Cappellari O, Cossu G (2013) Pericytes in development and pathology of skeletal muscle. Circ Res 113(3):341–347
Carlson BM (1981) Denervation, reinnervation, and regeneration of skeletal muscle. Otolaryngol Head Neck Surg 89(2):192–196
Chen G, Deng C, Li YP (2012) TGF-beta and BMP signaling in osteoblast differentiation and bone formation. Int J Biol Sci 8(2):272–288
Chen L et al (2016) Smooth muscle-alpha actin inhibits vascular smooth muscle cell proliferation and migration by inhibiting Rac1 activity. PLoS One 11(5):e0155726
Chen J et al (2017) CD146 coordinates brain endothelial cell-pericyte communication for blood-brain barrier development. Proc Natl Acad Sci U S A 114(36):E7622–E7631
Collins CA et al (2005) Stem cell function, self-renewal, and behavioral heterogeneity of cells from the adult muscle satellite cell niche. Cell 122(2):289–301
Cossu G et al (2015) Intra-arterial transplantation of HLA-matched donor mesoangioblasts in Duchenne muscular dystrophy. EMBO Mol Med 7(12):1513–1528
Costamagna D et al (2016) Smad1/5/8 are myogenic regulators of murine and human mesoangioblasts. J Mol Cell Biol 8(1):73–87
Crisan M et al (2008) A perivascular origin for mesenchymal stem cells in multiple human organs. Cell Stem Cell 3(3):301–313
Daneman R, Prat A (2015) The blood-brain barrier. Cold Spring Harb Perspect Biol 7(1):a020412
Dar A et al (2012) Multipotent vasculogenic pericytes from human pluripotent stem cells promote recovery of murine ischemic limb. Circulation 125(1):87–99
Deconinck AE et al (1997) Utrophin-dystrophin-deficient mice as a model for Duchenne muscular dystrophy. Cell 90(4):717–727
Delbono O (2003) Neural control of aging skeletal muscle. Aging Cell 2(1):21–29
Delbono O (2011) Expression and regulation of excitation-contraction coupling proteins in aging skeletal muscle. Curr Aging Sci 4(3):248–259
Dellavalle A et al (2007) Pericytes of human skeletal muscle are myogenic precursors distinct from satellite cells. Nat Cell Biol 9(3):255–267
Dellavalle A et al (2011) Pericytes resident in postnatal skeletal muscle differentiate into muscle fibres and generate satellite cells. Nat Commun 2:499
Duffield JS et al (2013) Host responses in tissue repair and fibrosis. Annu Rev Pathol 8:241–276
Dulauroy S et al (2012) Lineage tracing and genetic ablation of ADAM12(+) perivascular cells identify a major source of profibrotic cells during acute tissue injury. Nat Med 18(8):1262–1270
Dumont NA et al (2015) Satellite cells and skeletal muscle regeneration. Compr Physiol 5(3):1027–1059
Etchevers HC et al (2001) The cephalic neural crest provides pericytes and smooth muscle cells to all blood vessels of the face and forebrain. Development 128(7):1059–1068
Fan Y et al (1996) Rapid death of injected myoblasts in myoblast transfer therapy. Muscle Nerve 19(7):853–860
Farrington-Rock C et al (2004) Chondrogenic and adipogenic potential of microvascular pericytes. Circulation 110(15):2226–2232
Foster K et al (2008) Contribution of neural crest-derived cells in the embryonic and adult thymus. J Immunol 180(5):3183–3189
Galvez BG et al (2006) Complete repair of dystrophic skeletal muscle by mesoangioblasts with enhanced migration ability. J Cell Biol 174(2):231–243
Gannon FH et al (1997) Bone morphogenetic protein 2/4 in early fibromatous lesions of fibrodysplasia ossificans progressiva. Hum Pathol 28(3):339–343
Gautam J, Nirwane A, Yao Y (2017) Laminin differentially regulates the stemness of type I and type II pericytes. Stem Cell Res Ther 8(1):28
Gilbert PM et al (2010) Substrate elasticity regulates skeletal muscle stem cell self-renewal in culture. Science 329(5995):1078–1081
Goritz C et al (2011) A pericyte origin of spinal cord scar tissue. Science 333(6039):238–242
Grady RM et al (1997) Skeletal and cardiac myopathies in mice lacking utrophin and dystrophin: a model for Duchenne muscular dystrophy. Cell 90(4):729–738

Grenier G et al (2013) BMP-9 expression in human traumatic heterotopic ossification: a case report. Skelet Muscle 3(1):29
Hashimoto N et al (2004) Muscle reconstitution by muscle satellite cell descendants with stem cell-like properties. Development 131(21):5481–5490
Hellstrom M et al (1999) Role of PDGF-B and PDGFR-beta in recruitment of vascular smooth muscle cells and pericytes during embryonic blood vessel formation in the mouse. Development 126(14):3047–3055
Heslop L, Morgan JE, Partridge TA (2000) Evidence for a myogenic stem cell that is exhausted in dystrophic muscle. J Cell Sci 113(Pt 12):2299–2308
Humphreys BD et al (2010) Fate tracing reveals the pericyte and not epithelial origin of myofibroblasts in kidney fibrosis. Am J Pathol 176(1):85–97
James AW et al (2012) Perivascular stem cells: a prospectively purified mesenchymal stem cell population for bone tissue engineering. Stem Cells Transl Med 1(6):510–519
Jin S et al (2008) Notch signaling regulates platelet-derived growth factor receptor-beta expression in vascular smooth muscle cells. Circ Res 102(12):1483–1491
Juhas M, Bursac N (2013) Engineering skeletal muscle repair. Curr Opin Biotechnol 24(5):880–886
Korn J, Christ B, Kurz H (2002) Neuroectodermal origin of brain pericytes and vascular smooth muscle cells. J Comp Neurol 442(1):78–88
Kostallari E et al (2015) Pericytes in the myovascular niche promote post-natal myofiber growth and satellite cell quiescence. Development 142(7):1242–1253
Kragstrup TW, Kjaer M, Mackey AL (2011) Structural, biochemical, cellular, and functional changes in skeletal muscle extracellular matrix with aging. Scand J Med Sci Sports 21(6):749–757
Krueger M, Bechmann I (2010) CNS pericytes: concepts, misconceptions, and a way out. Glia 58(1):1–10
Kumar A et al (2017) Specification and diversification of Pericytes and smooth muscle cells from Mesenchymoangioblasts. Cell Rep 19(9):1902–1916
Levy MM et al (2001) Osteoprogenitor cells of mature human skeletal muscle tissue: an in vitro study. Bone 29(4):317–322
Lin SL et al (2008) Pericytes and perivascular fibroblasts are the primary source of collagen-producing cells in obstructive fibrosis of the kidney. Am J Pathol 173(6):1617–1627
Love FM, Thompson WJ (1999) Glial cells promote muscle reinnervation by responding to activity-dependent postsynaptic signals. J Neurosci 19(23):10390–10396
Lukjanenko L et al (2013) Genomic profiling reveals that transient adipogenic activation is a hallmark of mouse models of skeletal muscle regeneration. PLoS One 8(8):e71084
Majesky MW (2007) Developmental basis of vascular smooth muscle diversity. Arterioscler Thromb Vasc Biol 27(6):1248–1258
Majesky MW et al (2011) Vascular smooth muscle progenitor cells: building and repairing blood vessels. Circ Res 108(3):365–377
Mann CJ et al (2011) Aberrant repair and fibrosis development in skeletal muscle. Skelet Muscle 1(1):21
Mauro A (1961) Satellite cell of skeletal muscle fibers. J Biophys Biochem Cytol 9:493–495
Miller SR, Perera SN, Baker CV (2017) Constitutively active Notch1 converts cranial neural crest-derived frontonasal mesenchyme to perivascular cells in vivo. Biol Open 6(3):317–325
Montarras D et al (2005) Direct isolation of satellite cells for skeletal muscle regeneration. Science 309(5743):2064–2067
Morales MG et al (2013) Reducing CTGF/CCN2 slows down mdx muscle dystrophy and improves cell therapy. Hum Mol Genet 22(24):4938–4951
Moransard M et al (2011) NG2 expressed by macrophages and oligodendrocyte precursor cells is dispensable in experimental autoimmune encephalomyelitis. Brain 134(Pt 5):1315–1330
Motohashi N et al (2008) Muscle CD31(−) CD45(−) side population cells promote muscle regeneration by stimulating proliferation and migration of myoblasts. Am J Pathol 173(3):781–791
Muir AR, Kanji AH, Allbrook D (1965) The structure of the satellite cells in skeletal muscle. J Anat 99(Pt 3):435–444

Muller SM et al (2008) Neural crest origin of perivascular mesenchyme in the adult thymus. J Immunol 180(8):5344–5351

Negroni E et al (2009) In vivo myogenic potential of human CD133+ muscle-derived stem cells: a quantitative study. Mol Ther 17(10):1771–1778

Nehls V, Drenckhahn D (1991) Heterogeneity of microvascular pericytes for smooth muscle type alpha-actin. J Cell Biol 113(1):147–154

Neradil J, Veselska R (2015) Nestin as a marker of cancer stem cells. Cancer Sci 106(7):803–811

Nishimura R et al (2012) Regulation of bone and cartilage development by network between BMP signalling and transcription factors. J Biochem 151(3):247–254

Owens GK, Kumar MS, Wamhoff BR (2004) Molecular regulation of vascular smooth muscle cell differentiation in development and disease. Physiol Rev 84(3):767–801

Pagano AF et al (2015) Muscle regeneration with intermuscular adipose tissue (IMAT) accumulation is modulated by mechanical constraints. PLoS One 10(12):e0144230

Peault B et al (2007) Stem and progenitor cells in skeletal muscle development, maintenance, and therapy. Mol Ther 15(5):867–877

Pierantozzi E et al (2016) Tissue-specific cultured human Pericytes: perivascular cells from smooth muscle tissue have restricted mesodermal differentiation ability. Stem Cells Dev 25(9):674–686

Polito A, Reynolds R (2005) NG2-expressing cells as oligodendrocyte progenitors in the normal and demyelinated adult central nervous system. J Anat 207(6):707–716

Quan TE, Cowper SE, Bucala R (2006) The role of circulating fibrocytes in fibrosis. Curr Rheumatol Rep 8(2):145–150

Quattrocelli M et al (2014) Notch signaling regulates myogenic regenerative capacity of murine and human mesoangioblasts. Cell Death Dis 5:e1448

Que J et al (2008) Mesothelium contributes to vascular smooth muscle and mesenchyme during lung development. Proc Natl Acad Sci U S A 105(43):16626–16630

Ramirez DM et al (2014) Molecular and cellular mechanisms of heterotopic ossification. Histol Histopathol 29(10):1281–1285

Relaix F, Zammit PS (2012) Satellite cells are essential for skeletal muscle regeneration: the cell on the edge returns centre stage. Development 139(16):2845–2856

Ryall JG, Schertzer JD, Lynch GS (2008) Cellular and molecular mechanisms underlying age-related skeletal muscle wasting and weakness. Biogerontology 9(4):213–228

Sacco A et al (2008) Self-renewal and expansion of single transplanted muscle stem cells. Nature 456(7221):502–506

Sampaolesi M et al (2003) Cell therapy of alpha-sarcoglycan null dystrophic mice through intra-arterial delivery of mesoangioblasts. Science 301(5632):487–492

Sampaolesi M et al (2006) Mesoangioblast stem cells ameliorate muscle function in dystrophic dogs. Nature 444(7119):574–579

Shepro D, Morel NM (1993) Pericyte physiology. FASEB J 7(11):1031–1038

Shimono K et al (2013) The pathophysiology of heterotopic ossification: current treatment considerations in dentistry. Jpn Dent Sci Rev 50:1–8

Shore EM, Kaplan FS (2010) Inherited human diseases of heterotopic bone formation. Nat Rev Rheumatol 6(9):518–527

Shore EM et al (2006) A recurrent mutation in the BMP type I receptor ACVR1 causes inherited and sporadic fibrodysplasia ossificans progressiva. Nat Genet 38(5):525–527

Silini A et al (2012) Regulator of G-protein signaling 5 (RGS5) protein: a novel marker of cancer vasculature elicited and sustained by the tumor's proangiogenic microenvironment. Cell Mol Life Sci 69(7):1167–1178

Sims DE (1986) The pericyte—a review. Tissue Cell 18(2):153–174

Sims D et al (1994) Heterogeneity of pericyte populations in equine skeletal muscle and dermal microvessels: a quantitative study. Anat Histol Embryol 23(3):232–238

Sugiura Y, Lin W (2011) Neuron-glia interactions: the roles of Schwann cells in neuromuscular synapse formation and function. Biosci Rep 31(5):295–302

Suzuki S et al (2010) The neural stem/progenitor cell marker nestin is expressed in proliferative endothelial cells, but not in mature vasculature. J Histochem Cytochem 58(8):721–730

Takegahara Y et al (2014) Myotube formation is affected by adipogenic lineage cells in a cell-to-cell contact-independent manner. Exp Cell Res 324(1):105–114
Ten Broek RW, Grefte S, Von den Hoff JW (2010) Regulatory factors and cell populations involved in skeletal muscle regeneration. J Cell Physiol 224(1):7–16
Thompson LV (2009) Age-related muscle dysfunction. Exp Gerontol 44(1–2):106–111
Tilton RG, Kilo C, Williamson JR (1979) Pericyte-endothelial relationships in cardiac and skeletal muscle capillaries. Microvasc Res 18(3):325–335
Torrente Y et al (2004) Human circulating AC133(+) stem cells restore dystrophin expression and ameliorate function in dystrophic skeletal muscle. J Clin Invest 114(2):182–195
Uezumi A et al (2010) Mesenchymal progenitors distinct from satellite cells contribute to ectopic fat cell formation in skeletal muscle. Nat Cell Biol 12(2):143–152
Valadares MC et al (2014) Human adipose tissue derived pericytes increase life span in Utrn (tm1Ked) Dmd (mdx)/J mice. Stem Cell Rev 10(6):830–840
Verdijk LB et al (2014) Satellite cells in human skeletal muscle; from birth to old age. Age (Dordr) 36(2):545–547
Vettor R et al (2009) The origin of intermuscular adipose tissue and its pathophysiological implications. Am J Physiol Endocrinol Metab 297(5):E987–E998
Vezzani B, Pierantozzi E, Sorrentino V (2016) Not all pericytes are born equal: pericytes from human adult tissues present different differentiation properties. Stem Cells Dev 25(20)
von Tell D, Armulik A, Betsholtz C (2006) Pericytes and vascular stability. Exp Cell Res 312(5):623–629
Walston JD (2012) Sarcopenia in older adults. Curr Opin Rheumatol 24(6):623–627
Wang G et al (2015) Origin and differentiation of vascular smooth muscle cells. J Physiol 593(14):3013–3030
Willis BC, du Bois RM, Borok Z (2006) Epithelial origin of myofibroblasts during fibrosis in the lung. Proc Am Thorac Soc 3(4):377–382
Wilm B et al (2005) The serosal mesothelium is a major source of smooth muscle cells of the gut vasculature. Development 132(23):5317–5328
Wynn TA (2008) Cellular and molecular mechanisms of fibrosis. J Pathol 214(2):199–210
Yablonka-Reuveni Z (2011) The skeletal muscle satellite cell: still young and fascinating at 50. J Histochem Cytochem 59(12):1041–1059
Yao Y et al (2016) Laminin regulates PDGFRbeta(+) cell stemness and muscle development. Nat Commun 7:11415
Yin H, Price F, Rudnicki MA (2013) Satellite cells and the muscle stem cell niche. Physiol Rev 93(1):23–67
Zawadzka M et al (2010) CNS-resident glial progenitor/stem cells produce Schwann cells as well as oligodendrocytes during repair of CNS demyelination. Cell Stem Cell 6(6):578–590
Zeisberg EM et al (2007) Endothelial-to-mesenchymal transition contributes to cardiac fibrosis. Nat Med 13(8):952–961
Zhang X et al (2011) The Nell-1 growth factor stimulates bone formation by purified human perivascular cells. Tissue Eng Part A 17(19–20):2497–2509

Chapter 5
Pericytes in the Gut

Marta Ramirez, Nuria Pell, Marc Mejias, and Mercedes Fernandez

Abstract This review chapter describes the current knowledge about the nature of pericytes in the gut, their interaction with endothelial cells in blood vessels, and their pathophysiological functions in the setting of chronic liver disease. In particular, it focuses on the role of these vascular cell types and related molecular signaling pathways in pathological angiogenesis associated with liver disease and in the establishment of the gut-vascular barrier and the potential implications in liver disease through the gut-liver axis.

Keywords Chronic liver disease · Pathological angiogenesis · Gut-vascular barrier · Portosystemic collateral vessels · Mesenteric vascular bed · Intestinal circulation

5.1 Introduction

This chapter reviews our current understanding of the role of pericytes and endothelial cells in the gut vasculature during homeostasis and in disease. Specifically, we highlight the function and importance of these cell types as key components of the major vascular beds involved in the pathophysiology of chronic liver disease, which are leading causes of death and liver transplantation worldwide. These vascular beds

---

Marta Ramirez and Nuria Pell are Co-first authors.

M. Ramirez · N. Pell
Angiogenesis in Liver Disease Research Group, IDIBAPS Biomedical Research Institute, Hospital Clinic, University of Barcelona, Barcelona, Spain

M. Mejias · M. Fernandez (✉)
Angiogenesis in Liver Disease Research Group, IDIBAPS Biomedical Research Institute, Hospital Clinic, University of Barcelona, Barcelona, Spain

Biomedical Research Networking Center on Hepatic and Digestive Disease (CIBEREHD), Spanish National Institute of Health, Barcelona, Spain
e-mail: mercefernandez@ub.edu; https://www.mercedesfernandezlab.com

A. Birbrair (ed.), *Pericyte Biology in Different Organs*,
Advances in Experimental Medicine and Biology 1122,
https://doi.org/10.1007/978-3-030-11093-2_5

include not only the intrahepatic circulation but also the intestinal and mesenteric microvasculature, as well as the portosystemic collaterals and gastroesophageal varices. All these vessels are active structures affected by the dynamic interplay of distinct pathophysiological events and complex signals capable of increasing vasodilation, vascular remodeling, angiogenesis, vasculogenesis, and hemodynamic pressure stress in this pathological setting. We also describe here our current knowledge on the gut-vascular barrier and the potential implications in liver disease through the gut-liver axis. Another important focus of this chapter relates with the use of pericytes and endothelial cells as novel therapeutic targets to improve the outcome in patients suffering from chronic liver disease. This is of particular significance due to the lack of effective treatment options beyond liver transplantation, which is also not the cure since many patients die on the waiting list due to a shortage of organ donors.

5.2 Pericytes in Blood Vessels

Blood vessels are composed of two interacting cell types. Endothelial cells form the inner lining of the vessel wall, while pericytes envelop the surface of the vascular tube. For many years, pericytes have been thought to exclusively provide structural support to the vessel tube. However, we now know that these cells are functionally significant and play important roles in microvascular physiology, capillary blood flow regulation, and angiogenesis (Hirschi and D'Amore 1996; Bergers and Song 2005).

Pericytes are enveloped in a vascular basement membrane that is continuous with the endothelial basement membrane. At distinct points in the basement membrane, pericytes and endothelial cells form specialized junctions with each other. These sites are thought to play an anchoring role (Armulik et al. 2011). The number, distribution, and structure of pericytes vary among different organs and vascular beds, suggesting that they may have vascular bed- or tissue-specific roles (Hirschi and D'Amore 1996). Common pericyte molecular markers include platelet-derived growth factor receptor beta (PDGFRβ), neuron-glial antigen 2 (NG2), alpha smooth muscle actin (αSMA), or desmin, but no single entirely pericyte-specific marker is known given that expression of all of them can change depending on developmental states or pathological reactions (Armulik et al. 2011; Bergers and Song 2005).

5.2.1 Interaction with Endothelial Cells

Pericytes maintain active communication with the underlying endothelial cells, via direct contact or paracrine signaling. In addition, each pericyte can contact several endothelial cells at a time, even from different capillaries, and thus integrate coordinate signals between them (Armulik et al. 2011). Pericytes play an important role in regulation of endothelial cell proliferation and differentiation, contractility and tone, and stabilization and permeability.

Several molecular signaling pathways have been described to be involved in the cross talk between pericytes and endothelial cells. Platelet-derived growth factor-B (PDGFB), angiopoietins, and transforming growth factor beta (TGFβ) are the best described factors that paracrinely regulate the interactions between both cell types and will be further discussed below, but also other signaling pathways are implicated, like heparin-binding epidermal growth factor (HB-EGF), which is expressed by endothelial cells to recruit mural cells, or Notch and Ephrin signaling, which have been described to be implicated in mural cell association to the endothelium (Armulik et al. 2011; Gaengel et al. 2009).

5.2.2 *Functions*

The location of pericytes in capillaries and their muscle cell characteristics point at a role of these cells in the regulation of capillary blood flow through the contraction of the endothelium, but their functions are more diverse, including regulation of sprouting tubes for new capillary growth or maturation and stability of blood vessels (Bergers and Song 2005).

Direct in vivo evidence of pericyte contraction is limited since very subtle narrowing of capillaries would be sufficient to reduce blood flow and thus this contraction may be below the level of resolution of any morphological method. However, the presence of a contractile machinery in pericytes, which express αSMA, suggests that they could function in a way similar to vascular smooth muscle cells, being able to produce vasodilation and vasoconstriction. It has also been found that pericytes express tropomyosin and muscle myosin, besides cyclic-GMP-dependent protein kinase, which is involved in the contraction of smooth muscle cells. Receptors for both cholinergic and adrenergic responses are expressed in pericytes, and also molecules like angiotensin II and endothelin-1 have been reported to bind to pericytes and paracrinely regulate capillary vascular tone (Hirschi and D'Amore 1996; Armulik et al. 2011).

Pericytes also have an important role in angiogenesis and in the maturation of the newly formed vessels (Bergers and Song 2005; Stapor et al. 2014). In vessel sprouting, endothelial cells are stimulated by vascular endothelial growth factor (VEGF) and other angiogenic factors to degrade basal membrane, migrate, and proliferate (Gerhardt et al. 2003; Jakobsson et al. 2010). In this stage, pericytes contribute to the production of VEGF and to the establishment of VEGF gradients which guide the sprouting process (Bergers and Song 2005). After the microvessel forms a lumen, endothelial cells secrete growth factors, such as PDGFB, to attract pericytes, which express the PDGFRβ receptor (Potente et al. 2011). Pericyte attachment to the endothelial wall stabilizes the nascent vessel and induces endothelial differentiation and growth arrest once the angiogenic process is completed (Bergers and Song 2005; Armulik et al. 2011).

The angiopoietin-Tie2 axis has also been described to be important in the pericyte coating of blood vessels. The receptor Tie2 is expressed in the endothelial cells, and it induces pericyte recruitment and vessel stabilization, while its ligands, angiopoietin-1 and angiopoietin-2, are predominantly expressed in perivascular cells and endothelial cells, respectively. The former, Ang-1, is mainly an agonistic ligand of Tie2, so it is believed to promote vessel maturation, while the latter, Ang-2, is considered a destabilizing factor in blood vessels, and its expression is mainly associated with sites of vascular remodeling (Bergers and Song 2005; Armulik et al. 2011).

TGFβ has been described as an important factor in stabilization of newly formed blood vessels. Once pericyte precursors begin the coating of a forming endothelial tube, TGFβ is activated, leading to inhibition of endothelial cell proliferation and migration and induction of differentiation of perivascular precursors into pericytes (Bergers and Song 2005; Armulik et al. 2011).

Summarizing, pericytes are essential components of microvessels, not only giving structural support to endothelial cells but also contributing to capillary blood flow and angiogenesis through mechanical and signaling roles. Thus, many studies have pointed at the benefits of targeting both pericytes and endothelial cells as an angioregulatory therapy in different pathologies, including chronic liver disease, as described later on in this chapter (Kelly-Goss et al. 2014; Bergers et al. 2003; Fernandez et al. 2007).

5.3 Blood Supply of the Intestines

The intestines are the main components of the lower part of the gastrointestinal tract, and they have the essential function of digestion and absorption of all the ingested food and water. For that purpose, a sufficient amount of tissue oxygenation is mandatory. Therefore, the gastrointestinal circulation delivers the required oxygen for the secretory, absorptive, and motor activity of the intestines allowing the posterior delivery of nutrients and water to all the organs of the body using both the blood and the lymphatic vasculature. The anatomy and physiology of the intestinal tissue are complex and so is the blood supply that allows its homeostasis and vital functions, describing that complexity is the aim of this section.

5.3.1 Macroscopic Anatomical Considerations

The intestines are the last part of the continuous passageway that conforms the entire gastrointestinal tract, and they are divided into two main sections which are the small and the large intestine. The small intestine is subdivided into duodenum, jejunum, and ileum macroscopically but has also a microscopically division as named previously. The large intestine begins after the ileum and starts with the cecum and the vermiform appendix and continues with the ascending, transverse, descending, and sigmoid colon to end with the rectum and the anus (Gray and Lewis 2000) (Fig. 5.1).

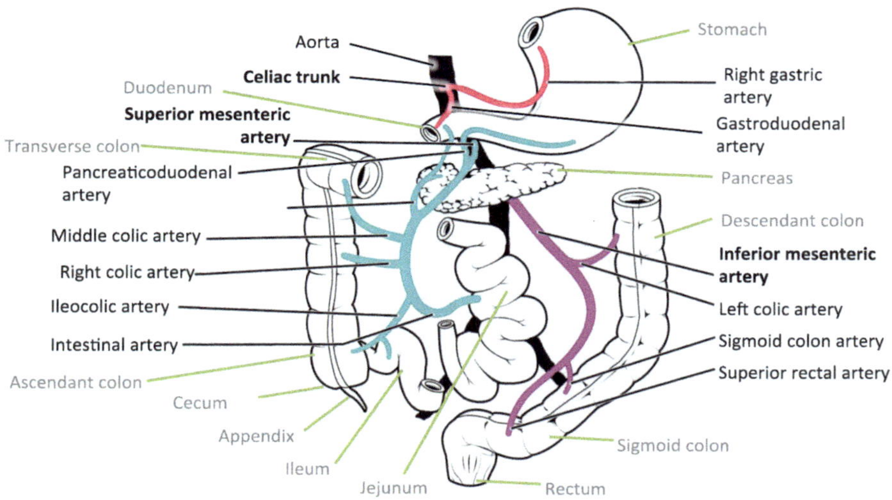

Fig. 5.1 Blood supply to the gastrointestinal tract. Intestinal tissues are written in gray, while vascular components are written in black. Gastrointestinal tract blood supply comes from three aortic branches called celiac trunk and superior and inferior mesenteric arteries. Each one of those branches has additional branches that allow blood irrigation of the whole tissue from the duodenum to the transverse colon. *Modified from Gray's Anatomy: superior mesenteric artery (2015)*

Blood supply for the intestines comes from different vessels, but all of them derive from the aorta which is divided into celiac trunk, superior mesenteric artery, and inferior mesenteric artery (Kachlik et al. 2010) (Fig. 5.1). The first branch is the celiac trunk, which subdivides into three branches and provide blood irrigation for the pylorus of the stomach and the duodenum by the branch named gastroduodenal artery. Below that ramification and also coming from the aorta, there is the superior mesenteric artery (Fig. 5.2), which branches into five major ramifications that provide blood to the small intestine and the proximal large intestine. The distal end of the duodenum is irrigated by the inferior pancreaticoduodenal artery, the smallest branch, and then there are the intestinal arteries divided into those providing blood to the ileum (ileal arteries) and the jejunum (jejunal arteries) (Matheson et al. 2000). All those interconnected branches are supported by the mesentery, and there are arches helping the continuous blood flow and preventing interruptions that would lead toward tissue death. Following those branches, there are the ileocolic artery feeding the terminal ileum, the cecum, and the appendix and the right colic artery providing for the ascending colon, and finally, the transverse colon receives blood from the middle colic artery (Fig. 5.1). The last aortic branch irrigating intestines is the inferior mesenteric artery which provides blood to the large intestine by the following branches: the left colic artery for the descending colon, the sigmoid artery for the sigmoid colon, and the superior rectal artery for the rectum. However, the blood flow oxygenating the anus comes from lower rectal arteries branched from the ileal arteries (Harper and Chandler 2016). Arterial vessels go into the villus giv-

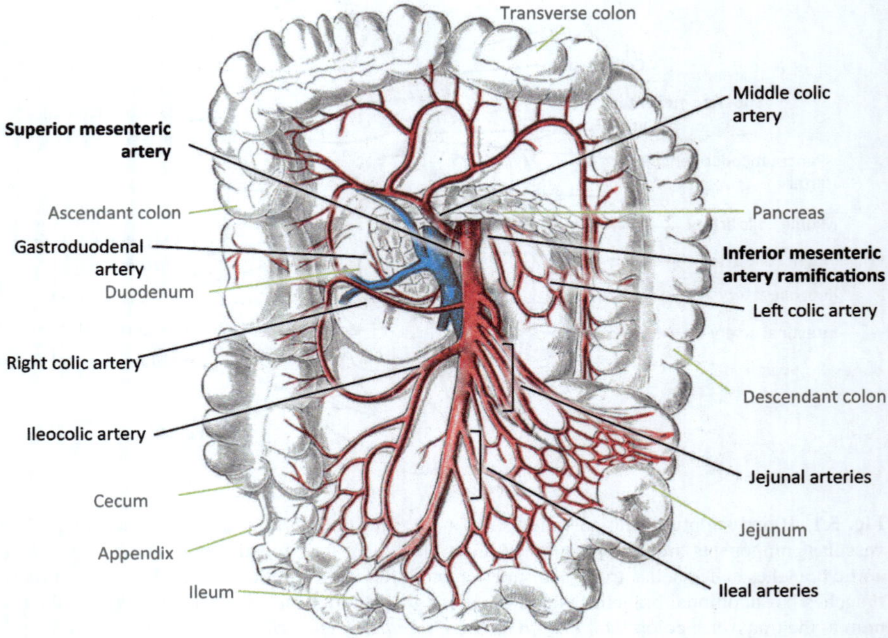

Fig. 5.2 Superior and inferior mesenteric arteries. Intestinal tissues are written in gray, while vascular components are written in black. The hand-fan-like structure of the mesentery and the vasculature going through it allows the formation of a very stable and organized mesh. Notice that most of the intestinal structures are supplied by the superior mesenteric artery ant its branches. The inferior mesenteric artery irrigates the last segment of the intestines beginning at the descendant colon. *Modified from Anatomy and Physiology web site: http://cnx.org/content/col11496/1.6/ (2013)*

ing rise to capillary networks that will drain intro the venous channel. The venous system draining the whole gastrointestinal tract is almost parallel to the blood supply system even at the arcade points. The superior and inferior mesenteric veins will drain into the portal vein sending it to the liver and subsequently back to the heart (Kvietys 2010).

5.3.2 Microscopic Anatomical Considerations

The intestinal layers are interconnected by its microvasculature. It conforms a branched network made of arterioles, venules, and capillaries that go all over the three main tissue layers: mucosa, muscularis, and submucosa (Granger et al. 2015). The blood supply enters the intestine by the surface of the serosa which is in contact with the mesentery (structure that gives support to the vasculature in the peritoneum). The first-order arterioles are found in the submucosa coming from the muscular layer and from them derive the second-order arterioles that will end up with the third order that will get into the mucosa and go down to the tip of the villi where they become capillaries and form a mesh-like structure (Thorburn et al. 2018) (Fig. 5.3). The morphology of the villi microcirculation at the small intestine is generally in an eccentrical manner having a single arteriole that goes up to the tip from which capillaries derive making up a network with numerous anastomoses. In contrast, the microvasculature at the colon is more similar to the stomach in which branches of arterioles and capillaries pass along the luminal surface of the mucosa and form a network around the glands being much closer to the epithelium than in the small intestine (Kvietys 2010). For the draining of the products of the chime and the collection for the posterior distribution of all the nutrients, there are the venules. They start at the villi and pierce the submucosa where they will become submucosal collecting venules and bring all the absorbed products to the portal circulation. Further details on this microvasculature and its physiological and pathological role as intestinal vascular barrier will be described later on in this chapter.

5.3.3 *Functional Considerations*

The blood passing through the intestines account for approximately 20–25% of the whole cardiac output in a non-fed state, but once the digestion is done and absorption takes place, the blood flow of each artery is increased up to 200% creating a

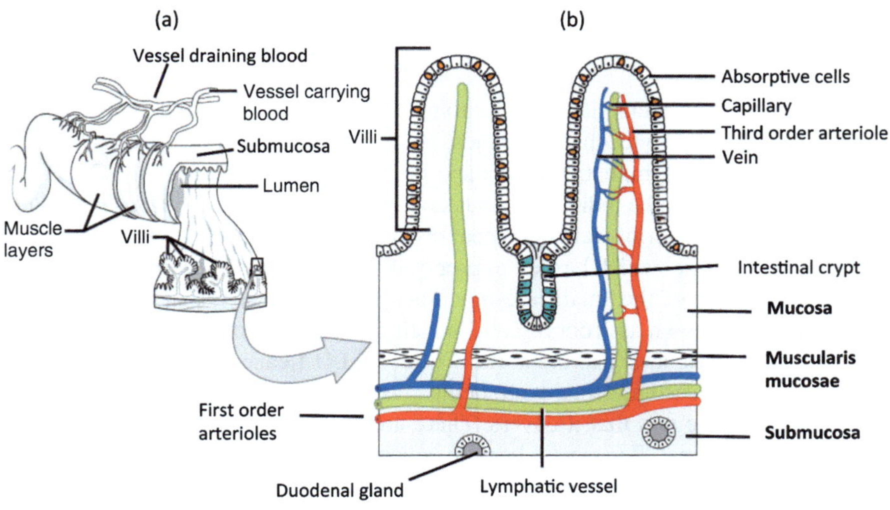

Fig. 5.3 The villi vasculature. (**a**) Cut of an intestine fragment with the lumen and the tissue layers visible: muscle, mucosa, and submucosa (inside de villi). Veins and arteries pierce the tissue in a parallel manner. (**b**) Villi amplification. The submucosa contains the first-order arterioles that will become second-order arterioles and subsequently capillaries that will carry on the absorption function. It also contains glands for mucus and enzyme secretions. *Modified from Anatomy and Physiology web site: http://cnx.org/content/col11496/1.6/ (2016)*

state of hyperemia as all the nutrients pass over the mucosal surface and enter inside the villi (Matheson et al. 2000). That increase will depend uniquely on the chime composition and will later decrease when interluminal pressure is raised (Granger et al. 2015). Depending on the functional importance of the tissue layer, it will receive more or less blood flow (Sparks 2011). The intestinal mucosa, which is full of capillaries and is the first layer of interplay between nutrients and vasculature as previously described, is essential for proper nutrient absorption and thus is the one receiving more blood input. The submucosa is the one preceding the submucosa and has the first-order arterioles that will allow the blood supply required at the villi. Besides, it contains glandular cells that produce serous and mucous secretions and immature enterocytes which will play a valuable role in nutrient absorption (Fig. 5.3). Finally, the muscular layers provide the ability to contract and thus allow the mixing and the movement of the chime, they are pierced by the vasculature coming from the mesentery and the serosa. Those two layers receive the rest of the blood addressed to the intestine.

There are several metabolic factors that maintain homeostasis of intestinal tissues such as decreased or increased PO_2, pH, osmolarity, or adenosine (Matheson et al. 2000). The gastrointestinal circulation contributes also to the defense against ingested toxins or noxious agents protecting intestine tissue from excessive tissue damage by a profound and complex vasoregulation (Granger et al. 2015). The nature of the blood flow in the gastrointestinal tract makes it essential for the proper perfusion and function of all the vital organs therefore becoming a compelling subject of study.

5.4 Blood Supply of the Mesentery

The mesentery is a double layer of peritoneum composed of mesothelium, connective tissue, and adipocytes that surrounds, holds, and gives support to the intestines, allowing blood and lymphatic vessels and nerves to supply them (Fig. 5.2). It was previously thought to be a discontinuous collection of discrete structures separately attached into the posterior wall, but recent research has found the mesentery to be one contiguous structure, which has led to proposals for its reclassification as an organ.

The mesentery of the small intestine emerges from the back of the abdominal cavity, the area known as "root region," which corresponds to the attachment of the superior mesenteric artery to the aorta (Coffey and O'Leary 2016). The mesentery spans the entire gastrointestinal tract, being classified into different areas depending on the intestinal region that it covers. There are six flexures in the gastrointestinal tract: duodenojejunal, ileocaecal, hepatic, splenic, and those between the descending and sigmoid colon and the sigmoid and rectum. These flexures are used to delimitate the different portions of the mesentery: small intestine mesentery; right, transverse, and left mesocolon; mesosigmoid area; and mesorectum, respectively (Coffey and O'Leary 2016). The mesentery is compactly folded within the peritoneal cavity, combining mobile regions, like the small intestinal mesentery, and other regions flattened against the abdominal wall. This structure prevents intestines collapsing into the pelvis by its attachment points to the posterior abdominal wall, specifically at the right and left mesocolon, and by its connection with the pelvic wall at the mesosigmoid regions (Coffey and O'Leary 2016).

As mentioned before, the mesentery is essential for blood supply to the intestines. It holds the superior and inferior mesenteric arteries (SMA and IMA) and veins (SMV and IMV) that will irrigate and drain the gastrointestinal tract, respectively (Drake et al. 2015) (Fig. 5.2). The SMA and IMA arise from the abdominal aorta and travel in the mesentery to irrigate the splanchnic organs, including the mesentery itself. The SMV and IMV, which both run alongside their associated arteries, drain the blood to the portal vein.

Focusing in the mesenteric microcirculation, the capillaries between arterioles and venules are responsible for the exchange of O_2 and nutrients with the parenchyma. Besides, the arterioles are responsible for most of the resistance to blood flow, so the regulation of blood flow in the mesentery takes place mainly on the walls of the arterioles through contraction or dilation (Jacobson 1982).

The mesentery is a key piece in chronic liver disease as it is affected by excessive vasodilatation and pathological angiogenesis (Fig. 5.4), which work together to promote and perpetuate the increased splanchnic blood flow and portal hypertension, as described below. However, it is also important in other abdominal and non-abdominal pathologies such as colorectal cancer, diverticular disease, cardiovascular disease, inflammatory bowel disease, obesity, and the metabolic syndrome (Coffey and O'Leary 2016).

5.5 Vascular Alterations in Chronic Liver Disease

Chronic liver diseases, including cirrhosis of the liver, are leading causes of death and liver transplantation worldwide. They are accompanied by profound circulatory disturbances that are not limited to the intrahepatic circulation but involve also the splanchnic and systemic vascular beds. These hemodynamic disturbances are responsible for the development of a hyperdynamic circulatory state and portal hypertension, which are major complications of chronic liver diseases.

5.5.1 Portal Hypertension

The portal hypertensive syndrome is characterized by a pathological increase in blood pressure in the portal vein. It is initiated by an increase in vascular resistance to portal blood flow at a presinusoidal (portal vein thrombosis), sinusoidal (cirrhosis of the liver, chronic hepatitis, alcoholic liver disease, and hepatic schistosomiasis), or postsinusoidal level (Budd-Chiari syndrome). The dominant cause of portal hypertension relates to liver cirrhosis, which increases resistance through the hepatic sinusoids due to distortion of the liver vascular architecture caused by fibrosis,

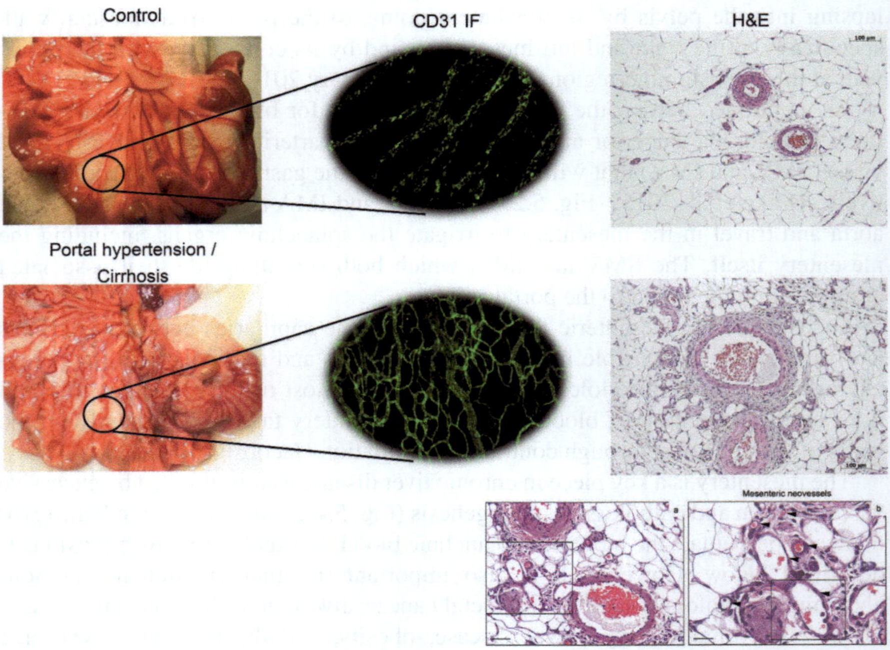

Fig. 5.4 The mesenteric vascular bed. Angiogenesis and vasodilation are markedly increased in the mesenteric vascular bed during portal hypertension and cirrhosis, as demonstrated using immunofluorescence for the endothelial cell marker CD31 and hematoxylin and eosin histological staining

scarring, and nodule formation, as well as by hepatic sinusoidal cellular alterations, with imbalance between vasodilators and vasoconstrictors, promoting sinusoidal constriction (Gupta et al. 1998; Rockey and Chung 1998; Schuppan and Afdhal 2008). Another major driver of portal hypertension locates extrahepatically and is an increase in blood flow in splanchnic organs, as described below.

5.5.2 *Increased Splanchnic Blood Flow*

The increase in blood flow in splanchnic organs draining into the portal vein and the subsequent increase in portal venous inflow represents a significant factor maintaining and worsening portal hypertension (Fig. 5.5). It perpetuates and exacerbates portal pressure elevation, promotes ascites and spontaneous bacterial peritonitis, and is associated with the formation of an extensive network of portosystemic collateral vessels (i.e., vascular channels linking portal and systemic venous circulations).

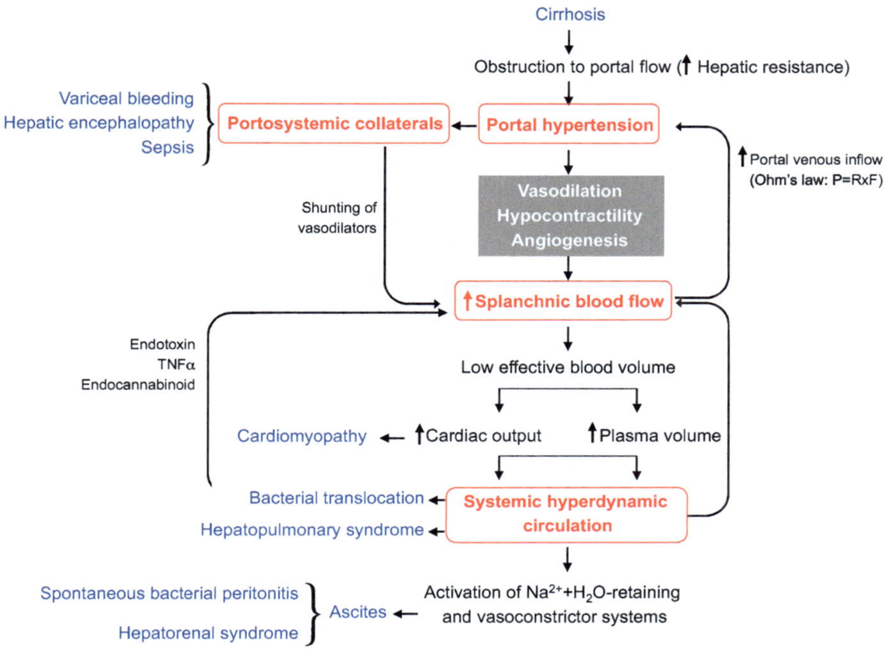

Fig. 5.5 Pathophysiology of portal hypertension in chronic liver disease. The portal hypertensive syndrome is characterized by a pathological increase in blood pressure in the portal vein. In cirrhosis, portal hypertension is initiated by an increased hepatic resistance to portal blood flow caused by the distortion of liver vascular architecture associated with fibrogenesis and angiogenesis and by an increased hepatic vascular tone due to intrahepatic vasoconstriction secondary to an imbalance between decreased endogenous dilators and increased vasoconstrictor stimuli. Portal hypertension is aggravated by an increased blood flow in splanchnic organs draining into the portal vein due to enhanced vasodilation and angiogenesis and hypocontractility to endogenous vasoconstrictors, with subsequent elevation in portal venous inflow. Such increased portal venous inflow is a significant factor maintaining and worsening portal pressure elevation. Increased splanchnic blood flow also leads to systemic hyperdynamic circulation. Another characteristic feature of portal hypertension is the formation of portosystemic collateral vessels responsible for life-threatening consequences like gastroesophageal variceal bleeding, portosystemic encephalopathy, and sepsis

The development of splanchnic hyperdynamic circulation with increased splanchnic blood flow and portal venous inflow in chronic liver disease is mainly due to mesenteric arteriolar vasodilation, decreased vascular responsiveness to endogenous vasoconstrictors, and neovascularization, including angiogenesis and vasculogenesis (Fernandez 2015). This feature also highlights the potential clinical relevance of applying combination therapies acting both on prevention/regression of new splanchnic vessels by antiangiogenic agents and on modulation of vasomotor dynamics by vasoactive substances.

5.5.3 *Portosystemic Collaterals*

The most common and clinically threatening collateral vessels are the gastroesophageal varices, which are fragile and particularly prone to leak blood and even rupture, causing upper gastrointestinal tract bleeding (Fig. 5.6). This hemorrhage is often torrential and difficult to staunch, and, despite many advances made in this field, it continues to be the most dramatic and lethal complication of portal hypertension (Garcia-Tsao et al. 2007; Garcia-Tsao and Bosch 2010; Sharara and Rockey 2001). Furthermore, because portosystemic venous shunts bypass the liver, several substances normally metabolized by the liver, such as drugs, toxins, hormones, and bacteria, can escape from collaterals to the systemic circulation, leading to other potentially lethal consequences, such as portosystemic encephalopathy, spontaneous bacterial peritonitis, or systemic infections (Fernandez 2015; Iwakiri et al. 2014).

The current understanding is that portosystemic collateralization is a highly complex multifactorial process that involves two different but complementary mechanisms: the opening, dilation, and remodeling of preexisting collaterals and

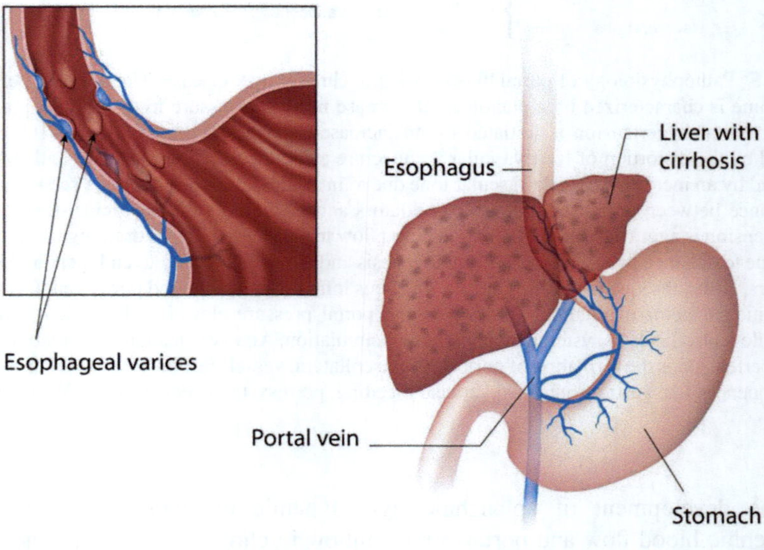

Fig. 5.6 Gastroesophageal varices. VEGF-dependent angiogenesis plays a crucial part in the formation of portosystemic collateral vessels and gastroesophageal varices. These varices are fragile and particularly prone to leak blood and even rupture, causing upper gastrointestinal tract bleeding. This hemorrhage is often torrential and difficult to staunch, and, despite many advances made in this field, it continues to be the cause of significant morbidity and mortality in patients. Furthermore, because portosystemic venous shunts bypass the liver, noxious substances that are normally metabolized by the liver can escape from collaterals to the central venous system, leading to other potentially lethal consequences, such as portosystemic encephalopathy, spontaneous bacterial peritonitis, or systemic infections

also the de novo formation and maturation of new collateral vessels, through active neoangiogenesis (Reiberger et al. 2009, 2013; Van Steenkiste et al. 2009; Fernandez et al. 2007). Both mechanisms may be driven by similar environmental variables and work in a coordinated manner with the final common goal of developing more abundant and more functional collateral vessels capable of conducting blood efficiently, to accommodate the sustained portal pressure elevation and the greatly increased portal venous flow. Worthy of note is that advanced portal hypertension is typically associated with an extensive network of high-flow portosystemic collaterals that may carry over even 90% of the blood entering the portal system.

5.5.4 *Pathological Angiogenesis*

As described above, angiogenesis represents a critical and clinically important pathological hallmark and driving force in the processes of hyperdynamic splanchnic circulation and portosystemic collateralization during chronic liver disease (Fernandez et al. 2007, 2009; Fernandez 2015) (Fig. 5.7). It also promotes the establishment and maintenance of the abnormal angioarchitecture distinctive of the cirrhotic liver, being functionally linked with fibrogenesis and inflammation (Tugues et al. 2007; Fernandez et al. 2009; Mejias et al. 2009; Fernandez 2015). Accordingly, growing evidence suggests that therapeutic inhibition of angiogenesis could be a valuable treatment strategy with multiple beneficial effects, reducing the development of new collaterals and ameliorating portal hypertension and liver cirrhosis (Fernandez et al. 2009; Fernandez 2015).

5.5.4.1 Vascular Endothelial Growth Factor

VEGF is the main proangiogenic growth factor implicated in angiogenesis during cirrhosis. VEGF and its signaling pathway are switched on very early during the natural history of chronic liver disease, preceding the increase in splanchnic blood flow and, therefore, occurring prior to shear stress augmentation (Fernandez et al. 2004, 2007; Abraldes et al. 2006; Calderone et al. 2016). Interestingly, we have recently demonstrated that this early VEGF overexpression in portal hypertension and cirrhosis is posttranscriptionally regulated by cytoplasmic polyadenylation element binding proteins (CPEB), which are in turn activated by Aurora kinase-A (Fig. 5.8), as described later on (Calderone et al. 2016).

VEGF signaling pathway promotes an extensive neovascularization in the cirrhotic liver and the mesenteric vascular bed during cirrhosis, increasing the splanchnic blood flow and contributing to the development of portosystemic collateral vessels, intrahepatic fibrosis, and inflammation (Fernandez et al. 2009; Fernandez 2015). An important consideration is that VEGF not only is involved in neovessel formation but also has an active role in hemodynamic processes, inducing systemic, splanchnic, and portocollateral hyperdynamic circulation by stimulation of

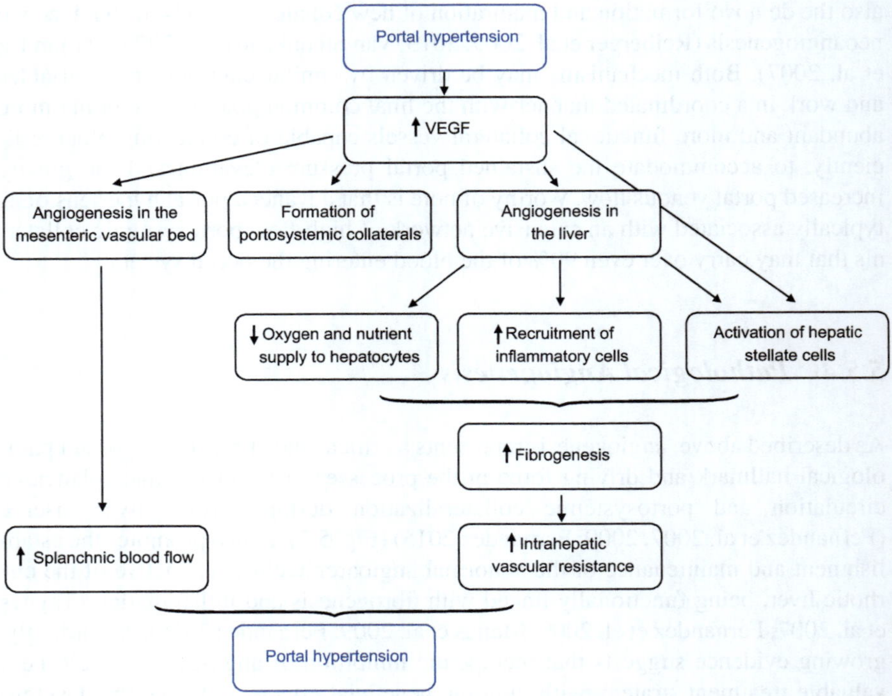

Fig. 5.7 Pathological angiogenesis. Pathological neovascularization stimulated mainly by VEGF represents a critical and clinically important hallmark in portal hypertension. Neovessel formation in the mesentery contributes to increase splanchnic blood flow draining into the portal vein, which, in turn, augments portal venous inflow. Such increased portal venous inflow perpetuates and exacerbates portal pressure elevation during chronic liver disease. Angiogenesis also plays a pivotal role in the development of an extensive network of portosystemic collateral vessels, which include gastroesophageal varices. These varices are particularly prone to rupture, causing massive, life-threatening gastroesophageal bleeding. Collateral vessels are also responsible for other major consequences of chronic liver disease, including portosystemic encephalopathy and sepsis. Neovascularization has also been implicated as a crucial player in the establishment and maintenance of the abnormal angioarchitecture distinctive of the cirrhotic liver, being intimately linked to the progression of fibrogenesis and inflammation and playing a major contribution to the aggravation of portal hypertension

vasodilating agents such as endothelial nitric oxide synthase (eNOS) (Abraldes et al. 2006). This hyperdynamic circulation, in turn, enhances vascular shear stress, further stimulating the production of VEGF and NO and perpetuating and aggravating the portal hypertensive syndrome.

Several studies have shown that the blockade of VEGF or its receptor (VEGFR2), using several angiogenesis inhibitors with different modes of action and prophylactic and therapeutic strategies, effectively decreases portal pressure and mesenteric neoangiogenesis, reduces splanchnic blood flow and portosystemic collateralization, and attenuates liver fibrosis (Fernandez et al. 2004, 2005, 2007; Mejias et al. 2009; Gallego et al. 2017) (Figs. 5.9 and 5.10).

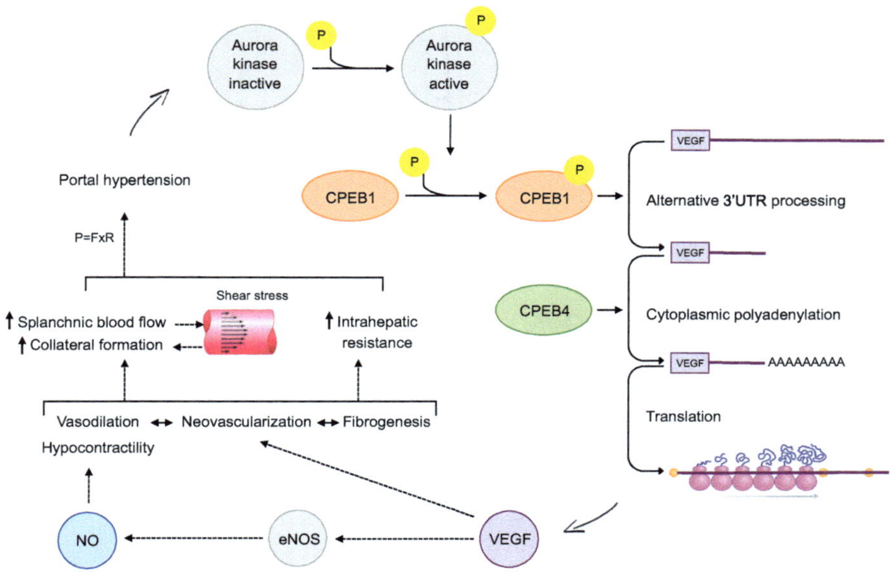

Fig. 5.8 Molecular pathophysiology. Portal hypertension activates Aurora kinase-A, which is a well-known activator of CPEB1. Then, CPEB1 and CPEB4 sequentially lead to increased VEGF protein, which induces angiogenesis and also stimulates nitric oxide (NO) production. In the cirrhotic liver, angiogenesis is intimately linked to liver fibrogenesis, contributing to increase intrahepatic vascular resistance. Extrahepatically, angiogenesis works together with vasodilation and hypocontractility to increase splanchnic blood flow and portosystemic collateral formation. And the combination of all these factors perpetuates portal hypertension

5.5.4.2 Platelet-Derived Growth Factor

Besides VEGF, the expression of platelet-derived growth factor (PDGF) and its cognate receptor PDGFRβ is also markedly upregulated during chronic liver disease and portal hypertension in mesentery and cirrhotic liver and plays a significant role in the excessive neovascularization of these vascular beds. Thus, during angiogenesis, endothelial cells secrete PDGF and, thereby, paracrinically stimulate the recruitment of PDGFRβ-positive pericytes, which, in turn, act as supportive vascular smooth muscle cells, intervening in neovessel stabilization and maturation, as described above. Binding of PDGF to its PDGFRβ is also an essential step in the conversion of quiescent hepatic stellate cell, a liver-specific pericyte, into myofibroblasts and the ensuing recruitment of these cells to sites of liver fibrosis.

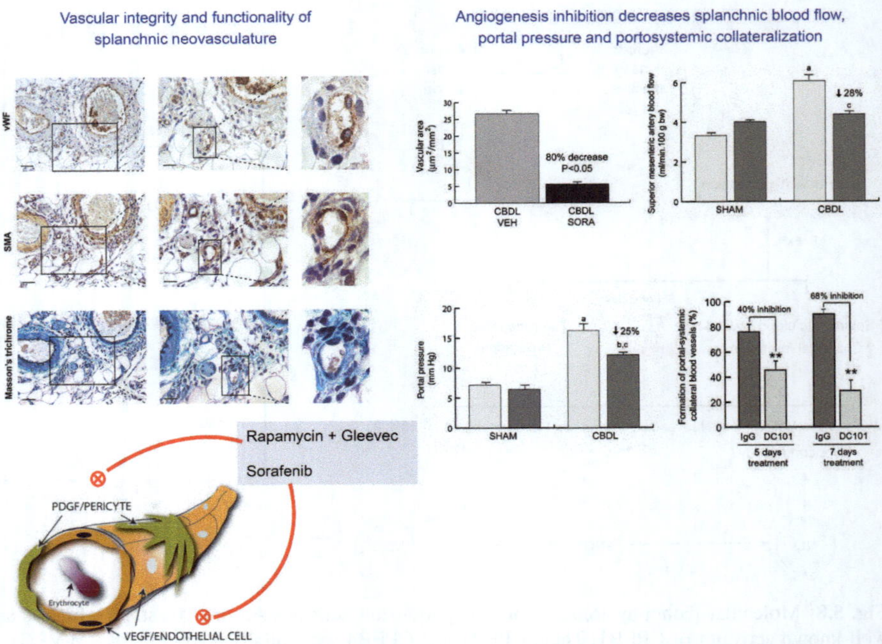

Fig. 5.9 Multiple beneficial effects of angiogenesis inhibition. Splanchnic neovasculature generated by angiogenesis during portal hypertension is functional, has vascular integrity, is perfused, and contributes to increase splanchnic blood flow. Accordingly, inhibition of angiogenesis using different strategies translates into marked decrease in splanchnic blood flow, portal pressure, and portosystemic collateral vessels. These beneficial effects are more efficient when using a combined antiangiogenic treatment directed against pericytes and endothelial cells. This is due to the fact that in the formation of a new blood vessel, there is first the production of an angiogenic factor, which activates endothelial cells; these cells proliferate, migrate, and form an endothelial tube. This tube is then covered by pericytes that provide stabilization and maturation to the newly formed vessel. The first steps of this angiogenic process are mainly modulated by VEGF, whereas the stabilization and maturation of the neovessel are mainly modulated by platelet-derived growth factor (PDGF). Therefore, simultaneous targeting of the VEGF signaling pathway, that is the endothelial cells, and the PDGF signaling pathway, that is the pericytes, gives a greater vascular destabilization and a better vascular regression than the targeting of either alone. This is what we found using a combination of rapamycin (to inhibit the VEGF pathway) and Gleevec (to inhibit the PDGF pathway) or a multikinase inhibitor such as sorafenib

5.5.4.3 Combined Antiangiogenic Treatment Directed against Pericytes and Endothelial Cells

As mentioned before, in the process of neovascularization, VEGF plays a predominant role in the initial stages of formation of new blood vessels, activating the proliferation of endothelial cells and the subsequent formation of an endothelial tubule, while maturation of the newly formed vessels is mainly modulated by the proangiogenic growth factor PDGF, which regulates the investiture of the endothelial tubule

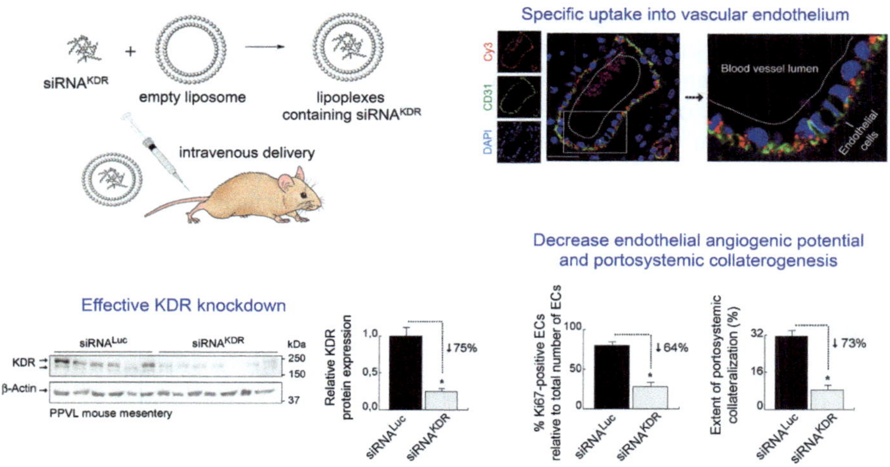

Fig. 5.10 Gene therapy with cell-targeted molecule-targeted liposomal siRNAs. We have recently used siRNA^{KDR} lipoplexes to efficiently and specifically target the kinase insert domain receptor KDR (also known as VEGF receptor-2) in vascular endothelial cells in vivo after systemic intravenous administration. This therapy markedly ameliorates the development of portosystemic collateral vessels and impairs the pathological angiogenic potential of endothelial cells in a murine model of portal hypertension

with mural cell and pericyte populations, thereby stabilizing the vascular architecture of the nascent vessel. Based on these considerations, it was hypothesized that the simultaneous targeting of the VEGF and PDGF signaling pathways, which is the simultaneous targeting of endothelial cells and pericytes, could provide a greater vascular destabilization and a better vascular regression than targeting either alone. Indeed, combined antiangiogenic treatment directed against endothelial cells and pericytes, using VEGF and PDGF inhibitors simultaneously, provides a synergistic benefit in reducing circulatory abnormalities in portal hypertension (Fernandez et al. 2007). This is due to the fact that removal of pericyte coverage by anti-PDGF molecules leads to exposed endothelial tubes, making endothelial cells much more susceptible to anti-VEGF treatment (Fig. 5.9). In this regard, concurrent targeting of VEGFR-2 and PDGFR-β signaling pathways using low doses of the multikinase inhibitor sorafenib, used in clinical practice for the treatment of hepatocellular carcinoma (Llovet and Bruix 2009), significantly reduces portosystemic collateralization, hyperdynamic splanchnic circulation, intrahepatic fibrosis, and portal pressure in experiments in rats with portal hypertension and cirrhosis (Reiberger et al. 2009; Mejias et al. 2009; Tugues et al. 2007), with potential beneficial effects also in humans (Pinter et al. 2012). In addition, the continued administration of the VEGF signaling inhibitor rapamycin plus the PDGF signaling inhibitor Gleevec markedly decreased the splanchnic neovascularization and the pericyte coverage of neovessels in portal hypertensive rats (Fernandez et al. 2007). This combined treatment also resulted in a virtually complete reversal of the increased portal pressure (40%

reduction) and the increased portal venous blood inflow of these animals. This is important since clinical studies have shown a dramatic reduction of the risk of portal hypertensive complications and improved survival in patients achieving a decrease in portal pressure of at least 20% under drug therapy. Notably, the magnitude of the effects of the combination treatment was superior than the addition of the effects of either drug alone, suggesting a synergistic regulatory interaction between the VEGF and PDGF signaling pathways in mediating the maintenance of the vascular and hemodynamic abnormalities observed in portal hypertensive rats. These findings further suggest that in the absence of proliferating pericytes (i.e., after PDGF signaling inhibition), the endothelium is more vulnerable to antiangiogenic therapies targeting endothelial cells, such as VEGF signaling blockade.

5.5.4.4 Endogenous Angiogenesis Inhibitors

Several lines of evidence support the critical role of endogenous angioinhibitors in the pathophysiology of chronic liver disease and the therapeutic benefit of interventions aimed at augmenting the expression of these naturally occurring molecules. Thus, we have found that the powerful endogenous angiogenesis inhibitor pigment epithelium-derived factor (PEDF) is unidirectionally upregulated together with VEGF, in an attempt to counteract the proangiogenic activity of VEGF (Fig. 5.11). Exogenous PEDF overexpression by adenovirus-mediated gene transfer moves the

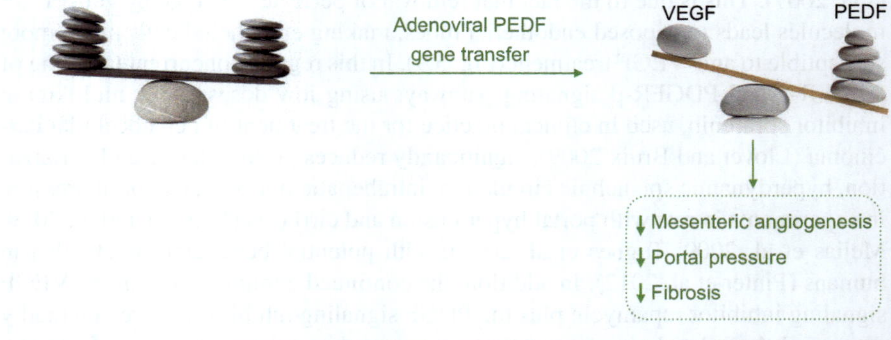

Fig. 5.11 Antiangiogenic and antifibrogenic activity of pigment epithelium-derived factor. Pigment epithelium-derived factor (PEDF) is an endogenous angiogenesis inhibitor that is unidirectionally upregulated together with VEGF in portal hypertension in an attempt to counteract the proangiogenic activity of VEGF. Exogenous overexpression of PEDF moves the balance in favor to inhibition of angiogenesis, translating into beneficial effects

balance between VEGF and PEDF in favor of inhibition and angiogenesis, translating into portal pressure decrease and partial correction of excessive angiogenesis in experimental portal hypertension (Mejias et al. 2015; Vespasiani-Gentilucci and Rombouts 2015).

Another novel approach for halting chronic liver disease progression could be disruption of the negative feedback loop between VEGF and vasohibin-1, which is a protein that is normally present in the organism and has antiangiogenic activity. This therapeutic strategy is a good way to bring VEGF levels back to normal, but not below normal, preserving the baseline VEGF levels necessary for homeostasis of healthy vessels and other angiogenic physiologic events and translating into amelioration of hemodynamic disturbances and reduction of liver fibrosis and portal pressure in portal hypertension (Coch et al. 2014; Chatterjee 2014) (Fig. 5.12).

5.5.4.5 Cytoplasmic Polyadenylation Element Binding Proteins

We have recently identified a new mechanism of regulation of pathologic VEGF expression and angiogenesis through sequential and nonredundant functions of two members of the family of cytoplasmic polyadenylation element binding proteins,

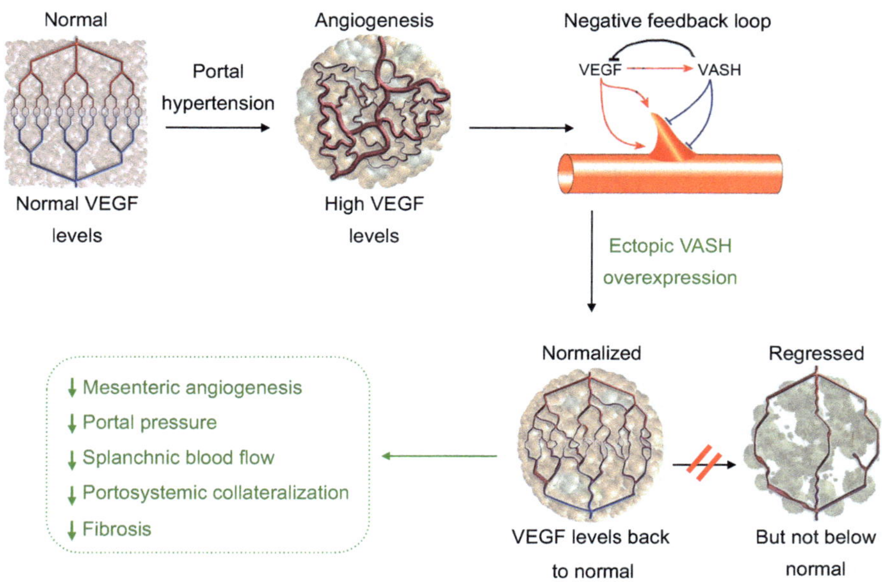

Fig. 5.12 Disruption of negative feedback loop between vasohibin-1 and VEGF decreases portal pressure, angiogenesis, and fibrosis in cirrhotic rats. Vasohibin-1 is an endogenous protein with antiangiogenic activity implicated in a negative feedback loop with VEGF. Disruption of this negative feedback loop in portal hypertension brings VEGF levels back to normal, but not below normal, preserving the physiological VEGF required for vascular homeostasis of healthy vessels and for physiologic angiogenic processes. And that translates into amelioration of several portal hypertension-associated abnormalities

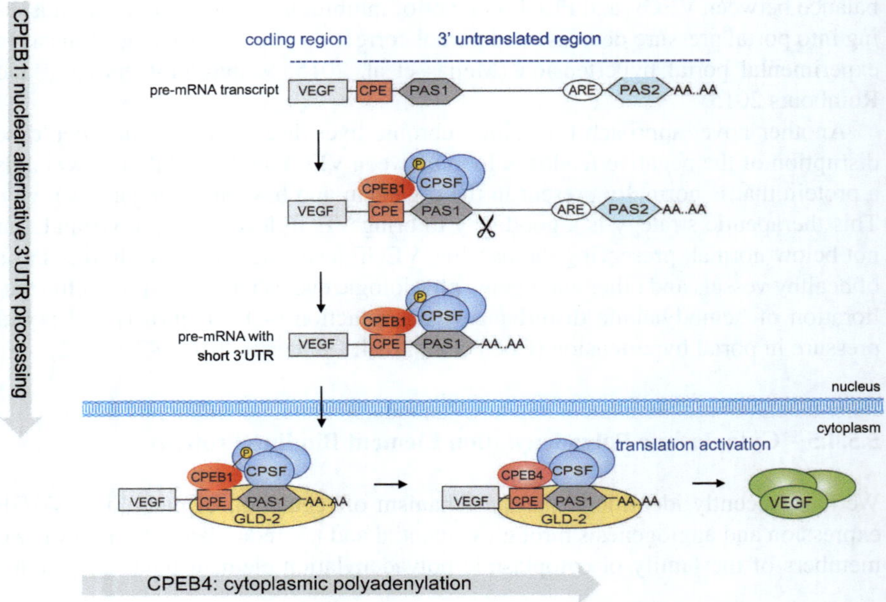

Fig. 5.13 Sequential functions of CPEB1 and CPEB4 regulate pathologic expression of VEGF and angiogenesis in chronic liver disease. CPEB proteins are RNA-binding proteins that bind to and regulate the expression of a specific group of mRNAs, which have, on their 3'untranslated regions, some sequences, call CPEs, cytoplasmic polyadenylation elements. One of these mRNAs is VEGF. The binding of CPEB1 to VEGF mRNA promotes alternative 3'UTR processing of VEGF mRNA in the nuclei. And subsequently, CPEB4 promotes the cytoplasmic polyadenylation of VEGF mRNA in the cytoplasm, which activates translation and generation of high levels of VEGF proteins. Importantly, this regulatory mechanism operates only in pathologic conditions, when CPEB1 and CPEB4 are overexpressed, as we have seen in portal hypertension

CPEB1 and CPEB4 (Figs. 5.8 and 5.13). Importantly, this regulatory mechanism operates only in pathologic conditions, when CPEB1 and CPEB4 are overexpressed, as we have seen in portal hypertension and cirrhosis (Calderone et al. 2016). CPEB proteins are RNA-binding proteins that bind to and regulate the expression of a specific group of mRNAs, which have, on their non-coding 3'-untranslated regions (3'UTR), some sequences named cytoplasmic polyadenylation elements (CPEs) (Bava et al. 2013; Fernandez-Miranda and Mendez 2012; Pique et al. 2008). We have found that one of these CPEB-regulated mRNAs is VEGF mRNA. Upon portal hypertension and cirrhosis induction, there is a rapid upregulation and activation by autophosphorylation of the serine/threonine kinase Aurora kinase-A in the mesentery and the liver. Activated Aurora kinase-A, in turn, phosphorylates and activates CPEB1 (Mendez et al. 2000a, b; Sarkissian et al. 2004). Activation of CPEB1 then promotes alternative nuclear processing within 3'UTRs of VEGF and CPEB4 mRNAs, resulting in deletion of translation repressor elements. The subsequent overexpression of CPEB4 promotes cytoplasmic polyadenylation of VEGF mRNA, increasing its translation and generating high levels of VEGF (Calderone et al. 2016). In addition to promoting pathologic angiogenesis and remodeling in splanchnic, portocollateral, and intrahepatic circulations (Fernandez et al. 2009; Fernandez 2015), excessive VEGF production also contributes to the splanchnic vasodilation and hypocontractility that characterizes portal hypertension through augmentation of NO production (Abraldes et al. 2006). These pathogenic processes synergistically contribute to increase splanchnic blood flow and portosystemic collateralization, sustaining and aggravating portal hypertension.

Importantly, this CPEB-mediated regulatory mechanism is essential for pathologic angiogenesis but dispensable for physiologic neovascularization. Thus, targeting CPEBs could lead to safer treatment outcomes by specifically reducing excessive pathologic VEGF production instead of indiscriminately perturbing both pathologic and physiologic VEGF synthesis, minimizing potential adverse side effects. Consistent with this notion, CPEB depletion reduced portosystemic collateralization and mesenteric neovascularization and attenuated the progression of the portal hypertensive syndrome in portal vein-ligated mice, without affecting the normal vasculature or physiological angiogenesis (Calderone et al. 2016). It is worth mentioning that CPEBs could also play a role in pathological angiogenesis during cancer (Ortiz-Zapater et al. 2011), and we have recently demonstrated an important role of these RNA-binding proteins in hepatic steatosis as well (Maillo et al. 2017).

5.5.5 *Pathological Postnatal Vasculogenesis*

Most research in the neovascularization field has focused on angiogenesis, the formation of neovessels from activation and proliferation of mature endothelial cells in existing vasculature. However, it is now evident that alternative vascularization mechanisms may occur postnatally, including vasculogenesis, the de novo formation of vessels out of vascular stem/progenitor cells, which historically was thought to occur exclusively during embryological development. Indeed, recent studies from our group have identified vascular stem/progenitor cells residing dormant in the vascular wall of postnatal mesenteric vessels under normal conditions but possessing sphere-forming ability and proliferative potential, producing large numbers of daughter cells (i.e., proliferative progenitors or transit-amplifying cells) that differentiate toward either endothelial cells or smooth muscle cells when activated by injury stimuli, such as upon portal hypertension and cirrhosis induction (Garcia-Pras et al. 2017). Importantly, these cells structurally and functionally contribute to abnormal neovessel formation (Fig. 5.14), indicating that abnormal neovascularization in this pathological setting might conceivably be a heterogeneous process, arising through a combination of both angiogenesis and vasculogenesis. Mechanistically, we also found that the RNA-binding protein CPEB4 is an important factor responsible for the proliferative activity of stem/progenitor cell progeny, adding therefore

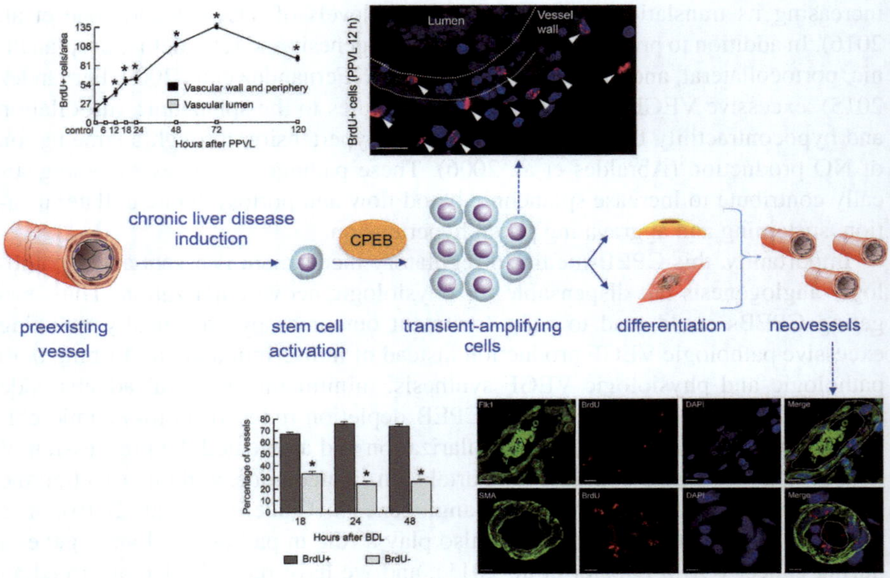

Fig. 5.14 Vascular stem cells: new culprits uncovered. Recent studies from our lab highlight the functional significance of pathologic neovascularization derived from vascular stem/progenitor cells as an important mechanism of formation of new blood vessels in adults, in the setting of chronic liver disease, and identify these stem cells as potential new therapeutic targets. Thus, we have demonstrated the existence in the vascular wall of adult mesenteric blood vessels of a distinctive population of vascular stem/progenitor cells. These cells display some of the most widely accepted criteria for stem cell recognition, including quiescence and slow-cycling properties, high proliferative potentiality, capability of growing as cellular spheres in suspension culture, expression of specific biomarkers of stem cells, and the ability to self-renew and generate daughter cells in response to liver disease induction. This vascular stem cell progeny is able to differentiate into endothelial and smooth muscle cell lineages and readily contribute physically and functionally to neovascularization in vivo during chronic liver disease. Hence, chronic liver disease-associated abnormal neovascularization might conceivably be a heterogeneous process, arising through a combination of both neoangiogenesis and neovasculogenesis. Accordingly, therapeutic targeting of both vascular stem cell-derived neovascularization (vasculogenesis) and new vessel growth mechanisms that utilize non-stem cell constituents (angiogenesis) may effectively block abnormal neovessel formation and improve antiangiogenic therapeutics

another facet to the "proangiogenic" activity of CPEB4, namely, regulation of cell proliferation in vascular stem/progenitor cell descendants, which could coordinately act with the recently demonstrated VEGF-dependent function (Calderone et al. 2016). These findings may also have direct translational implications. Thus, therapeutic targeting of both vascular stem cell-derived neovascularization (vasculogenesis) and new vessel growth mechanisms that utilize non-stem cell constituents (angiogenesis) may effectively block abnormal neovessel formation and improve antiangiogenic therapeutics.

5.6 Gut-Vascular Barrier in Chronic Liver Disease

In the intestine, the epithelial lining separates internal organs from the enteric environment loaded with various foreign substances, including microbiota and its metabolic products as well as nutrients and wastes (Garrett et al. 2010) (Fig. 5.3). The lamina propria lying beneath the enterocytes in the intestinal villi especially that in the lower part houses a largest pool of macrophages for maintaining mucosal homeostasis against the gut microbiota and for the constant need of epithelial renewal (Bain and Mowat 2014). The lamina propria contains microvessels as well as a central lacteal (lymph vessel) and lymphoid tissue, in addition to immune cells. The portal venous system ensures that substances absorbed in the intestine transit first through the liver, where they can be further metabolized and detoxified.

5.6.1 Gut-Liver Axis

It was in the early 1950s when a relationship between liver disease and intestine was first recognized, after hepatic coma was associated with the absorption of nitrogenous metabolites from the intestine (Phillips et al. 1952). Following it, Volta et al. (1987) described for the first time the so-called gut-liver axis being critical in the progression of chronic liver disease. From there on, many advances have been done in clarifying the role of the gut in chronic liver disease.

Both quantitative and qualitative changes in the intestinal microbiota have been associated with liver disease, even though the causes and/or consequence is still unknown (Schnabl and Brenner 2014; Usami et al. 2015; Bhat et al. 2016). For instance, in cirrhotic patients, gut dysbiosis is characterized by an overgrowth of pathogenic bacteria that would favor an increase of bacterial translocation, systemic inflammation, and systemic infection characteristic of liver cirrhosis. Furthermore, it is widely known that the portal venous system is another key component in ensuring that substances absorbed in the intestine transit through the liver. In this regard, many studies have also supported a role of the endotoxemia from gut-derived bacterial translocation in the hyperdynamic circulation characteristic of cirrhotic patients. In a study conducted by Guarner et al. in the early 1990s, endotoxemia was correlated with nitrite serum levels of cirrhotic patients (Guarner et al. 1993). After treatment with colistin, cirrhotic patients had both nitrite and endotoxin levels reduced suggesting a lineal relationship between the intestine and liver. A series of recent studies have further demonstrated that translocation of bacteria and their products across the intestinal barrier is a common place in patients with liver disease (Tsiaoussis et al. 2015).

The mainly growing attention in the last years on the gut-liver axis has been in the microbiota recirculation through the lymphatics during liver pathologies, focusing on the epithelium barrier and leaving rather unexplored the possible role of blood circulation in this scenario. The normal intestinal protection system against pathogens depends on both non-specific mechanisms and specific immunological responses (Jankowski et al. 1994; Alverdy 1990). As it has been extensively reported, gut epithelium acts as the first barrier against the bacterial translocation of gut microbiota into the lymphatics (Artis 2008; Backhed et al. 2005), together with luminal and submucosal factors. When normal microbiota from the gut, present in the mucus layer, suffers alterations both in number and composition, viable bacteria from the lumen go through the intestinal wall and translocate to extraintestinal sites promoting immunological responses. Upon liver disease, tight and adherent junctions present in attaching epithelial cells get deregulated, allowing selective migration of microorganisms and bacterial products. Thus, how bacteria participate or are excluded from the blood circulation was completely unexplored for many years even though the location of the intestinal capillaries under the epithelial layer suggested an important role.

5.6.2 Gut-Vascular Barrier

In the year 2015, Spadoni et al. described for the first time a new anatomical structure in the murine and human intestines, called the gut-vascular barrier (GVB) (Spadoni et al. 2015). The GVB plays a critical role in health and disease by limiting systemic dissemination of microbes and toxins while allowing nutrients to access the circulation. Therefore, in addition to the epithelial barrier, the GVB presents a second, independent barrier that regulates the translocation of luminal bacteria and their ligands, as well as innocuous food antigens.

The GVB shares many key morphological and functional characteristics with the well-known blood-brain barrier, which separates circulating blood from the brain's extracellular fluid. It is composed of closely interacting intestinal vascular endothelial cells, pericytes, and enteroglial cells. Endothelial cells in the GVB develop elaborate junctional complexes that include tight junction and adherens junction, which strictly control paracellular trafficking of solutes and fluids. Pericytes associated with GVB microvasculature seem to be necessary for the stabilization and maintenance of the vascular barrier (Armulik et al. 2010; Daneman et al. 2010).

Spadoni et al. proposed how bacteria can disrupt the GVB, allowing their passage into the bloodstream (portal vein) and spread to the liver, spleen, or other peripheral tissues (Spadoni et al. 2015) (Fig. 5.15). They found that the expression of the plasmalemma vesicle-associated protein-1 (PV1) was upregulated in blood capillaries upon infection with bacterial pathogens, such as *Salmonella typhimurium*. This correlated with increased barrier permeability, with dissemination of the bacterium to the liver and the spleen, and with liver damage. In addition to PV1, the Wnt/β-catenin signaling pathway seems to play a role in regulating GVB and preventing the translocation of bacteria under homeostatic and pathological conditions (Liebner et al. 2008; Spadoni et al. 2016). Other pathways found upregulated in a transcriptome analysis (Spadoni et al. 2016), such as SHH signaling pathways, could also be altered after pathogen infection.

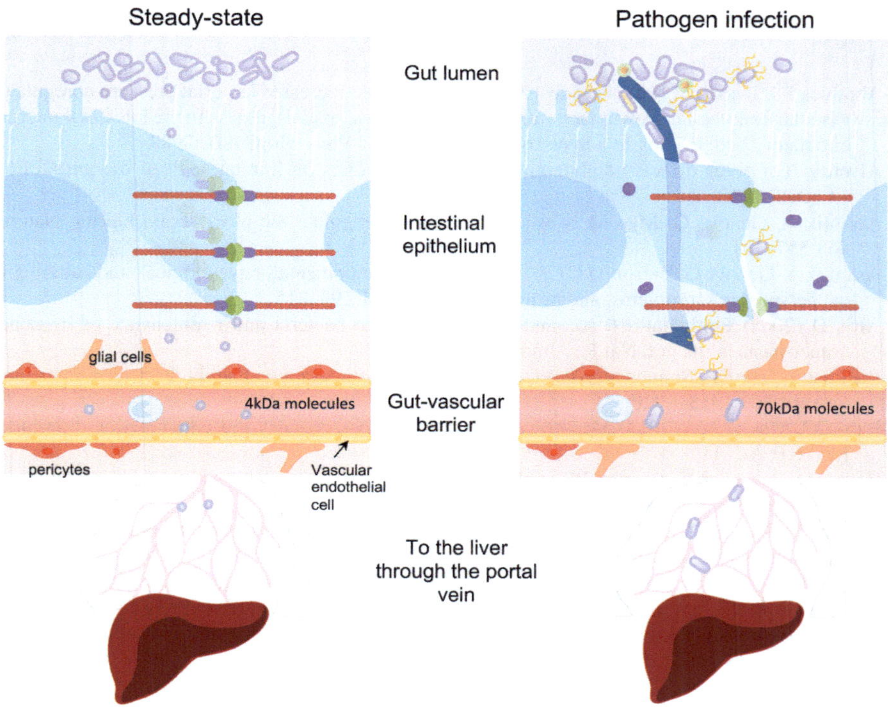

Fig. 5.15 Gut-vascular barrier. In a steady state, pericytes and glial cell surround the vascular endothelial cells forming the barrier that regulates the diffusion of molecules of 4 kD from the intestine to the bloodstream and liver. Under bacterial infection, both epithelium and vascular barriers become leaky, allowing and promoting bacterial spreading through the portal vein

Importantly, disruption of the GVB in the human gut may lead to liver damage, as observed in patients with celiac disease (Spadoni et al. 2016). The microvascular endothelium of the gut barrier also plays a crucial role during inflammation in inflammatory bowel disease. Understanding mechanistically how alterations of the GVB may disrupt systemic immune homeostasis and promote liver damage warrants investigation in different pathological settings. This knowledge may also open a new window of therapeutic treatments to block the entrance of undesirable pathogens.

Grant Support This is supported by grants from the Spanish Ministry of Economy and Competitiveness (MINECO; SAF2014–55473-R and BES2015–071399); the Spanish Ministry of Science, Innovation and Universities (SAF2017–87988-R); the European Union FEDER funds; the Spanish Association Against Cancer (AECC); the Worldwide Cancer Research Foundation; and the CERCA Programme (Generalitat de Catalunya, Spain). CIBERehd is an initiative from the Instituto de Salud Carlos III.

Disclosures All authors declare no competing financial interests.

References

Abraldes JG, Iwakiri Y, Loureiro-Silva M et al (2006) Mild increases in portal pressure upregulate vascular endothelial growth factor and endothelial nitric oxide synthase in the intestinal microcirculatory bed, leading to a hyperdynamic state. Am J Phys 290:G980–G987

Alverdy JC (1990) Effects of glutamine-supplemented diets on immunology of the gut. JPEN 14:109S–113S

Armulik A, Genove G, Mäe M et al (2010) Pericytes regulate the blood–brain barrier. Nature 468:557–561

Armulik A, Genove G, Betsholtz C (2011) Pericytes: developmental, physiological, and pathological perspectives, problems, and promises. Dev Cell 21:193–215

Artis D (2008) Epithelial-cell recognition of commensal bacteria and maintenance of immune homeostasis in the gut. Nat Rev Immunol 8:411–420

Backhed F, Ley RE, Sonnenburg JL et al (2005) Host-bacterial mutualism in the human intestine. Science 307:1915–1920

Bain CC, Mowat AM (2014) Macrophages in intestinal homeostasis and inflammation. Immunol Rev 260:102–117

Bava FA, Eliscovich C, Ferreira PG et al (2013) CPEB1 coordinates alternative 3'-UTR formation with translational regulation. Nature 495:121–125

Bergers G, Song S (2005) The role of pericytes in blood-vessel formation and maintenance. Neuro-Oncology 7:452–464

Bergers G, Song S, Meyer-Morse M et al (2003) Benefits of targeting both pericytes and endothelial cells in the tumor vasculature with kinase inhibitors. J Clin Invest 111:1287–1295

Bhat M, Arendt BA, Bhat V et al (2016) Implication of the intestinal microbiome in complications of cirrhosis. World J Hepatol 8:1128–1136

Calderone V, Gallego J, Fernandez-Miranda G et al (2016) Sequential functions of CPEB1 and CPEB4 regulate pathologic expression of VEGF and angiogenesis in chronic liver disease. Gastroenterology 150:982–997

Chatterjee S (2014) Reversal of vasohibin-driven negative feedback loop of vascular endothelial growth factor/angiogenesis axis promises a novel antifibrotic therapeutic strategy for liver diseases. Hepatology 60:458–460

Coch L, Mejias M, Berzigotti A et al (2014) Disruption of negative feedback loop between vasohibin-1 and VEGF decreases portal pressure, angiogenesis and fibrosis in cirrhotic rats. Hepatology 60:633–647

Coffey JC, O'Leary DP (2016) The mesentery: structure, function, and role in disease. Lancet Gastroenterol Hepatol 1:238–247

Daneman R, Zhou L, Kebede AA et al (2010) Pericytes are required for blood–brain barrier integrity during embryogenesis. Nature 468:562–566

Drake RL, Vogl AW, Mitchell A (2015) Gray's anatomy for students, 3rd edn. Churchill Livingstone/Elsevier, Philadelphia, PA, pp 271–275

Fernandez M (2015) Molecular pathophysiology of portal hypertension. Hepatology 61:1406–1415

Fernandez M, Vizzutti F, Garcia-Pagan JC et al (2004) Anti-VEGF receptor-2 monoclonal antibody prevents portal-systemic collateral vessel formation in portal hypertensive mice. Gastroenterology 126:886–894

Fernandez M, Mejias M, Angermayr B et al (2005) Inhibition of VEGF receptor-2 decreases the development of hyperdynamic splanchnic circulation and portal-systemic collateral vessels in portal hypertensive rats. J Hepatol 43:98–103

Fernandez M, Mejias M, Garcia-Pras E et al (2007) Reversal of portal hypertension and hyperdynamic splanchnic circulation by combined vascular endothelial growth factor and platelet-derived growth factor blockade in rats. Hepatology 46:1208–1217

Fernandez M, Semela D, Bruix J et al (2009) Angiogenesis in liver disease. J Hepatol 50:604–620

Fernandez-Miranda G, Mendez R (2012) The CPEB-family of proteins, translational control in senescence and cancer. Ageing Res Rev 11:460–472

Gaengel K, Genove G, Armulik A et al (2009) Endothelial-mural cell signaling in vascular development and angiogenesis. Arterioscler Thromb Vasc Biol 29:630–638

Gallego J, Garcia-Pras E, Mejias M et al (2017) Therapeutic siRNA targeting endothelial KDR decreases portosystemic collateralization in portal hypertension. Sci Rep 7:14791

Garcia-Pras E, Gallego J, Coch L et al (2017) Role and therapeutic potential of vascular stem/progenitor cells in pathological neovascularisation during chronic portal hypertension. Gut 66:1306–1320

Garcia-Tsao G, Bosch J (2010) Management of varices and variceal hemorrhage in cirrhosis. N Engl J Med 362:823–832

Garcia-Tsao G, Sanyal AJ, Grace N et al (2007) Prevention and management of gastroesophageal varices and variceal hemorrhage in cirrhosis. Hepatology 46:922–938

Garrett WS, Gordon JI, Glimcher LH (2010) Homeostasis and inflammation in the intestine. Cell 140:859–870

Gerhardt H, Golding M, Fruttiger M et al (2003) VEGF guides angiogenic sprouting utilizing endothelial tip cell filopodia. J Cell Biol 161:1163–1177

Granger DN, Holm L, Kvietys PR (2015) The gastrointestinal circulation: physiology and pathophysiology. In: Terjung R (ed) Comprehensive physiology. Wiley, Hoboken, NJ. https://doi.org/10.1002/cphy.c150007

Gray H, Lewis WH (2000) Gray's anatomy of the human body, 20th edn. Bartleby, New York, NY

Guarner C, Soriano G, Tomas A et al (1993) Increased serum nitrite and nitrate levels in patients with cirrhosis: relationship to endotoxemia. Hepathology 18:1139–1143

Gupta TK, Toruner M, Chung MK et al (1998) Endothelial dysfunction and decreased production of nitric oxide in the intrahepatic microcirculation of cirrhotic rats. Hepatology 28:926–931

Harper D, Chandler B (2016) Splanchnic circulation. BJA Education 16:66–71

Hirschi K, D'Amore PA (1996) Pericytes in the microvasculature. Cardiovasc Res 32:687–698

Iwakiri Y, Shah V, Rockey DC (2014) Vascular pathobiology in chronic liver disease and cirrhosis - current status and future directions. J Hepatol 61:912–924

Jacobson ED (1982) Physiology of the mesenteric circulation. Physiologist 25:439–443

Jakobsson L, Franco CA, Bentley K et al (2010) Endothelial cells dynamically compete for the tip cell position during angiogenic sprouting. Nat Cell Biol 12:943–953

Jankowski JA, Goodlad RA, Wright NA (1994) Maintenance of normal intestinal mucosa: function, structure, and adaptation. Gut 35(Suppl 1):S1–S4

Kachlik D, Baca V, Stingl J (2010) The spatial arrangement of the human large intestinal wall blood circulation. J Anat 216:335–343

Kelly-Goss MR, Sweat RS, Stapor PC et al (2014) Targeting pericytes for angiogenic therapies. Microcirculation 21:345–357

Kvietys PR (2010) The gastrointestinal circulation (chap. 2: anatomy). Morgan & Claypool Life Sciences, San Rafael, CA

Liebner S, Corada M, Bangsow T et al (2008) Wnt/−catenin signaling controls development of the blood—brain barrier. J Cell Biol 183:409–417

Llovet JM, Bruix J (2009) Testing molecular therapies in hepatocellular carcinoma: the need for randomized phase II trials. J Clin Oncol 27:833–835

Maillo C, Martin J, Sebastian D et al (2017) Circadian- and UPR-dependent control of CPEB4 mediates a translational response to counteract hepatic steatosis under ER stress. Nat Cell Biol 19:94–105

Matheson PJ, Wilson MA, Garrison RN (2000) Regulation of intestinal blood flow. J Surg Res 93:182–196

Mejias M, Garcia-Pras E, Tiani C et al (2009) Beneficial effects of sorafenib on splanchnic, intrahepatic, and portocollateral circulations in portal hypertensive and cirrhotic rats. Hepatology 49:1245–1256

Mejias M, Coch L, Berzigotti A et al (2015) Antiangiogenic and antifibrogenic activity of pigment epithelium-derived factor (PEDF) in bile duct-ligated portal hypertensive rats. Gut 64:657–666

Mendez R, Hake LE, Andresson T et al (2000a) Phosphorylation of CPE binding factor by Eg2 regulates translation of c-Mos mRNA. Nature 404:302–307

Mendez R, Murthy KG, Ryan K et al (2000b) Phosphorylation of CPEB by Eg2 mediates the recruitment of CPSF into an active cytoplasmic polyadenylation complex. Mol Cell 6:1253–1259
Ortiz-Zapater E, Pineda D, Martínez-Bosch N et al (2011) Key contribution of CPEB4-mediated translational control to cancer progression. Nat Med 18:83–90
Phillips GB, Schwartz R, Gabuzda GJ Jr et al (1952) The syndrome of impending hepatic coma in patients with cirrhosis of the liver given certain nitrogenous substances. N Engl J Med 247:239–246
Pinter M, Sieghart W, Reiberger T et al (2012) The effects of sorafenib on the portal hypertensive syndrome in patients with liver cirrhosis and hepatocellular carcinoma-a pilot study. Aliment Pharmacol Ther 35:83–91
Pique M, Lopez JM, Foissac S et al (2008) A combinatorial code for CPE-mediated translational control. Cell 132:434–448
Potente M, Gerhardt H, Carmeliet P (2011) Basic and therapeutic aspects of angiogenesis. Cell 146:873–887
Reiberger T, Angermayr B, Schwabl P et al (2009) Sorafenib attenuates the portal hypertensive syndrome in partial portal vein ligated rats. J Hepatol 51:865–873
Reiberger T, Payer BA, Schwabl P et al (2013) Nebivolol treatment increases splanchnic blood flow and portal pressure in cirrhotic rats via modulation of nitric oxide signalling. Liver Int 33:561–568
Rockey DC, Chung JJ (1998) Reduced nitric oxide production by endothelial cells in cirrhotic rat liver: endothelial dysfunction in portal hypertension. Gastroenterology 114:344–351
Sarkissian M, Mendez R, Richter JD (2004) Progesterone and insulin stimulation of CPEB-dependent polyadenylation is regulated by Aurora a and glycogen synthase kinase-3. Genes Dev 18:48–61
Schnabl B, Brenner DA (2014) Interactions between the intestinal microbiome and liver diseases. Gastroenterology 146:1513–1524
Schuppan D, Afdhal NH (2008) Liver cirrhosis. Lancet 371:838–851
Sharara AI, Rockey DC (2001) Gastroesophageal variceal hemorrhage. N Engl J Med 345:669–681
Spadoni I, Zagato E, Bertocchi A et al (2015) A gut-vascular barrier controls the systemic dissemination of bacteria. Science 350:830–834
Spadoni I, Pietrelli A, Pesole G et al (2016) Gene expression profile of endothelial cells during perturbation of the gut vascular barrier. Gut Microbes 7:540–548
Sparks HV (2011) Effect of local metabolic factors on vascular smooth muscle. Supplement 7: handbook of physiology, the cardiovascular system, vascular smooth muscle, pp 475–513
Stapor PC, Sweat RS, Dashti DC et al (2014) Pericyte dynamics during angiogenesis: new insights from new identities. J Vasc Res 51:163–174
Thorburn T, Aali M, Lehmann C (2018) Immune response to systemic inflammation in the intestinal microcirculation. Front Biosci (Landmark Ed) 23:782–795
Tsiaoussis GI, Assimakopoulos SF, Tsamandas AC et al (2015) Intestinal barrier dysfunction in cirrhosis: current concepts in pathophysiology and clinical implications. World J Hepatol 7:2058–2068
Tugues S, Fernandez-Varo G, Muñoz-Luque J et al (2007) Antiangiogenic treatment with sunitinib ameliorates inflammatory infiltrate, fibrosis, and portal pressure in cirrhotic rats. Hepatology 46:1919–1926
Usami M, Miyoshi M, Yamashita H (2015) Gut microbiota and host metabolism in liver cirrhosis. World J Gastroenterol 21:11597–11608
Van Steenkiste C, Geerts A, Vanheule E et al (2009) Role of placental growth factor in mesenteric neoangiogenesis in a mouse model of portal hypertension. Gastroenterology 137:2112–2124
Vespasiani-Gentilucci U, Rombouts K (2015) Boosting pigment epithelial-derived factor: a promising approach for the treatment of early portal hypertension. Gut 64:523–524
Volta U, Bonazzi C, Bianchi FB et al (1987) IgA antibodies to dietary antigens in liver cirrhosis. Ric Clin Lab 17:235–242

Chapter 6
Pericytes in Bone Marrow

Yuya Kunisaki

Abstract Bone marrow environments are composed of multiple cell types, most of which are thought to be derived from mesenchymal stem cells. In mouse bone marrow, stromal cells with $CD45^-$ $Tie2^-$ $CD90^-$ $CD51^+$ $CD105^+$ phenotype, Nestin-GFP^+, CXCL12-abundant reticular (CAR) cells, $PDGFR\alpha^+$ $Sca\text{-}1^+$ or $CD51^+$ $PDGFR\alpha^+$, and Prx-1-derived $CD45^-$ $Ter119^-$ $PDGFR\alpha^+$ $Sca\text{-}1^+$ populations select for MSC activity. There is evidence that these stromal cell populations display some significant overlap with each other and comprise important cellular constituents of the hematopoietic stem cell niche. Moreover, these mesenchymal cell populations share characteristics in their location as they all are found around bone marrow vessels (can be called "pericytes"). In this chapter, with reviewing the recent literatures, how the pericytes relate to physiological and pathological hematopoiesis is argued.

Keywords Pericytes · Perivascular cells · Hematopoietic stem cells · Niche · Mesenchymal stem cells · Skeletal progenitor · Microenvironments · Bone marrow vessels · Leukemia · Cancer · Cancer microenvironments · Cytokine niche

6.1 Introduction

Somatic stem cells self-renew to maintain tissue homeostasis for the lifetime of organisms through tightly controlled proliferation and differentiation (Orkin and Zon 2008; Li and Clevers 2010; Hsu et al. 2011). Hematopoietic stem cells (HSCs) are essentially required for the hematopoietic homeostasis. Therefore, they do not only ensure lifelong replenishment of all blood lineages but also keep their pool constant. Recent studies have highlighted the importance of bone marrow microenvironments that regulate HSC functions (HSC niches) (Kunisaki and Frenette 2012; Morrison and Scadden 2014). In the HSC biology, there has been a considerable interest and debate regarding whether or not quiescence and proliferation of HSCs

---

Y. Kunisaki (✉)
Kyushu University Hospital, Center for Cellular and Molecular Medicine, Fukuoka, Japan
e-mail: kunisaki@cancer.med.kyushu-u.ac.jp

A. Birbrair (ed.), *Pericyte Biology in Different Organs*,
Advances in Experimental Medicine and Biology 1122,
https://doi.org/10.1007/978-3-030-11093-2_6

are regulated by distinct niches (Orford and Scadden 2008). Recent studies have revealed that bone marrow perivascular cells, "pericytes," have mesenchymal stem cell (MSC) activity and that two types of vessels, arterioles and sinusoids, are accompanied with distinct pericytes that are associated with quiescent and proliferative HSCs, respectively (Kunisaki et al. 2013; Asada et al. 2017).

6.2 What Does Bone Marrow Do?

Bone marrow is the main organ that maintains the number of circulating leukocytes and HSCs by controlling their release and recruitment. HSCs self-renew and differentiate into all blood types in response to various demands during life (Orkin and Zon 2008). Functions of HSCs are finely regulated by specialized microenvironments commonly referred to as "niches" in bone marrow (Kunisaki and Frenette 2012; Morrison and Scadden 2014; Mendelson and Frenette 2014). In addition, HSCs are constitutively present at low level in the circulation and constantly released into peripheral blood and home to bone marrow during homeostasis (Laird et al. 2008).

HSCs are released into peripheral blood in a circadian manner, which is controlled by the sympathetic nervous system through the regulation of CXCL12 levels in bone marrow (Mendez-Ferrer et al. 2008; Scheiermann et al. 2013). Vessels border blood circulation in all organs and are important to recruit leukocytes to the specific organs by expressing adhesion molecules and secreting chemokines according to the situations (Druzd et al. 2017; Scheiermann et al. 2012). Endothelial selectins (P-selectin and E-selectin) and vascular cell adhesion molecule-1 (VCAM-1) are constitutively expressed in bone marrow and have been shown to be important for the rolling and homing process of the HSCs (Scheiermann et al. 2012). E-selectin, an adhesion molecule constitutively expressed in certain BM sinusoid microdomains, promotes HSC proliferation, and blockade of E-selectin protects HSC following chemotherapy or irradiation, suggesting that sinusoids represent a proliferative niche for HSCs (Winkler et al. 2012). Therefore, vessels are structurally and functionally an important constituent of bone marrow.

6.3 Bone Marrow Vascular Structure

In bone marrow, there is an even distribution of the sinusoidal network that occupies ~30% of bone marrow volume, and individual sinusoidal vessels are regularly spaced and draining into the central vein. In addition to sinusoids, there are small-caliber (~20 μm) arterioles, which comprise a much smaller volumetric fraction ~1% of bone marrow. The vessels are distinguished from sinusoids in respect to their pronounced Tie-2-GFP and CD31 expression (Li et al. 2009), absence of in vivo staining with the sinusoid-specific Dil-Ac-LDL (Li et al. 2009), and strong staining with the artery-specific dye Alexa Fluor 633 (Shen et al. 2012). Bone marrow vasculature, hence, is composed of distinct vascular components, arterioles and sinusoids (Fig. 6.1a, b) (Kunisaki et al. 2013).

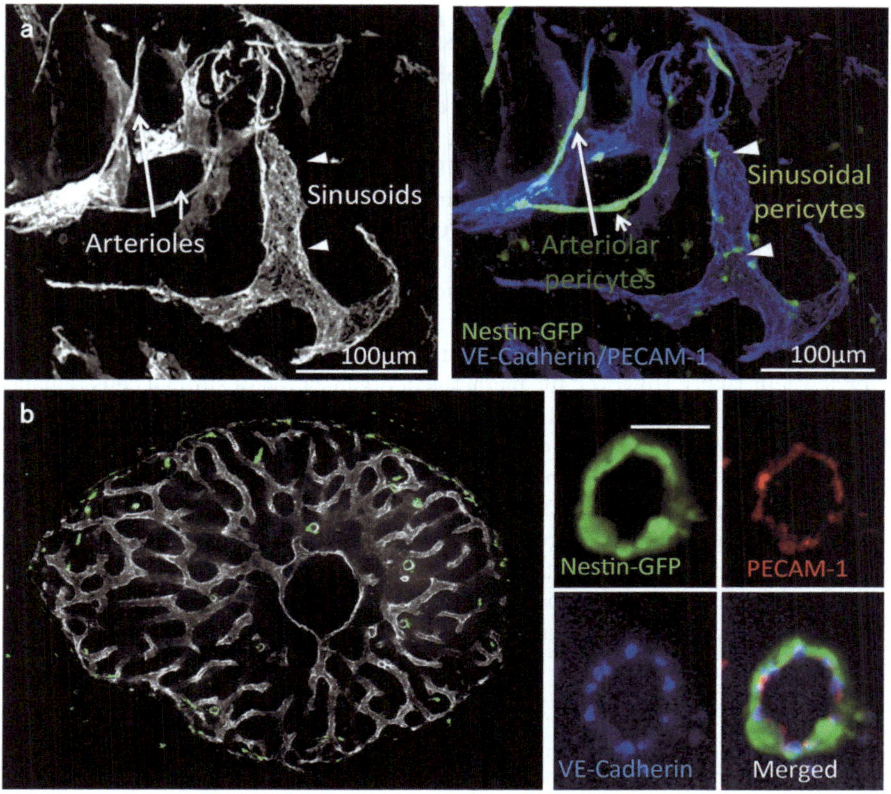

Fig. 6.1 Bone marrow vessels and pericytes. (**a**) Immunofluorescence images of mouse sternal bone marrow stained with anti-VE cadherin and PECAM1 antibodies (left). Two vessel types, arterioles and sinusoids, are ensheathed by distinct pericytes, Nestin-GFP^{high} and Nestin-GFP^{dim} cells, respectively (right). Scale bar: 100 μm. (**b**) Immunofluorescence images of cross-sectioned mouse femoral bone marrow stained with anti-VE cadherin and PECAM1 antibodies (left, right). Nestin-GFP^{dim} cells are not appreciated due to the contrast to Nestin-GFP^{high} cells. Nestin-GFP^{high} cells exist an outer layer wrapping endothelial cells forming a vascular lumen (right)

6.4 Mesenchymal Precursor and Osteoprecursor Cells as Bone Marrow Components

Recent studies have highlighted the critical importance of bone marrow stromal cells as essential constituents for the HSC niche through the production of factors such as CXCL12 and SCF. Bone marrow stromal cells are derived from mesenchymal stem cells (MSCs) (Caplan 1991; Rodda and McMahon 2006; Frenette et al. 2013). Stromal progenitor activity in bone marrow was initially isolated from clonal populations of fibroblastic colony-forming units (CFU-F) that exhibit self-renewal and capacity to differentiate into the major mesenchymal lineages (Zhou et al.

2010). In mouse bone marrow, the cell populations exhibiting the mesenchymal stem/progenitor activity are isolated as Nestin-GFP$^+$ (Mendez-Ferrer et al. 2010), CD45$^-$ Tie2$^-$ CD90$^-$ CD51$^+$ CD105$^+$ phenotype (Chan et al. 2009), CXCL12-abundant reticular (CAR) cells (Omatsu et al. 2010; Sugiyama et al. 2006), leptin receptor-positive cells (Ding et al. 2012), PDGFRα^+ Sca-1$^+$ (Morikawa et al. 2009) or CD51$^+$ PDGFRα^+ (Pinho et al. 2013), and Prx-1-derived CD45$^-$ Ter119$^-$ PDGFRα^+ Sca-1$^+$ populations (Greenbaum et al. 2013). Recent lineage tracing analyses using tamoxifen-inducible Osx-cre (Osx-cre^{ERT2}) have revealed that expression of Osx is not restricted in osteoblasts and that Osx-expressing cells at the neonatal stage are bone marrow stromal precursors (Mizoguchi et al. 2014). Owen and Friedenstein predicted 25 years ago that the differentiation tree of stromal cells of the bone marrow is as complex as their hematopoietic counterparts (Owen and Friedenstein 1988). However, the hierarchical organization of stromal cells remains unknown. In addition, stromal cells marked by various genetic reporter mouse strains, e.g., Prx-1-cre, Osterix-cre, NG2-cre^{ERTM}, leptin receptor-cre, and NG2-cre, have been reported as important niche constituents (Kunisaki et al. 2013; Omatsu et al. 2010; Greenbaum et al. 2013; Ding and Morrison 2013). There is evidence that these stromal cell populations display some significant overlap with each other and comprise important cellular constituents of the HSC niche, suggesting that MSCs organize the bone marrow environment. Moreover, these mesenchymal cell populations share characteristics in their location as they all are found around bone marrow vessels (can be called "pericytes").

6.5 Developmental Origins of the Pericytes

Developmental origins of pericytes in bone marrow have still been under investigations. There are several studies reporting that mesenchymal cells arise from the neural crest or mesoderm. The neural crest stem cells (NCSCs), which exhibit the potentials to self-renew and differentiate into neurons, glial cells, and myofibroblasts (Morrison et al. 1999), have been described in various adult tissues. Okano's group report the presence of NCSCs in bone marrow (Nagoshi et al. 2008) and suggest the possibility that NCSCs might have overlap MSCs in bone marrow. Another study from Yamazaki's group examined the origins of bone marrow stromal cells by using the genetically labeled mouse models as Mesp1-cre for mesoderm-derived cells and P0-cre for neural crest-derived cells and showed their presence in bone marrow (Komada et al. 2012). These studies demonstrate that bone marrow NCSCs exhibit mesenchymal activity as they differentiate into osteo-, chondro-, and adipo-lineages in vitro as do bone marrow MSCs, suggesting that there are two stromal cell types exhibiting MSC activity that are derived from distinct origins in bone marrow. However, whether NCSCs overlap the pericytes, such as Nestin-GFP+, LepR+, and CAR cells that comprise the HSC niches, is yet known.

6.6 Arterioles and Sinusoids Are Ensheathed by Distinct Mesenchymal Pericytes

One study using the genetic mouse model, where GFP is expressed under nestin promoter (Nestin-GFP), shows anatomical relationships of pericytes and distinct vascular structures, arterioles and sinusoids (Kunisaki et al. 2013). Nestin-GFP$^+$ pericytes contain all bone marrow mesenchymal stem cell activity (Mendez-Ferrer et al. 2010) and two distinct subsets that are associated with distinct vessel types: rare Nestin-GFP cells expressing the NG2 antigen (~10% of Nestin-GFP$^+$ cells) but not leptin receptor are exclusively associated with arterioles, whereas NG2$^-$ and LEPR$^+$ Nestin-GFP cells (~90% of Nestin-GFP$^+$ cells) are associated with sinusoids (Fig. 6.1a, b) (Kunisaki et al. 2013).

6.7 Arteriolar Pericytes Are Quiescent and Protected from Myeloablation

The Nestin-GFP$^+$ pericytes associated with arterioles are more quiescent than the sinusoidal pericytes, as expression of the nuclear proliferation markers, Ki-67 and proliferation cell nuclear antigen (PCNA), in arteriolar pericytes was significantly lower than in sinusoidal pericytes and other stromal cells. As the arteriolar pericytes are largely preserved structurally and numerically as compared to the sinusoidal pericytes after 5-fluorouracil (5FU), a drug that kills cycling cells, the quiescent status of the arteriolar pericytes is functionally demonstrated (Kunisaki et al. 2013).

6.8 HSC Niches

The "niche" concept has been validated by numerous studies since Schofield postulated its presence in 1978 (Schofield 1978). Most of the HSCs divide infrequently and are quiescent in the niche (Kiel et al. 2007). They, however, are reversibly activated in response to hematopoietic stresses (Wilson et al. 2008). Cell cycle quiescence is a key behavior of stem cells, which protects them from being exhausted by exogenous insults and is also assumed to prevent them from acquiring genetic mutations that potentially result in consequent malignant transformations (Lobo et al. 2007).

The identification of cellular constituents of the HSC niche has recently been the subject of intense investigations. Osteoblasts have been proposed to promote HSC quiescence via direct contact (Sugimura et al. 2012) and the secretion of angiopoietin-1 (Arai et al. 2004) or osteopontin (Nilsson et al. 2005; Stier et al. 2005). On the other hand, an emerging role of bone marrow vasculature has recently gained support and interest (Kiel et al. 2007, 2005; Takakura 2012) as other studies have found that most HSCs are localized adjacent to blood vessels (sinusoids), near perivascular cell populations characterized as CXCL12-abundant reticular (CAR) cells (Omatsu et al. 2010; Sugiyama et al. 2006), Nestin-GFP$^+$ mesenchymal progenitors (Mendez-Ferrer et al. 2010), and leptin receptor (LEPR)$^+$ cells (Ding et al. 2012). The proximity of HSCs to sinusoidal vessels has been reported in many previous studies. The relationship between HSCs and arteries also has been well documented in the emergence of HSCs during development as definitive hematopoiesis is beginning in the aorta-gonad-mesonephros (AGM) region (Mikkola and Orkin 2006).

6.9 Assessment of Spatial Relationships by Computational Simulation

Unbiased assessment by computational simulation is useful to evaluate the significance of the association between HSCs and bone marrow structures observed in situ (Kunisaki et al. 2013; Chen et al. 2016; Acar et al. 2015). In a null model, HSCs are randomly placed on the unoccupied regions of binary spatial maps of bone marrow structures, sinusoids, arterioles, and bone lining osteoblasts, defined from the images of whole-mount prepared bone, and Euclidean distance of the random HSCs to the structures is measured. The means of 1000 simulations define a distribution of mean distances one would observe for non-preferentially localized HSCs in relation to the respective structures. If the in situ distance measurements are not statistically different from those obtained by a random placement of HSCs on the same structures in silica, this will indicate a non-preferential spatial HSC distribution.

In one study using this modeling (Kunisaki et al. 2013), the mean distance observed in situ to sinusoids could not be statistically differentiated from that of randomly placed HSCs. By contrast, the observed mean distance to arterioles was highly statistically different from that of randomly placed HSCs. The analyses using the combination of high-quality imaging and the computational simulation have revealed that quiescent HSCs are specifically associated with small-caliber arterioles (arteriolar niches) (Fig. 6.2).

6.10 Contributions by Distinct Pericytes to HSC Maintenance

There are two types of Nestin-GFP$^+$ pericytes expressing different surface markers, nerve/glial antigen 2 (NG2) and leptin receptor (Lepr) that are associated with arterioles and sinusoid, respectively, in bone marrow (Kunisaki et al. 2013). Both arteriolar and sinusoidal pericytes show high gene expression of cytokines essential for HSC maintenance such as CXCL12 and stem cell factor (SCF) (Asada et al. 2017).

To investigate contributions of arteriolar and sinusoidal pericytes, we can utilize genetic mouse models with a cre/flox targeted gene deleting system in which

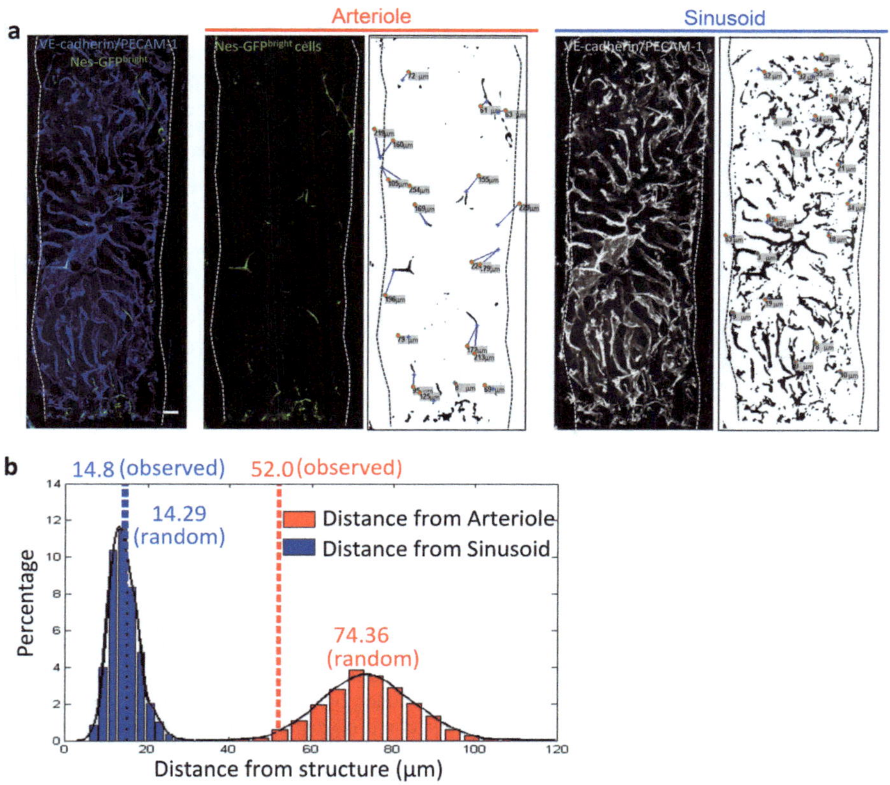

Fig. 6.2 A computational modeling of the HSC localization in relation to bone marrow arterioles or sinusoids. (**a**) Computational simulation of randomly distributed HSCs on images of whole-mount prepared sterna. To establish the null-model, binary spatial maps of the sinusoids, arterioles were defined from the images of whole-mount prepared sterna. To simulate a null model in which HSCs are not preferentially localized in the marrow, we randomly placed 20 HSCs (to reflect the mean HSCs/sternum observed in situ) on the unoccupied regions of the spatial maps and measured the Euclidean distance of HSCs to the nearest vascular structure. Scale bar: 100 μm. (**b**) The means of 1000 simulations defined a distribution of mean distances one would observe for non-preferentially (random) localized HSCs in relation to the respective structures. If the in situ distance measurements are not statistically significantly different from those obtained by a random placement of HSCs on the same structures in silica, this would indicate a non-preferential spatial HSC distribution. The cumulative probability of observing the in situ mean was calculated based on the normal distribution obtained from our simulation on a map of each bone marrow structure

CXCL12 or SCF (e.g., CXCL12-flox or SCF-flox animals) can be deleted in specific cell types (e.g., NG-2 cre^{ERTM} or leptin receptor cre) (Asada et al. 2017; Ding et al. 2012; Greenbaum et al. 2013; Ding and Morrison 2013). CXCL12 deletions in sinusoidal pericytes by using Lepr-cre/$Cxcl12^{fl/-}$ mice mobilize HSCs but have no effect on bone marrow HSC numbers (Asada et al. 2017; Ding and Morrison 2013). To examine further a role of CXCL12 produced by $NG2^+$ arteriolar pericytes on HSC maintenance, tamoxifen-inducible NG2-cre^{ERTM}/$Cxcl12^{fl/-}$ mice are used. Deletion of CXCL12 postnatally in $NG2^+$ arteriolar niche cells significantly reduces

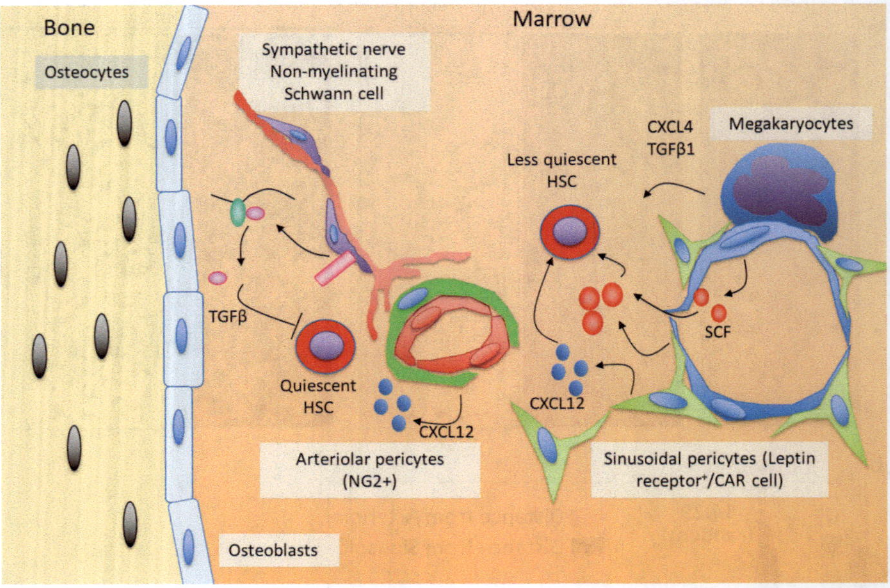

Fig. 6.3 Hematopoietic stem cell niches in bone marrow. Hematopoietic stem cells, HSCs, reside in a specialized microenvironment called the niche. A variety of cells, osteoblasts, endothelial cells, pericytes, non-myelinating Schwann cells, and megakaryocytes, have been reported as niche components. Arterioles and sinusoids that are ensheathed by distinct perivascular NG2$^+$ Nestin-GFP^{bright} or Lepr$^+$ Nestin-GFP^{dim} cells, respectively, comprise distinct cytokine niches for HSCs as the arteriolar niche promotes quiescence of HSCs, whereas sinusoids form another niche, which play major roles in HSC proliferation

the number of HSCs in bone marrow. Deletion of SCF in Lepr-cre targeted cells shows a significant reduction of HSC numbers in bone marrow (Ding et al. 2012), whereas no significant change is observed in NG2-cre^{ERTM}/SCF$^{fl/-}$ mice, suggesting that Lepr$^+$ vascular niches rather than NG2$^+$ arteriolar niches are the most important source of SCF in bone marrow. These results highlight distinct contributions of pericytes primarily located in separate vascular niches, arteriolar and sinusoidal, in HSC maintenance (Fig. 6.3).

6.11 Bone Marrow Pericytes in Pathological Conditions

6.11.1 Bone Marrow Microenvironments Remodeled by Leukemia Cells

In recent years, there have been several studies reporting that leukemia cells remodel bone marrow microenvironments to favor leukemia progression. These changes suppress normal hematopoiesis that could lead to myelosuppression in leukemia.

Passegue's group reported that BCR-/ABL-positive leukemia cells in the mouse chronic myelogenous leukemia model promote bone marrow pericytes (=mesenchymal progenitor cells) to differentiate into bone lineages that suppress normal hematopoiesis (Schepers et al. 2013). Hong's group, likewise, demonstrated that leukemia precursor cells reconstruct bone marrow that helps the leukemia cells escape from chemotherapy (Duan et al. 2014). In the study from Frenette's group, Nestin-GFP-positive mesenchymal pericytes are induced to proliferate and differentiate into osteo-lineages in mice transplanted with MLL-AF9 AML cells, resulting in altered distribution of normal HSCs apart from the arteriolar niche with reduction in their number (Hanoun et al. 2014).

Furthermore, the group of Mendez-Ferrer reported that in the myeloproliferative disorder model mice induced by the human JAK2 (V617F) mutant gene, tumor cells impair nerves by producing IL-1β and transform bone marrow microenvironments that promote tumor development. These changes induce apoptosis of pericytes, resulting in suppressed HSC functions (Arranz et al. 2014). Nowak et al. also reported that human myelodysplastic syndrome cells facilitate patient-derived xenografts in mice by transplanting simultaneously with mesenchymal stromal cells derived from the same patient (Medyouf et al. 2014). These results collectively support the notion that cancer cells hijack bone marrow microenvironments to support cancer growth by involving the pericytes.

6.12 Leukemia Development and Pericytes

Studies using genetic modification mouse models show that hematopoietic tumors can occur by dysfunction of bone marrow microenvironments. Two studies from the groups of Orkin et al. and Purton et al. demonstrated that deficiency of the retinoic acid receptor γ gene or Rb oncogene causes myeloproliferative disorders (Walkley et al. 2007a, b). More recently, using the tissue cell-specific mouse genetic modification model, Scadden's group shows that spontaneous development of myelodysplastic syndrome by deletion of a microRNA processing enzyme, Dicer 1, specifically in bone precursor cells (Raaijmakers et al. 2010). In addition, the group of Kousteni reported that acute myeloid leukemia develops by constitutively activating β-catenin specifically in osteoblasts (Kode et al. 2014).

6.13 Metastatic Bone Cancer and Pericytes

Recent studies have revealed that environmental cues regulate not only primary tumor growth but also the establishment and progression of metastasis (Yoneda and Hiraga 2005). Bones are preferred for metastases in various malignancies including malignant melanoma and breast and prostate cancer. The metastatic processes highly depend on the vasculature that provides a route for cancer cell to disseminate into the sites and access oxygen and nutrients. Tumor vessels are also embedded with pericytes, which are reported to play important roles in capturing circulating tumor cells at the metastasized tissues (Caplan 2017). One study from Caplan's group shows that bone marrow pericytes attract the circulating melanoma cells by releasing their "fingers" through the fenestrations in the endothelial layers. These processes are enhanced by platelet-derived growth factor (PDGF)-BB released from broken platelets(Correa et al. 2016). Collectively, the vasculature wrapped with "pericytes" from metastatic niches for cancer not only for leukemia in bone marrow.

6.14 Concluding Remarks

"Pericytes," synonymous to perivascular mesenchymal progenitors in bone marrow, border vascular endothelial cells and bone marrow cavities and play important roles in supporting hematopoiesis as niches as well as in organ regeneration and angiogenesis.

HSCs are essentially required for hematopoietic homeostasis, for that signals from the HSC niche are critical. Studies herein report that such selected microenvironment exists and highlight the possibility of heterogeneity among niche factor-producing pericytes. As oligopotent HSCs are identified (Notta et al. 2016; Sanjuan-Pla et al. 2013; Yamamoto et al. 2013; Pinho et al. 2018), further studies will determine the extent by which HSC heterogeneity is matched by niche heterogeneity.

Recently, cellular heterogeneity within a tumor has been highlighted, and among aggressively proliferating tumor cells, a small fraction is found in quiescent status, termed "cancer stem cells," causing refractoriness to anticancer therapy and relapsed diseases. Even in cancer cells that appear to proliferate in an unlimited manner, environmental cues are believed to control their cell cycle status, quiescence, or proliferation. Therefore, "decision" whether symmetric or asymmetric division is essential for hematologic tumors such as leukemia to maintain "leukemic stem cell (LSC)" clones. The anatomical and functional interactions between hematopoietic cells or leukemia and their microenvironments are essential to efficiently control normal and pathological hematopoiesis (Fig. 6.4). The outcome that further studies will achieve with the niche is expected to add new concept as "anticancer niche" to cancer therapy.

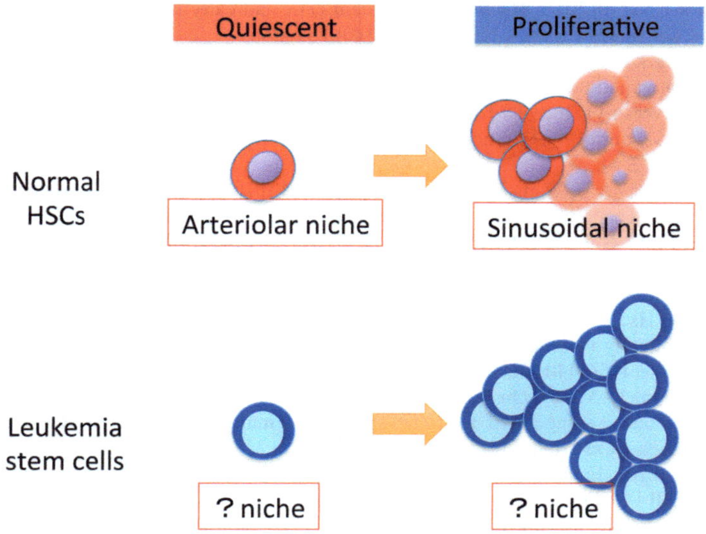

Fig. 6.4 Leukemia microenvironments as a therapeutic target. In steady state, the fate of HSCs, quiescence, proliferation, or differentiation, is determined by their special microenvironments (HSC niche). Arteriolar niches keep HSCs to be quiescent, whereas sinusoidal niches rather enforce them to be actively proliferated or mobilized. With gaining our knowledge about environmental cues for pathological hematopoiesis, a new concept "anticancer niche therapy" will be emerging

References

Acar M, Kocherlakota KS, Murphy MM, Peyer JG, Oguro H, Inra CN et al (2015) Deep imaging of bone marrow shows non-dividing stem cells are mainly perisinusoidal. Nature 526(7571):126–130

Arai F, Hirao A, Ohmura M, Sato H, Matsuoka S, Takubo K et al (2004) Tie2/angiopoietin-1 signaling regulates hematopoietic stem cell quiescence in the bone marrow niche. Cell 118(2):149–161

Arranz L, Sanchez-Aguilera A, Martin-Perez D, Isern J, Langa X, Tzankov A et al (2014) Neuropathy of haematopoietic stem cell niche is essential for myeloproliferative neoplasms. Nature 512(7512):78–81

Asada N, Kunisaki Y, Pierce H, Wang Z, Fernandez NF, Birbrair A et al (2017) Differential cytokine contributions of perivascular haematopoietic stem cell niches. Nat Cell Biol 19(3):214–223

Caplan AI (1991) Mesenchymal stem cells. J Orthop Res 9(5):641–650

Caplan AI (2017) New MSC: MSCs as pericytes are sentinels and gatekeepers. J Orthop Res 35(6):1151–1159

Chan CK, Chen CC, Luppen CA, Kim JB, DeBoer AT, Wei K et al (2009) Endochondral ossification is required for haematopoietic stem-cell niche formation. Nature 457(7228):490–494

Chen JY, Miyanishi M, Wang SK, Yamazaki S, Sinha R, Kao KS et al (2016) Hoxb5 marks long-term haematopoietic stem cells and reveals a homogenous perivascular niche. Nature 530(7589):223–227

Correa D, Somoza RA, Lin P, Schiemann WP, Caplan AI (2016) Mesenchymal stem cells regulate melanoma cancer cells extravasation to bone and liver at their perivascular niche. Int J Cancer 138(2):417–427

Ding L, Morrison SJ (2013) Haematopoietic stem cells and early lymphoid progenitors occupy distinct bone marrow niches. Nature 495(7440):231–235

Ding L, Saunders TL, Enikolopov G, Morrison SJ (2012) Endothelial and perivascular cells maintain haematopoietic stem cells. Nature 481(7382):457–462

Druzd D, Matveeva O, Ince L, Harrison U, He W, Schmal C et al (2017) Lymphocyte circadian clocks control lymph node trafficking and adaptive immune responses. Immunity 46(1):120–132

Duan CW, Shi J, Chen J, Wang B, Yu YH, Qin X et al (2014) Leukemia propagating cells rebuild an evolving niche in response to therapy. Cancer Cell 25(6):778–793

Frenette PS, Pinho S, Lucas D, Scheiermann C (2013) Mesenchymal stem cell: keystone of the hematopoietic stem cell niche and a stepping-stone for regenerative medicine. Annu Rev Immunol 31:285–316

Greenbaum A, Hsu YM, Day RB, Schuettpelz LG, Christopher MJ, Borgerding JN et al (2013) CXCL12 in early mesenchymal progenitors is required for haematopoietic stem-cell maintenance. Nature 495(7440):227–230

Hanoun M, Zhang D, Mizoguchi T, Pinho S, Pierce H, Kunisaki Y et al (2014) Acute myelogenous leukemia-induced sympathetic neuropathy promotes malignancy in an altered hematopoietic stem cell niche. Cell Stem Cell 15(3):365–375

Hsu YC, Pasolli HA, Fuchs E (2011) Dynamics between stem cells, niche, and progeny in the hair follicle. Cell 144(1):92–105

Kiel MJ, Yilmaz OH, Iwashita T, Terhorst C, Morrison SJ (2005) SLAM family receptors distinguish hematopoietic stem and progenitor cells and reveal endothelial niches for stem cells. Cell 121(7):1109–1121

Kiel MJ, He S, Ashkenazi R, Gentry SN, Teta M, Kushner JA et al (2007) Haematopoietic stem cells do not asymmetrically segregate chromosomes or retain BrdU. Nature 449(7159):238–242

Kode A, Manavalan JS, Mosialou I, Bhagat G, Rathinam CV, Luo N et al (2014) Leukaemogenesis induced by an activating beta-catenin mutation in osteoblasts. Nature 506(7487):240–244

Komada Y, Yamane T, Kadota D, Isono K, Takakura N, Hayashi S et al (2012) Origins and properties of dental, thymic, and bone marrow mesenchymal cells and their stem cells. PLoS One 7(11):e46436

Kunisaki Y, Frenette PS (2012) The secrets of the bone marrow niche: enigmatic niche brings challenge for HSC expansion. Nat Med 18(6):864–865

Kunisaki Y, Bruns I, Scheiermann C, Ahmed J, Pinho S, Zhang D et al (2013) Arteriolar niches maintain haematopoietic stem cell quiescence. Nature 502(7473):637–643

Laird DJ, von Andrian UH, Wagers AJ (2008) Stem cell trafficking in tissue development, growth, and disease. Cell 132(4):612–630

Li L, Clevers H (2010) Coexistence of quiescent and active adult stem cells in mammals. Science 327(5965):542–545

Li XM, Hu Z, Jorgenson ML, Slayton WB (2009) High levels of acetylated low-density lipoprotein uptake and low tyrosine kinase with immunoglobulin and epidermal growth factor homology domains-2 (Tie2) promoter activity distinguish sinusoids from other vessel types in murine bone marrow. Circulation 120(19):1910–1918

Lobo NA, Shimono Y, Qian D, Clarke MF (2007) The biology of cancer stem cells. Annu Rev Cell Dev Biol 23:675–699

Medyouf H, Mossner M, Jann JC, Nolte F, Raffel S, Herrmann C et al (2014) Myelodysplastic cells in patients reprogram mesenchymal stromal cells to establish a transplantable stem cell niche disease unit. Cell Stem Cell 14(6):824–837

Mendelson A, Frenette PS (2014) Hematopoietic stem cell niche maintenance during homeostasis and regeneration. Nat Med 20(8):833–846

Mendez-Ferrer S, Lucas D, Battista M, Frenette PS (2008) Haematopoietic stem cell release is regulated by circadian oscillations. Nature 452(7186):442–447

Mendez-Ferrer S, Michurina TV, Ferraro F, Mazloom AR, Macarthur BD, Lira SA et al (2010) Mesenchymal and haematopoietic stem cells form a unique bone marrow niche. Nature 466(7308):829–834

Mikkola HK, Orkin SH (2006) The journey of developing hematopoietic stem cells. Development 133(19):3733–3744
Mizoguchi T, Pinho S, Ahmed J, Kunisaki Y, Hanoun M, Mendelson A et al (2014) Osterix marks distinct waves of primitive and definitive stromal progenitors during bone marrow development. Dev Cell 29(3):340–349
Morikawa S, Mabuchi Y, Kubota Y, Nagai Y, Niibe K, Hiratsu E et al (2009) Prospective identification, isolation, and systemic transplantation of multipotent mesenchymal stem cells in murine bone marrow. J Exp Med 206(11):2483–2496
Morrison SJ, Scadden DT (2014) The bone marrow niche for haematopoietic stem cells. Nature 505(7483):327–334
Morrison SJ, White PM, Zock C, Anderson DJ (1999) Prospective identification, isolation by flow cytometry, and in vivo self-renewal of multipotent mammalian neural crest stem cells. Cell 96(5):737–749
Nagoshi N, Shibata S, Kubota Y, Nakamura M, Nagai Y, Satoh E et al (2008) Ontogeny and multipotency of neural crest-derived stem cells in mouse bone marrow, dorsal root ganglia, and whisker pad. Cell Stem Cell 2(4):392–403
Nilsson SK, Johnston HM, Whitty GA, Williams B, Webb RJ, Denhardt DT et al (2005) Osteopontin, a key component of the hematopoietic stem cell niche and regulator of primitive hematopoietic progenitor cells. Blood 106(4):1232–1239
Notta F, Zandi S, Takayama N, Dobson S, Gan OI, Wilson G et al (2016) Distinct routes of lineage development reshape the human blood hierarchy across ontogeny. Science 351(6269):aab2116
Omatsu Y, Sugiyama T, Kohara H, Kondoh G, Fujii N, Kohno K et al (2010) The essential functions of adipo-osteogenic progenitors as the hematopoietic stem and progenitor cell niche. Immunity 33(3):387–399
Orford KW, Scadden DT (2008) Deconstructing stem cell self-renewal: genetic insights into cell-cycle regulation. Nat Rev Genet 9(2):115–128
Orkin SH, Zon LI (2008) Hematopoiesis: an evolving paradigm for stem cell biology. Cell 132(4):631–644
Owen M, Friedenstein AJ (1988) Stromal stem cells: marrow-derived osteogenic precursors. Ciba Found Symp 136:42–60
Pinho S, Lacombe J, Hanoun M, Mizoguchi T, Bruns I, Kunisaki Y et al (2013) PDGFRalpha and CD51 mark human nestin+ sphere-forming mesenchymal stem cells capable of hematopoietic progenitor cell expansion. J Exp Med 210(7):1351–1367
Pinho S, Marchand T, Yang E, Wei Q, Nerlov C, Frenette PS (2018) Lineage-biased hematopoietic stem cells are regulated by distinct niches. Dev Cell 44(5):634–641.e4
Raaijmakers MH, Mukherjee S, Guo S, Zhang S, Kobayashi T, Schoonmaker JA et al (2010) Bone progenitor dysfunction induces myelodysplasia and secondary leukaemia. Nature 464(7290):852–857
Rodda SJ, McMahon AP (2006) Distinct roles for hedgehog and canonical Wnt signaling in specification, differentiation and maintenance of osteoblast progenitors. Development 133(16):3231–3244
Sanjuan-Pla A, Macaulay IC, Jensen CT, Woll PS, Luis TC, Mead A et al (2013) Platelet-biased stem cells reside at the apex of the haematopoietic stem-cell hierarchy. Nature 502(7470):232–236
Scheiermann C, Kunisaki Y, Lucas D, Chow A, Jang JE, Zhang D et al (2012) Adrenergic nerves govern circadian leukocyte recruitment to tissues. Immunity 37(2):290–301
Scheiermann C, Kunisaki Y, Frenette PS (2013) Circadian control of the immune system. Nat Rev Immunol 13(3):190–198
Schepers K, Pietras EM, Reynaud D, Flach J, Binnewies M, Garg T et al (2013) Myeloproliferative neoplasia remodels the endosteal bone marrow niche into a self-reinforcing leukemic niche. Cell Stem Cell 13(3):285–299
Schofield R (1978) The relationship between the spleen colony-forming cell and the haemopoietic stem cell. Blood Cells 4(1–2):7–25

Shen Z, Lu Z, Chhatbar PY, O'Herron P, Kara P (2012) An artery-specific fluorescent dye for studying neurovascular coupling. Nat Methods 9(3):273–276
Stier S, Ko Y, Forkert R, Lutz C, Neuhaus T, Grunewald E et al (2005) Osteopontin is a hematopoietic stem cell niche component that negatively regulates stem cell pool size. J Exp Med 201(11):1781–1791
Sugimura R, He XC, Venkatraman A, Arai F, Box A, Semerad C et al (2012) Noncanonical Wnt signaling maintains hematopoietic stem cells in the niche. Cell 150(2):351–365
Sugiyama T, Kohara H, Noda M, Nagasawa T (2006) Maintenance of the hematopoietic stem cell pool by CXCL12-CXCR4 chemokine signaling in bone marrow stromal cell niches. Immunity 25(6):977–988
Takakura N (2012) Guest editorial: mutual relationship between vascular biology and hematology. Int J Hematol 95(2):117–118
Walkley CR, Olsen GH, Dworkin S, Fabb SA, Swann J, McArthur GA et al (2007a) A microenvironment-induced myeloproliferative syndrome caused by retinoic acid receptor gamma deficiency. Cell 129(6):1097–1110
Walkley CR, Shea JM, Sims NA, Purton LE, Orkin SH (2007b) Rb regulates interactions between hematopoietic stem cells and their bone marrow microenvironment. Cell 129(6):1081–1095
Wilson A, Laurenti E, Oser G, van der Wath RC, Blanco-Bose W, Jaworski M et al (2008) Hematopoietic stem cells reversibly switch from dormancy to self-renewal during homeostasis and repair. Cell 135(6):1118–1129
Winkler IG, Barbier V, Nowlan B, Jacobsen RN, Forristal CE, Patton JT et al (2012) Vascular niche E-selectin regulates hematopoietic stem cell dormancy, self renewal and chemoresistance. Nat Med 18(11):1651–1657
Yamamoto R, Morita Y, Ooehara J, Hamanaka S, Onodera M, Rudolph KL et al (2013) Clonal analysis unveils self-renewing lineage-restricted progenitors generated directly from hematopoietic stem cells. Cell 154(5):1112–1126
Yoneda T, Hiraga T (2005) Crosstalk between cancer cells and bone microenvironment in bone metastasis. Biochem Biophys Res Commun 328(3):679–687
Zhou X, Zhang Z, Feng JQ, Dusevich VM, Sinha K, Zhang H et al (2010) Multiple functions of Osterix are required for bone growth and homeostasis in postnatal mice. Proc Natl Acad Sci U S A 107(29):12919–12924

Chapter 7
Cochlear Capillary Pericytes

Martin Canis and Mattis Bertlich

Abstract Capillary pericytes in the cochlea of mammals are—compared to pericytes in other tissues, like the CNS—relatively poorly researched. To begin with, there is still a considerable debate as to whether the very last precapillary arterioles should—due to their contractile properties—may be considered to be pericytes.

However, cochlear capillary pericytes have shifted into the center of attention in the past decade. Most mammals show a considerable number of pericytes in the stria vascularis of the cochlea—up to 1300 in a mouse alone. This high number may be explained by the observation that cochlear capillary pericytes may be differentiated into different subgroups, depending on the immune markers that are expressed by them. Corresponding with these subpopulations, cochlear pericytes fulfill three core functions in the physiology of the cochlea:

- Formation of the intrastrial blood-fluid barrier—Pericytes monitor the ion, fluid, and nutrient household and aid in the homeostasis thereof.
- Regulation of cochlear blood flow—By contraction on relaxation, pericytes contribute to the regulation of cochlear blood flow, a paramount function parameter of the cochlea.
- Immune response—Pericytes actually contribute to the immune response in inflammation of the cochlea.

Due to these central roles in the physiology of the cochlea, pericytes actually play a major role in numerous cochlear pathologies, including, but not limited to, sudden sensorineural hearing loss, acoustic trauma, and inflammation of the cochlea.

Keywords Capillary pericytes · Cochlea · Cochlear blood flow · Cochlear blood flow regulation · Strial blood-fluid barrier · Immune response · Fluid homeostasis · Cochlear pathology

M. Canis · M. Bertlich (✉)
The Department of Otorhinolaryngology, Head and Neck Surgery, University Hospital, Munich, Federal Republic of Germany
e-mail: Mattis.Bertlich@med.uni-muenchen.de

A. Birbrair (ed.), *Pericyte Biology in Different Organs*,
Advances in Experimental Medicine and Biology 1122,
https://doi.org/10.1007/978-3-030-11093-2_7

7.1 Introduction

Pericytes in the cochlea are generally considered to be those cells that adhere to the outer wall of capillaries of the stria vascularis or the spiral ligament. While pericytes in many other tissues have been characterized relatively well, the function of cochlear capillary pericytes remains widely unclear.

In the cochlea, there are two distinct subpopulations of pericytes—those that adhere to the capillaries of the stria vascularis and those that adhere to the capillaries of the spiral ligament. Both show remarkable differences in terms of cellular markers these express—while the pericytes of the stria vascularis are very rich in the structural protein desmin, those of the spiral ligament show a rich expression of contractile proteins such as smooth muscle actin and tropomyosin. It is probable that while those pericytes that are rich in desmin fulfill a structural purpose, those rich in contractile proteins may take an active part in regulating cochlear microcirculation.

In the following chapter, we therefore discuss the role of cochlear capillary pericytes in various pathologies of the cochlea, like physical or acoustic trauma, bacterial inflammation, or sudden sensorineural hearing loss. Moreover, we discuss the signaling pathways involved and the role each pericyte subpopulation plays in these pathologies.

Finally, we will discuss the pharmacological properties of pericytes in the cochlea and give an outlook as to where research on cochlear capillary pericytes may head in the following years.

7.2 Main Text

7.2.1 What Is a Cochlear Pericyte?

Generally speaking, pericytes are cells that adhere to the external walls of capillaries and postcapillary venules; they appear in most tissues when energy demand is high and a dense net of capillaries is present, e.g., the central nervous system (CNS) or the kidneys. Knowing this, it is not surprising that pericytes also appear in the stria vascularis as well as the spiral ligament of the cochlea.

However, as the recent debate about the pericytes of the CNS has pointed out (Hill et al. 2015; Attwell et al. 2016), there is no black-and-white line as to what is already or still a pericyte and what is a functionally or morphologically similar cell, but not yet a pericyte. For example, while some authors have decided to already consider the precapillary arteriolar cells to be pericytes (Attwell et al. 2016), many authors do not agree with this division (Hill et al. 2015; Fernandez-Klett et al. 2010). However, this is not so much the case in the cochlea, where a relatively common

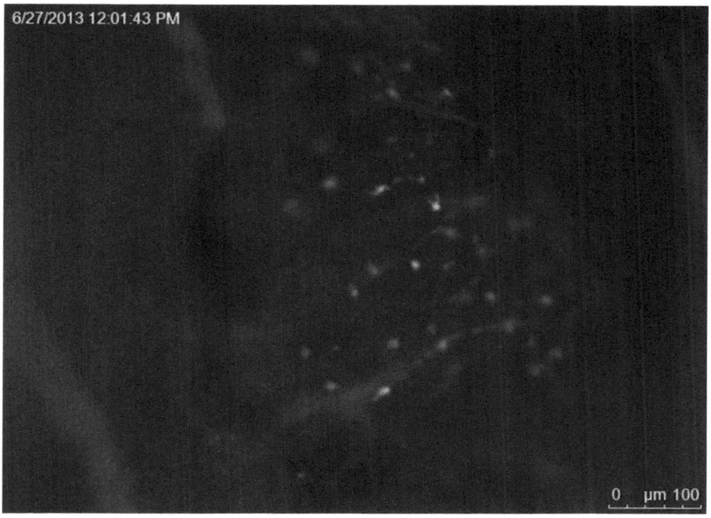

Fig. 7.1 Capillary pericytes in the stria vascularis of a guinea pig, visualized by in vivo fluorescence microscopy; pericytes were marked with topical application diaminofluorescein-2-diacetate; vessels of the stria vascularis were contrasted by i.v. injection of fluorescein-isothiocyanate-dextran

definition has been agreed upon: pericytes are commonly considered to be those cells that explicitly adhere to the outer walls of the capillaries of the stria vascularis or the spiral ligament and express a defined set of cellular markers (Fig. 7.1) (Dai et al. 2009; Shi et al. 2008).

7.2.2 Blood Supply of the Cochlea

Blood supply to the cochlea is of paramount importance for the function of the organ; in particular the endocohlear potential (Lamm and Arnold 2000) and the subsequent production of endolymph directly correlate to steady blood flow in the stria vascularis (Shi 2011). The entire blood supply to the cochlea makes up 10^{-7} of the entire cardiac output (Nakashima et al. 2003) and derives from the anterior inferior cerebellar artery, from which the spiral modiolar artery branches off, radiating over the scala vestibuli and across the spiral lamina. During its course, the spiral modiolar artery has numerous radial branches that eventually form the capillaries of the stria vascularis and the spiral ligament (Shi 2011).

7.2.3 Morphology and Cellular Markers

Cochlear pericytes are found numerously (up to 1300 in the cochlea of a C57/BL6 mouse) (Neng et al. 2015) and exhibit a very heterogeneous range in terms of morphology and function in the cochlea. Generally speaking, pericytes of the cochlea commonly express platelet-derived growth factor receptor ß, desmin, neural proteoglycan 2, and CD90. However, those pericytes that are commonly found in the spiral ligament typically express connexin 40, a gap junction protein, α-smooth muscle actin (α-SMA), and tropomyosin, while those present at capillaries of the stria vascularis regularly show a rich expression of desmin without any expression of α-SMA or tropomyosin (Shi 2011). This is a defining difference between these different subpopulations of cochlear capillary pericyte and may be indicative of different functions of the cochlear pericytes.

7.2.4 Physiological Functions of Cochlear Capillary Pericytes

Cochlear pericytes form part of the intrastrial blood-fluid barrier that controls the exchange between the intrastitial space and blood. The main objective of this highly specialized network is to shield intrastitial space from any toxins and other potentially harmful substances that might be dissolved in the blood, just like the blood-brain barrier. Precise regulation of the transport into and out of the intrastrial space is also paramount for maintaining exact composition of the inner ear fluids (Juhn et al. 2001) and therefore, ultimately, the ability to hear. Impairment of the integrity of the blood labyrinth barrier has been associated with numerous inner ear pathologies, including suppurative labyrinthitis (Zhang et al. 2015), acoustic trauma (Shi 2009), and age-related hearing loss (Neng et al. 2015).

The intrastrial blood-fluid barrier is mainly formed by endothelial cells, which are connected by tight junctions, and the underlying basement membrane. However, pericytes do contribute in a large number to the formation of the barrier (Shi et al. 2008; Shi 2009). By means of close biochemical and anatomical linkage, pericytes monitor the ion, fluid, and nutrient household of the stria vascularis and may respond accordingly (Peppiatt et al. 2006; Hall et al. 2014; Shi 2016).

Additionally, cochlear pericytes seem to be directly involved in the regulation of cochlear microcirculation in the stria vascularis. Since these pericytes evidently show the ability to contract (Dai et al. 2009) and may even do so upon presentation of a physiological stimulus (Dai et al. 2009; Bertlich et al. 2017a) as well as expressing contractile proteins(Dai et al. 2009), a role in the local regulation of blood flow in the stria vascularis seems very likely. Fittingly, a very similar phenomenon has been reported with the capillary pericytes of the central nervous system, where pericytes may locally increase or decrease cochlear blood flow according to local demand (Peppiatt et al. 2006).

Additionally, pericytes may actually contribute to the immune response in the stria vascularis. Pericytes have been reported to show very close interlinking with perivascular-resident macrophage-like melanocytes (Zhang et al. 2013), cells that contribute to the immune reaction of the stria vascularis. Since pericytes, like the perivascular-resident macrophage-like melanocytes, show a considerable reaction after stimulation with bacterial lipopolysaccharide (Zhang et al. 2013, 2015), a contribution of pericytes to the immune response seems likely.

7.2.5 Pathophysiology of Cochlear Pericytes

It has been established that cochlear capillary pericytes contribute to the regulation of cochlear blood flow; however, (permanent) impairment of cochlear blood flow has repeatedly been discussed as the final common pathological pathway in numerous entities. Among those are acoustic trauma, where longtime exposure to noise may lead to increased oxygen demand of the cochlea, eventually causing a relative lack of oxygen (Arpornchayanon et al. 2013). A longer-lasting lack of oxygen eventually activates the TNF pathway, which is known to cause pericytes to contract and eventually reduce cochlear blood flow (Bertlich et al. 2017a; Arpornchayanon et al. 2013).

In addition to this, the pericytes alter their physical configuration after longer-lasting exposure to loud noise—they migrate from the endothelial wall and express significantly elevated levels of the structural protein desmin (Shi 2009). This migration of pericytes from the endothelial cells is known to decrease the integrity of the blood-fluid barrier of the cochlea (Shi 2016).

In addition to this, impairment of cochlear blood flow has been discussed to be the leading cause for sudden sensorineural hearing loss (SSNHL). It has been observed that fibrinogen, a serum protein that is part of the plasmatic coagulation system, is both a risk factor for and a potential treatment target for SSNHL (Suckfull 2002). Since increased fibrinogen levels have been shown to impair hearing thresholds as well as microcirculation (Ihler et al. 2012a) [and thus partial oxygen pressure (Lamm and Arnold 2000)], a contribution of cochlear capillary pericytes is very probable.

Fittingly, it has been shown that capillary pericytes of the CNS typically exhibit a rigor mortis-like state after continuous insufficient partial oxygen pressures and thus continuously impair cochlear blood flow (Hall et al. 2014). Assuming that the cochlear capillary pericytes exhibit similar properties, this could also explain why when causal treatment is administered, hearing thresholds may return to basal values without any permanent alterations (Weiss et al. 2017).

Cochlear capillary pericytes also play a major role in bacterial suppurative labyrinthitis, a severe complication of inflammation of the middle ear. Inflammation of the middle ear is commonly caused by gram-positive bacteria such as *Streptococcus pneumoniae* or *Streptococcus pyogenes*—bacteria which are known to secrete exotoxins called streptolysin or pneumolysin. In addition to this, the gram-negative bacteria that regularly cause otitis media, like *Escherichia coli*, *Haemophilus influenzae*, and *Pseudomonas aeruginosa*, are known to contain an endotoxin, lipopolysaccharide, which is known to be set free during an immune response. In suppurative labyrinthitis, a transition of these toxins into the cochlea has repeatedly been hypothesized (Ishihara et al. 2016; Buckiova et al. 2012). Since both the exotoxin and the endotoxin are well known to activate the tumor necrosis factor alpha pathway, impairment of cochlear blood flow (Ihler et al. 2013) by pericyte contraction is most likely. Moreover, there is also a second role which pericytes play in suppurative labyrinthitis. Capillary pericytes significantly contribute to the upkeep of the blood-fluid barrier of the inner ear. In otitis media that has been caused by injection of lipopolysaccharide into the tympanic cavity, pericytes show significant migration from the vessel wall, their original site, and thus destabilize the blood-fluid barrier (Zhang et al. 2015). This observation is similar to those made in acoustic trauma (Shi 2009) as well as other tissues like the retina (Pfister et al. 2008).

Finally, cochlear capillary pericytes probably also contribute in the immune response to physical trauma to the cochlea. This is of particular importance since the insertion of an electrode during the surgical implantation of a cochlear implant is a procedure that is becoming more and more common in otorhinolaryngology (Wang et al. 2014). Surgical placement of the electrode of the cochlear implant inside the cochlea commonly causes direct damage to the basilar membrane (Roland and Wright 2006) as well as causing a local, abacterial inflammation. Together, both pose a risk for residual hearing in patients undergoing cochlear implants. However, it has been shown that application of etanercept, a fusion protein inhibiting the TNF pathway, is effective in preserving residual hearing under cochlear electrode implantation in an animal model (Ihler et al. 2014). Thus, an involvement of cochlear pericytes seems very probable. Fittingly, similar changes in pericyte morphology have been described in the central nervous system reacting to physical trauma (Dore-Duffy et al. 2000).

Overall, it seems evident that both the impairment of cochlear blood flow and the breakdown of the cochlear blood-fluid barrier play a major role in the pathophysiology of numerous pathologies of the inner hear. However, it is unlikely that both mechanisms are separate reactions to different stimuli. It is much more likely that both reactions—impairment of cochlear blood flow and breakdown of the blood-fluid barrier of the cochlea—are distinct reactions to inflammatory stimuli. Due to the considerable heterogeneity of the pericytes (Dai et al. 2009; Shi et al. 2008), it seems probably that different subpopulations that are present in the stria vascularis or the spiral ligament react to the same stimulus in a different manner, according to each subpopulation specification. This would be in line with the different immunohistochemical markers that have been described for these populations and would make both reactions that have been described different sides of the same medal.

Bearing this in mind, it is more than likely that pericytes are involved in numerous more pathologies of the inner ear, like radiation-induced hearing loss (Mujica-Mota et al. 2014), chemotoxicity (Sheth et al. 2017; Jeong et al. 2007), and also possibly hereditary hearing impairments.

7.2.6 Pharmacological Features of Cochlear Capillary Pericytes

The aforementioned properties make pericytes valid targets for pharmacotherapy of numerous inner ear pathologies. For example, the neutralization of tumor necrosis factor alpha by the use of etanercept seems to be a promising treatment for many of the aforementioned diseases; in addition to this, testing of various other remedies is still undergoing both in vitro (Jeong et al. 2007) and the animal model (Bertlich et al. 2017b; Sharaf et al. 2016). In addition to this, it has been hypothesized that the blood flow-promoting effect of betahistine on cochlear microcirculation (Ihler et al. 2012b; Bertlich et al. 2014) is mediated by pericytes. However, it has been shown that this is not the case, as this effect seems to be mediated by precapillary arterioles (Bertlich et al. 2017c). However, several authors do already consider these cell populations to be pericytes in other tissues (Attwell et al. 2016).

7.3 Outlook and Future Trends

When it comes to where research in cochlear capillary pericytes is heading, the most important findings are the differences but more importantly the similarities to the capillary pericytes of the central nervous system. Many functions of the pericytes of the central nervous have been described, and it will be crucial to see of the pericytes of the cochlea behave the same or in a different manner.

The most important aspect in this respect will be the role of pericytes in the damage to the stria vascularis and the spiral ligament. This includes proliferation and migration in response to injury—in particular in the long term—as well as regenerative and in particular stem cell properties of pericytes, since all these are considerably better described in pericytes of the central nervous system. A better understanding of the role of pericytes in the (patho)physiology of the cochlea will lead to a better understanding of cochlear homeostasis and function overall.

In addition to this, the selective pharmacological targeting of relevant structures within the cochlea—and this explicitly includes pericytes—will also see a steady increase.

References

Arpornchayanon W, Canis M, Ihler F, Settevendemie C, Strieth S (2013) TNF-alpha inhibition using etanercept prevents noise-induced hearing loss by improvement of cochlear blood flow in vivo. Int J Audiol 52:545–552. https://doi.org/10.3109/14992027.2013.790564

Attwell D, Mishra A, Hall CN, O'Farrell FM, Dalkara T (2016) What is a pericyte? J Cereb Blood Flow Metab 36:451–455

Bertlich M, Ihler F, Sharaf K, Weiss BG, Strupp M, Canis M (2014) Betahistine metabolites, Aminoethylpyridine, and Hydroxyethylpyridine increase cochlear blood flow in Guinea pigs in vivo. Int J Audiol 53:753–759. http://www.ncbi.nlm.nih.gov/pubmed/25014609

Bertlich M, Ihler F, Weiss BG, Freytag S, Strupp M, Canis M (2017a) Cochlear Pericytes are capable of reversibly decreasing capillary diameter in vivo after tumor necrosis factor exposure. Otol Neurotol 38:e545–e550

Bertlich M, Ihler F, Weiss BG, Freytag S, Jakob M, Strupp M et al (2017b) Fingolimod (FTY-720) is capable of reversing tumor necrosis factor induced decreases in Cochlear blood flow. Otol Neurotol 38:1213–1216

Bertlich M, Ihler F, Weiss BG, Freytag S, Strupp M, Jakob M et al (2017c) Role of capillary pericytes and precapillary arterioles in the vascular mechanism of betahistine in a Guinea pig inner ear model. Life Sci 187:17–21

Buckiova D, Ranjan S, Newman TA, Johnston AH, Sood R, Kinnunen PKJ et al (2012) Minimally invasive drug delivery to the cochlea through application of nanoparticles to the round window membrane. Nanomedicine (Lond) 7:1339–1354

Dai M, Nuttall A, Yang Y, Shi X (2009) Visualization and contractile activity of cochlear pericytes in the capillaries of the spiral ligament. Hear Res 254:100–107

Dore-Duffy P, Owen C, Balabanov R, Murphy S, Beaumont T, Rafols JA (2000) Pericyte migration from the vascular wall in response to traumatic brain injury. Microvasc Res 60:55–69

Fernandez-Klett F, Offenhauser N, Dirnagl U, Priller J, Lindauer U (2010) Pericytes in capillaries are contractile in vivo, but arterioles mediate functional hyperemia in the mouse brain. Proc Natl Acad Sci U S A 107:22290–22295

Hall CN, Reynell C, Gesslein B, Hamilton NB, Mishra A, Sutherland BA et al (2014) Capillary pericytes regulate cerebral blood flow in health and disease. Nature 508:55–60. https://doi.org/10.1038/nature13165

Hill RA, Tong L, Yuan P, Murikinati S, Gupta S, Grutzendler J (2015) Regional blood flow in the Normal and ischemic brain is controlled by arteriolar smooth muscle cell contractility and not by capillary Pericytes. Neuron 87:95–110

Ihler F, Strieth S, Pieri N, Gohring P, Canis M (2012a) Acute hyperfibrinogenemia impairs cochlear blood flow and hearing function in Guinea pigs in vivo. Int J Audiol 51:210–215. https://doi.org/10.3109/14992027.2011.622302

Ihler F, Bertlich M, Sharaf K, Strieth S, Strupp M, Canis M (2012b) Betahistine exerts a dose-dependent effect on cochlear stria vascularis blood flow in Guinea pigs in vivo. PLoS One 7:e39086

Ihler F, Sharaf K, Bertlich M, Strieth S, Reichel CA, Berghaus A et al (2013) Etanercept prevents decrease of cochlear blood flow dose-dependently caused by tumor necrosis factor alpha. Ann Otol Rhinol Laryngol 122:468–473

Ihler F, Pelz S, Coors M, Matthias C, Canis M (2014) Application of a TNF-alpha-inhibitor into the scala tympany after cochlear electrode insertion trauma in Guinea pigs: preliminary audiologic results. Int J Audiol 53:810–816. https://doi.org/10.3109/14992027.2014.938369

Ishihara H, Kariya S, Okano M, Zhao P, Maeda Y, Nishizaki K (2016) Expression of macrophage migration inhibitory factor and CD74 in the inner ear and middle ear in lipopolysaccharide-induced otitis media. Acta Otolaryngol 136:1011–1016

Jeong H-J, Kim S-J, Moon P-D, Kim N-H, Kim J-S, Park R-K et al (2007) Antiapoptotic mechanism of cannabinoid receptor 2 agonist on cisplatin-induced apoptosis in the HEI-OC1 auditory cell line. J Neurosci Res 85:896–905

Juhn SK, Hunter BA, Odland RM (2001) Blood-labyrinth barrier and fluid dynamics of the inner ear. Int Tinnitus J 7:72–83

Lamm K, Arnold W (2000) The effect of blood flow promoting drugs on cochlear blood flow, perilymphatic pO(2) and auditory function in the normal and noise-damaged hypoxic and ischemic Guinea pig inner ear. Hear Res 141:199–219

Mujica-Mota MA, Lehnert S, Devic S, Gasbarrino K, Daniel SJ (2014) Mechanisms of radiation-induced sensorineural hearing loss and radioprotection. Hear Res 312:60–68

Nakashima T, Naganawa S, Sone M, Tominaga M, Hayashi H, Yamamoto H et al (2003) Disorders of cochlear blood flow. Brain Res Brain Res Rev 43:17–28
Neng L, Zhang J, Yang J, Zhang F, Lopez IA, Dong M et al (2015) Structural changes in thestrial blood-labyrinth barrier of aged C57BL/6 mice. Cell Tissue Res 361:685–696
Peppiatt CM, Howarth C, Mobbs P, Attwell D (2006) Bidirectional control of CNS capillary diameter by pericytes. Nature 443:700–704
Pfister F, Feng Y, vom Hagen F, Hoffmann S, Molema G, Hillebrands J-L et al (2008) Pericyte migration: a novel mechanism of pericyte loss in experimental diabetic retinopathy. Diabetes 57:2495–2502
Roland PS, Wright CG (2006) Surgical aspects of cochlear implantation: mechanisms of insertional trauma. Adv Otorhinolaryngol 64:11–30
Sharaf K, Ihler F, Bertlich M, Reichel CA, Berghaus A, Canis M (2016) Tumor necrosis factor-induced decrease of Cochlear blood flow can be reversed by Etanercept or JTE-013. Otol Neurotol 37:e203–e208
Sheth S, Mukherjea D, Rybak LP, Ramkumar V (2017) Mechanisms of Cisplatin-induced ototoxicity and Otoprotection. Front Cell Neurosci 11:338
Shi X (2009) Cochlear pericyte responses to acoustic trauma and the involvement of hypoxia-inducible factor-1alpha and vascular endothelial growth factor. Am J Pathol 174:1692–1704
Shi X (2011) Physiopathology of the cochlear microcirculation. Hear Res 282:10–24
Shi X (2016) Pathophysiology of the cochlear intrastrial fluid-blood barrier (review). Hear Res 338:52–63
Shi X, Han W, Yamamoto H, Tang W, Lin X, Xiu R et al (2008) The cochlear pericytes. Microcirculation 15:515–529
Suckfull M (2002) Fibrinogen and LDL apheresis in treatment of sudden hearing loss: a randomised multicentre trial. Lancet (London, England) 360:1811–1817
Wang JT, Wang AY, Psarros C, Da Cruz M (2014) Rates of revision and device failure in cochlear implant surgery: a 30-year experience. Laryngoscope 124:2393–2399
Weiss BG, Bertlich M, Bettag SA, Desinger H, Ihler F, Canis M (2017) Drug-induced Defibrinogenation as new treatment approach of acute hearing loss in an animal model for inner ear vascular impairment. Otol Neurotol 38:648–654
Zhang F, Zhang J, Neng L, Shi X (2013) Characterization and inflammatory response of perivascular-resident macrophage-like melanocytes in the vestibular system. J Assoc Res Otolaryngol 14:635–643
Zhang J, Chen S, Hou Z, Cai J, Dong M, Shi X (2015) Lipopolysaccharide-induced middle ear inflammation disrupts the cochlear intra-strial fluid-blood barrier through down-regulation of tight junction proteins. PLoS One 10:e0122572

Chapter 8
Pericytes in the Placenta: Role in Placental Development and Homeostasis

Rodrigo S. N. Barreto, Patricia Romagnolli, Andressa Daronco Cereta, Leda M. C. Coimbra-Campos, Alexander Birbrair, and Maria Angelica Miglino

Abstract The placenta is the most variable organ, in terms of structure, among the species. Besides it, all placental types have the same function: production of viable offspring, independent of pregnancy length, litter number, or invasion level. The angiogenesis is a central mechanism for placental functionality, due to proper maternal-fetal communication and exchanges. Much is known about the vasculature structure, but little is known about vasculature development and cellular interactions. Pericytes are perivascular cells that were described to control vasculature stability and permeability. Nowadays there are several new functions discovered, such as lymphocyte modulation and activation, macrophage-like phagocytic properties, tissue regenerative and repair processes, and also the ability to modulate stem cells, majorly the hematopoietic. In parallel, placental tissues are known to be a particularly immune microenvironment and a rich stem cell niche. The pericyte function plethora could be similar in the placental microenvironment and could have a central role in placental development and homeostasis.

Keywords Capillary system · Maternal-fetal communication · Placental vascularization · Placentation · Perivascular cell

R. S. N. Barreto · P. Romagnolli · A. D. Cereta · M. A. Miglino (✉)
School of Veterinary Medicine and Animal Sciences, University of São Paulo, Butantã, Sao Paulo, Brazil
e-mail: miglino@usp.br

L. M. C. Coimbra-Campos
Department of Pathology, Federal University of Minas Gerais, Pampulha, Belo Horizonte, Brazil

A. Birbrair
Department of Radiology, Columbia University Medical Center, New York, NY, USA

Department of Pathology, Federal University of Minas Gerais, Pampulha, Belo Horizonte, Brazil

A. Birbrair (ed.), *Pericyte Biology in Different Organs*, Advances in Experimental Medicine and Biology 1122,
https://doi.org/10.1007/978-3-030-11093-2_8

8.1 Introduction

The eutherian mammals had a common ancestor that had the placenta with hemotrophic and histiotrophic bipotential (Vogel 2005). From this ancestor, all phylogenetical clades evolved. In the lower clades (Afrotheria and Xenarthra), there are only mammals with hemotrophic placentation or the ones whose mothers have low or no control of blood support to the fetus (Murphy et al. 2001; Vogel 2005; Carter and Mess 2007). However, in the upper clades (Euarchontoglires and Laurasiatheria), there are also mammals with histiotrophic placentation, where maternal blood vessels remain intact and the mother controls the blood supply (Murphy et al. 2001; Vogel 2005; Carter and Mess 2007). The appearance of less invasive placentas in upper clades, an evolutionary response, maybe to produce a more efficient placenta, by means of maternal blood control (Mess and Carter 2007), maternal-fetal transfer (Leiser and Kaufmann 1994), and increasing immune system barrier (Moffett and Loke 2006). The mechanisms that lead to the eutherian diversification are still not completely known; however, the genes that control the early embryo and placental development are largely phylogenetically conserved (Knox and Baker 2008). Regarding placenta-specific gene expression, human and cattle are more genetically similar than human and mouse, even with several morphological differences in between (Barreto et al. 2011). The plethora of the placenta's gross and fine morphological characteristics is distributed in mammals, although in all of them the same general function is achieved: production of healthy offspring. Then, in this chapter, we will discuss the placental diversity and speculate some possible roles of pericytes in placental development.

8.2 Placental Variability

In several mammals, placental morphology was thoroughly revised by Mossman (1987) and more recently by Wooding and Burton (2008a); also an actualization of nomenclature was made by Leiser and Kaufmann (1994). Herein, we will focus on some general morphological differences and afterward, in more detail, on the placental vascularization.

The placenta could be classified by the (1) arrangement of extraembryonic membranes, (2) gross morphological shape of maternal-fetal communication, (3) fine arrangement of maternal-fetal interdigitation, (4) number of tissue layers between maternal and fetal bloodstreams, and (5) arrangement of maternal and fetal vessels (Leiser and Kaufmann 1994).

The four extraembryonic membranes could be in contact with the uterus and arranged in several ways. The amnion, the non-vascularized and in close contact with the fetus. The yolk sac is transitory in some species and well developed in others, arising from the midgut as an endothelial and hematopoietic stem cell niche. The allantois arises from the hindgut and is the analogous extraembryonic urinary bladder, being also highly vascularized. And the chorion, which is derived from the trophectoderm, initially has its own capillary system, later fusing with allantois vessels, in several species, and it is the outer membrane being the first barrier (Leiser and Kaufmann 1994).

The arrangement of those four membranes could produce four models:

1. Chorionic placenta, present in early phases of all eutherians, with the chorion in major contact with the uterus.
2. Choriovitelline placenta, showing the chorion connected to vitelline vascularization, i.E., parts of Perissodactyla and carnivore placenta.
3. Chorioallantoic placenta, where the chorion is connected to allantoic vascularization, i.E., the predominant area in several eutherians.
4. Vitelline placenta, with the fetal circulation connected to the vitelline vessels, i.E., the area of rodent placenta, fish, and amphibia (Mossman 1987; Leiser and Kaufmann 1994; Wooding and Flint 1994; Wooding and Burton 2008a).

The maternal-fetal communication units could be spread across all the uterine wall extension or restricted to some area(s), forming an intense and deep contact or only a tenuous apposition of tissues. In the diffuse placenta, the maternal-fetal communication is spread (or diffused) by the entire uterine wall and just with apposition of maternal-fetal tissues, i.e., in pigs, dolphins, Perissodactyla, and some lower primates. In the cotyledonary placenta, the communication is organized in specialized areas intercalated by smooth areas and distributed along almost all the uterine wall, i.e., ruminants. In the zonary placenta, the communication is organized in a belt surrounding all the uterine wall inner circumference, i.e., most carnivores. Being more restricted, in the bidiscoidal placenta, the communication is organized in only two discs in close contact, i.e., some lower and higher primates. The highest concentration of communication is reached with the discoidal placenta, where only one disc is apparent, i.e., rodents, lagomorphs, great apes, and humans (Mossman 1987; Leiser and Kaufmann 1994; Wooding and Flint 1994; Wooding and Burton 2008a).

Microscopically, the maternal-fetal communication units could be arranged as:

1. Folded, when the uterine wall and chorion are folded, the simplest one, i.e., marsupials and some lower primates.
2. Lamellar, with the uterine wall and chorion forming some branched folds, i.e., carnivores.
3. Villous, in which the chorion forms inverted tree shape inside the uterine wall, i.e., ruminants, horses, higher primates, and humans.
4. Trabecular, a mixture of lamellar and villous types, i.e., some new-world monkeys.
5. Labyrinthine, most complex, where the chorion is bathed in channels or lacunas of the maternal bloodstream, i.e., rodents, lagomorphs, and some lower primates (Mossman 1987; Leiser and Kaufmann 1994; Wooding and Flint 1994; Wooding and Burton 2008a).

In the maternal-fetal communication, six tissue layers could be present, three on the fetal side (trophoblast, mesenchyme, and fetal endothelium) and three others on the maternal side (endometrial epithelium, endometrial stroma, and maternal endothelium). Moreover, different invasive deepness also happens between placental types, resulted by losing maternal tissue layers. In the noninvasive epitheliochorial placenta, all six tissue layers are preserved forming a complete barrier, i.e., lower primates, horses, and pigs. A specialization of this type, synepitheliochorial, presents migratory cells that detach from trophoblast and fuse with endometrial epithelium, conferring tenuous invasive behavior (Pereira et al. 2013), i.e., ruminants. Going deeper, in the endotheliochorial placenta, the chorion stays in contact with maternal endothelium, i.e., carnivores. The most intimate contact is made by hemochorial placenta, in which the chorion is directly batched by maternal blood, although one to three trophoblast layers could be present in hemomonochorial, i.e., caviomorph rodents, great apes, and humans at term; hemodichorial, i.e., lagomorphs and human; and hemotrichorial, i.e., some rodents, respectively (Mossman 1987; Leiser and Kaufmann 1994; Wooding and Flint 1994; Wooding and Burton 2008a).

The arrangement of maternal and fetal blood at arterial and venous levels (macroflow) is countercurrent, with some rare exceptions, where maternal and fetal vessels are in parallel and flow in opposite directions. However, at capillary level (microflow), the arrangement of maternal and fetal capillaries varies. It could be also countercurrent, efficient for passive diffusion, and present in small placentas, i.e., rodents and lagomorphs. Also, two other arrangements are possible, the crosscurrent flow, in which the maternal and fetal capillaries are arranged in transversal directions, i.e., carnivores and lower primates, and the multivillous flow which is a mixture of countercurrent and crosscurrent flows, i.e., ruminants, higher primates, and humans (Mossman 1987; Leiser and Kaufmann 1994; Wooding and Flint 1994; Wooding and Burton 2008a).

8.3 Placental Blood Flow

As discussed above, the placental morphology has a plethora of differences among species; consequently, the structure of maternal and fetal blood capillaries in the communication units varies in accordance with other morphological variations to maintain the efficiency of gas, nutrient, and waste exchanges (Mossman 1987; Leiser and Kaufmann 1994; Wooding and Flint 1994; Wooding and Burton 2008a). On the fetal side, the capillary bed is formed by non-fenestrated capillaries supplied by a fetal arteriole (Wooding and Burton 2008a). However, the phylogenetical evolution of eutherians modified majorly the maternal tissue arrangement, generating four different contact levels between trophoblast and maternal blood, resulting in significant differences in maternal blood flow organization and control. From the most distant to the closest contact, the epitheliochorial placenta is formed by two complex capillary beds (Enders et al. 1998). In the synepitheliochorial and endotheliochorial, there is modification or elimination of the endometrial epithelium, whereas in the hemochorial, observed in humans and rodents, the endometrial epithelium, the connective tissue, and endothelium are no longer present, and maternal blood is delivered in maternal blood spaces (Heinrich et al. 1988; Kaufmann et al. 1988).

In addition, in hemochorial, maternal arteries are reminiscent; however, their endothelium is replaced by migratory and invasive trophoblast cells. In humans, those arteries end abruptly at maternal blood spaces without channels to carry to the fetal surface (Wooding and Flint 1994). In hemochorial placentas, no exact control of maternal blood flow is possible, due to the absence of maternal arterial precapillary sphincters regulated by vasoactive compounds (i.e., estrogen, NO, prostacyclin, relaxin, and calcitonin) that are present in other placental types (Wooding and Flint 1994; Thornburg et al. 2006). Also, as other placental types, no neural innervation is present to control vasoconstriction or dilatation (Lachenmayer 1971a; Nikolov and Schiebler 1973; Myatt 1992), maybe by myometrial axon degeneration mediated by estrogen (Mónica Brauer and Smith 2015). Then, metabolic factors probably control the vascular walls (Lachenmayer 1971b; Heinrich et al. 1988), and myometrial contractility could reduce the blood flow in intervillous spaces in hemochorial placentas (Ramsey et al. 1967).

Throughout the pregnancy, both the placenta and fetus increase in size, and, as a result, more nutrients are needed for development and maintenance of those tissues (Baur 1977). Firstly, the mother increases her blood volume, blood flow, and cardiac output to be able to perfuse the developing placenta (Wooding and Burton 2008a). In humans, the blood volume increases around 1.5 to 2.0 L, and the cardiac output is around 50% (Thornburg et al. 2006); however, hemochorial placenta presents lower maternal blood rates than epithelio- or synepitheliochorial placentas (Wooding and Burton 2008a). As another mechanism, in some primates at late pregnancy, anastomoses are present between ovarian and uterine arteries to increase the maternal placental blood influx (Wehrenberg et al. 1977), such as in viscacha, a South American rodent (unpublished data). On the other side, umbilical blood flow needs to increase to supply the blood channels that occupy around 50% of fetal placental volume (Wooding and Flint 1994), but no relation of umbilical flow rate with placental type is observed (Wooding and Burton 2008a). The villous surface-placental volume ratio also varies (i.e., from 300 $cm^2 kg^{-1}$ in the mare to 10,400 $cm^2 kg^{-1}$ in the cat) and is lower in large animals than in small ones and higher in species with compact placentas (cattle, rats, cats, humans) than in diffuse ones (horses, pigs) (Baur 1977).

Altogether, placental blood flow, transport capacity, villous surface area, and several other parameters vary greatly between eutherian species, and none of them have direct influence in fetal growth rate (Wooding and Flint 1994), indicating that each type of placenta has its own control mechanisms.

8.4 Placental Vascular Architecture

In comparative aspects, the vascular architecture is not well explored in the literature. Also, in the capillary system, the formation during placental development and establishment is performed by different mechanisms and involves different placental cell population and regions (Mossman 1987; Leiser and Kaufmann 1994; Wooding and Flint 1994; Wooding and Burton 2008a). Herein, we will discuss the vascular architecture and formation of the capillary systems, or microvasculature, in four different placental models: human chorioallantoic, discoidal, villous, hemomonochorial, and multivillous flow; mouse choriovitellinic, discoidal, labyrinthine, hemotrichorial, and countercurrent flow; bovine chorioallantoic, cotyledonary, villous, synepitheliochorial, and multivillous flow; and feline chorioallantoic, zonary, labyrinthine, endotheliochorial, and crosscurrent flow.

8.4.1 Human Model: Hemochorial Villous Placenta

A detailed description of the fetal vasculature in the human placenta was made by Kaufmann et al. (1988) and could be observed in Fig. 8.1. The two umbilical arteries ramify in several chorionic arteries and enter the chorionic plate. From the chorionic plate, the fetal villous tree shape of communication is formed. The stem villi (supplied by chorionic arterioles) are ramified in intermediated villi (supplied by arterioles) and finally in terminal villi; the last have around 30 to 80 μm and are supplied only by fetal capillaries that occupy half of their stroma (Kaufmann et al. 1988). In the intermediated villi, the arterioles are surrounded by only one or two smooth muscle cell layers; however, some venules are surrounded only by pericytes (Rhodin 1968; Kaufmann et al. 1988). Also, precapillary sphincters, or structures with similar function, are absent, and then the arteries continue to capillaries by reducing their luminal diameters and losing the smooth muscle cell layers. The capillaries in the terminal villi form loops and sinusoids (only at mature placenta) near to villi tip; those sinusoids possess continuous endothelial wall and basement membrane, without fenestrations or pores. Functionally, those sinusoids decrease the blood flow, favoring the fetal-maternal exchange (Arts 1961; Kaufmann et al. 1988). As terminal villi development is sensitive to oxygen concentrations, in some hypoxia conditions (i.e., preeclampsia, intrauterine growth restriction, diabetes, etc.), there is an increase of development of those villi and, consequently, sinusoids (Firth and Leach 1996; Macara et al. 1996; Zygmunt et al. 2003; Kaufmann et al. 2004; van der Heijden et al. 2005; Resta et al. 2006; Burton et al. 2009; Dubova et al. 2013). In summary, the fetal microvasculature is formed by branching of umbilical arteries until the sinusoid in the tip of terminal villi.

However, the maternal microvasculature formation is more complex and involves different cell populations, such as from the maternal immune system and

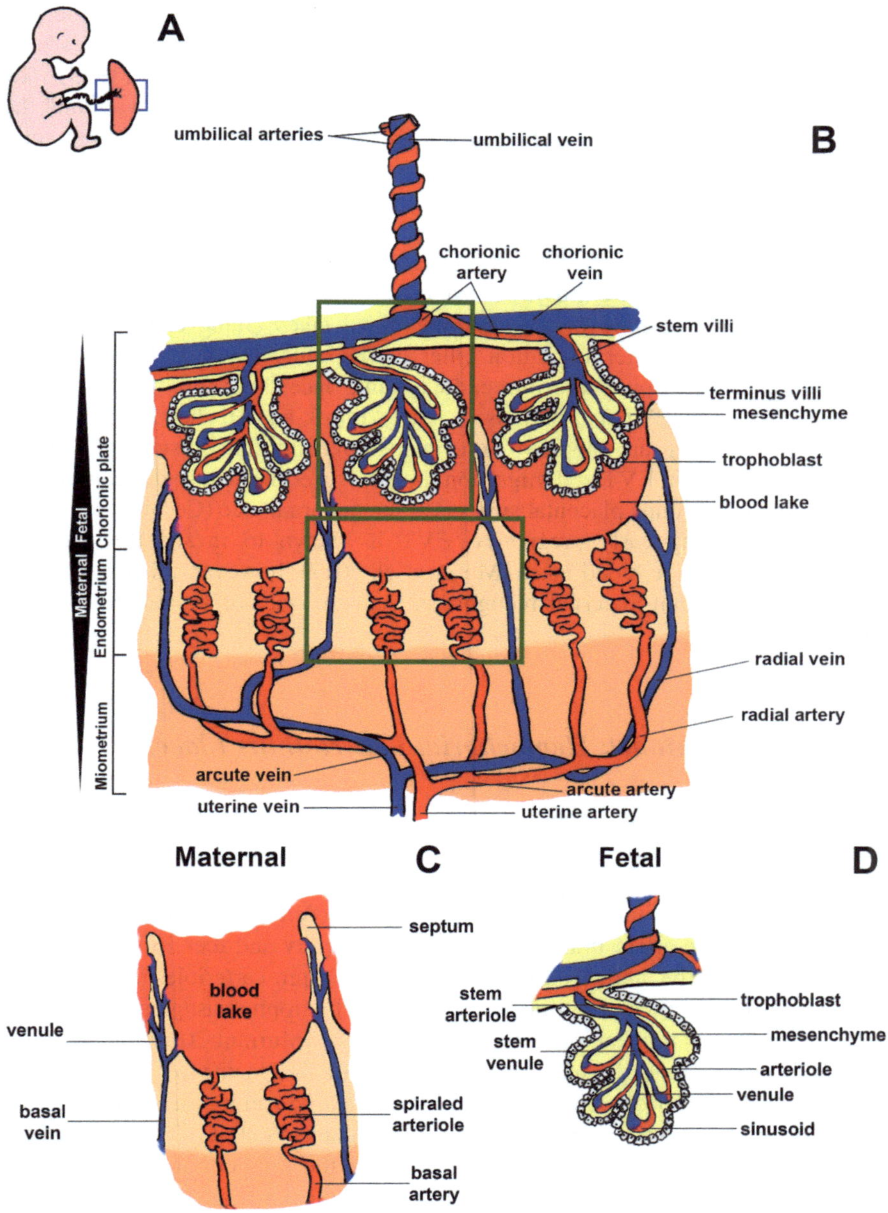

Fig. 8.1 Schema of human placental vascularization. In (**a**), human fetus and placenta. In (**b**), squared area of (**a**), showing layers and maternal and fetal vascularization. In (**c**), squared area of (**b**), showing maternal microvascularization. In (**d**), squared area of (**b**), showing fetal microvascularization

even fetal trophoblast cells. Firstly, branches of uterine artery enter the uterine wall to form the arcuate artery in the myometrium, branching again the radial artery, and basal artery enters the endometrium to form the spiral arterioles (1955). The spiral arterioles have no muscular and elastic tissue throughout the wall, and the lumen is dilated and tortuous. The blood flow from spiral arterioles goes directly to the placental bed, forming the blood lakes that are drained by openings in the basal plate and flow by the venous system until the uterine vein (Kaufmann et al. 1988). During the progress of the pregnancy, the spiral arteriole wall is remodeled by fetal cells with the interference of maternal immune system cells. Some migratory trophoblast cells, the extravillous trophoblast (EVT), secrete human leukocyte antigen (HLA)-G directly in the maternal blood lakes (Jiang et al. 2015) that attracts and activates uterine natural killer (uNK) cells. Then, uNK attract EVT cells to replace the spiral arteriole endothelium; once there, this mechanism of chemoattraction of EVT and uNK is increased (Arck and Hecher 2013; Tilburgs et al. 2015), having an important role in maternal immune system regulation. Therefore, defects in EVT cell migration to spiral arterioles cause deficiency both in vascular and immune placental systems at the maternal side (Cecati et al. 2011). In addition, during early pregnancy, EVT is known to invade the maternal venules that drain the blood lakes (Moser et al. 2017); however, this migration is more related to fetal-maternal recognition than to vascular system remodeling (He et al. 2017).

8.4.2 Mouse Model: Hemochorial Labyrinthine Placenta

The general structure of the mouse placenta was reviewed by Croy et al. (2015). Differently from human, no villous tree is formed, and then the fetal umbilical artery enters the chorionic plate, branched in innumerous chorionic arterioles, newly branching in the labyrinth into a dense capillary bed (Adamson et al. 2002; Murrant 2014). The uterine artery branches in the arcuate artery and then in several radial arteries that enter the uterine wall and branch now in spiral arterioles in the decidua (Murrant 2014). In mice, the intravascular invasion by trophoblast giant cells (TGC) is more discrete than what was observed in humans (Murrant 2014). Then, in the central junctional zone, the spiral arterioles are connected to one to four trophoblast tubes, named as arteriolar channels, that branch again forming the sinusoid meshwork layered by labyrinthine trophoblast cells (Adamson et al. 2002). Finally, the maternal blood in the labyrinthine sinusoids bath the trophoblast in the maternal spaces of the labyrinth and are drained through several venulous channels located in the junctional zone until the decidual venous sinusoids, a conditioned chronic hypoxia region (Leno-Durán et al. 2010). This condition may explain the glycogen-rich cells in the junctional zone, probably to deliver the substrate to permit glycogenolysis in nearby cells (Abrahamsohn and Zorn 1993). Almost similarly to humans, uNK cells also have a role in trophoblast invasion into maternal vessels (Yamada et al. 2014).

8.4.3 Cattle Model: Synepitheliochorial Villous Placenta

The umbilical artery is branched in several arteries to form the allantochorionic arterial system. Each branch enters the cotyledon and branches again into stem artery to irrigate each stem villus (Leiser et al. 1997b). Then, from the stem artery branches terminal arteriole to follow the villi ramification until the capillaries in the terminal villi of the arcade zone. The capillaries are initially tightly arranged, and then convolutions are made by capillaries coiling and anastomosing to create the sinusoids (Leiser et al. 1997a; b). After capillarization, terminal venules rise to form a tubelike venous system in the stem villi. This tubelike venous system converges to the allantochorionic venous system intended to the umbilical vein (Leiser et al. 1998).

Branches of the uterine artery penetrate the uterine wall, and in the caruncular stalk, they become vigorously spiraled, just after branching to form the less spiraled septal arterioles, which finally ramify to form the capillary complexes in the arcade zone (Leiser et al. 1997a). After capillarization, venules are formed, converging successively to create the stem veins and ultimately the uterine vein (Fig. 8.2).

8.4.4 Feline Model: Endotheliochorial Lamellar Placenta

The umbilical artery enters the chorioallantoic membrane in the placental girdle region and branches in stem arteries, which in turn branch in arterioles that overlay the funnel-like maternal system. Those arterioles branch to form the lamellar capillary system, from which are formed venules, which converge to the umbilical vein (Leiser and Kohler 1984).

The uterine artery penetrates the myometrium stablishing a network-like pattern. From this network, only one branch of the stem artery enters the endometrium in the girdle region that is larger than other stem arteries that irrigate the para- or interplacental regions. In direction of endometrial lamellae, the stem artery branches around five more times, forming a funnel-like system of arterioles. From those arterioles, one or two branches form the septal capillary network. The vascular connections in the funnel-like systems are performed by anastomosis with peripheral arterioles of each system. The capillary network continues forming venules and then converges to the stem vein in the junctional zone until the umbilical vein (Leiser and Kohler 1983; Dantzer 2002; Wooding and Burton 2008a) (Fig. 8.3).

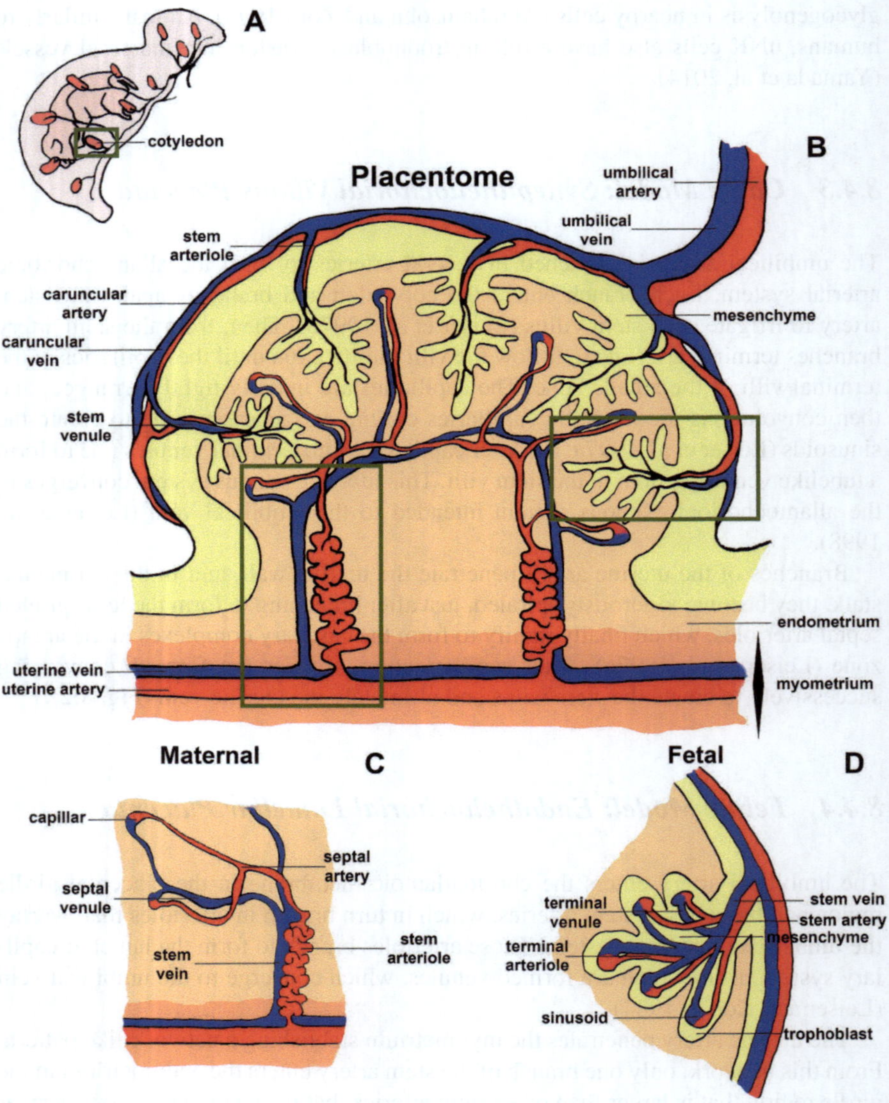

Fig. 8.2 Schema of bovine placental vascularization. In (**a**), bovine fetus and placenta. In (**b**), squared area of (**a**), showing layers and maternal and fetal vascularization inside the placentome. In (**c**), squared area of (**b**), showing maternal (or caruncular) microvascularization. In (**d**), squared area of (**b**), showing fetal (or cotyledonary) microvascularization

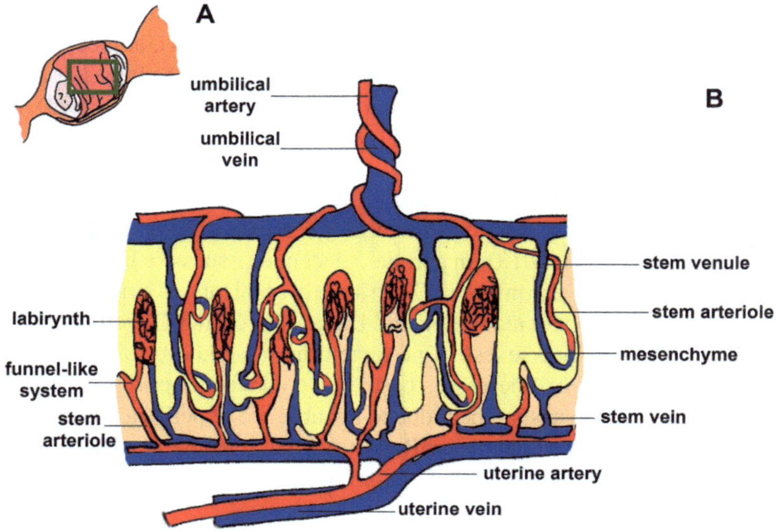

Fig. 8.3 Schema of cat placental vascularization. In (**a**), cat fetus and placenta. In (**b**), squared area of (**a**), showing layers and maternal and fetal macro- and microvascularization inside the placental girdle

8.5 Pericytes in the Placenta

As discussed in other chapters of this book, pericytes were firstly defined by their anatomical localization, i.e., adhered in the endothelium around the blood vessel wall with a general function related to vascular stability (Rouget 1873). However, other pericyte functions have been described, such as vascular permeability and blood flow control (Pallone et al. 1998, 2003; Pallone and Silldorff 2001); vascular development, maturation, and remodeling (Levéen et al. 1994; Soriano 1994; Lindahl et al. 1997; Hellström et al. 2001; Enge et al. 2002); blood-brain barrier integrity (Cuevas et al. 1984; Nakagawa et al. 2007; Nakamura et al. 2008; Shimizu et al. 2008; Krueger and Bechmann 2010; Bell et al. 2010; Armulik et al. 2010; Daneman et al. 2010; Al Ahmad et al. 2011; Kamouchi et al. 2011; Thanabalasundaram et al. 2011); and regulation of blood coagulation (Bouchard et al. 1997; Kim et al. 2006; Fisher 2009). In addition, in the last decade, new functions related to the immune system (Armulik et al. 2011; Hellerbrand 2013; Pfister et al. 2013; Nees et al. 2013; Hurtado-Alvarado et al. 2014; Pan et al. 2014) have been also described, i.e., lymphocyte modulation and activation (Fabry et al. 1993; Verbeek et al. 1995; Balabanov et al. 1999; Tu et al. 2011), macrophage-like phagocytic properties (Jeynes 1985; Hasan and Glees 1990; Balabanov et al. 1996; Thomas 1999; Castejón 2011), and dendritic cell modulation (Krautler et al. 2012). Besides those functions, the pericytes could also participate in several regenerative tissue and repair processes, i.e., formation of skeletal muscle cells (Crisan et al. 2008; Dellavalle et al. 2011), dental tissues (Feng et al. 2011; Zhao et al. 2014), adipocytes (Crisan et al. 2008), cartilage (Farrington-Rock et al. 2004), and bone (James et al. 2012), increasing heart function after infarct (Katare et al. 2011; Chen et al. 2013), healing (Zebardast et al. 2010), and fibrous tissue formation (Dulauroy et al. 2012; Duffield et al. 2013).

Pericytes can also modulate stem cells (Crisan et al. 2008; Bandeira et al. 2017), majorly the hematopoietic (Méndez-Ferrer et al. 2010; Ding et al. 2012; Corselli et al. 2013; Kunisaki et al. 2013). In parallel, placental tissues are known to be a rich stem cell niche, and those stromal cells have several applications similar as described for pericytes [reviewed by Antoniadou and David (2016) and Oliveira and Barreto-Filho (2015)]. So, this diverseness of pericyte function could be similar in the placental microenvironment. Therefore, pericytes could be important for placental development and homeostasis. Then, due to the absence of pericyte function knowledge in placenta-specific tissue, we will discuss what is known and speculate some possible functions including hypothesis from deficient or diseased models.

8.5.1 Major Genes Involved in Pericyte Angiogenesis Signaling

During the vasculogenesis in early human pregnancy (around the third week), it is possible to find two different cell groups directly derived from mesenchymal cells. Most of those cells are flattened, they stay in contact with the luminal surface of the forming vessel wall, and during this period, the basement membrane is absent. The second group of cells is in contact with the outer surface that is more polygonal and arranged in a weblike shape. Those cells are in contact with other mesenchymal cells, and they are surrounding the forming vessel wall and had been considered pericytes. Instead of intimate contact with flattened luminal cells (endothelium), the processes do not reach the lumen (Demir et al. 1989). This second group of cells, pericytes, could be differentiated directly from mesenchymal stem cells, or neural crest, by tumor growth factor (TGF)-β1 signaling (Chen and Lechleider 2004; Ding et al. 2004). In more mature placenta, it is possible to find decidual (Maier and Pober 2011) and placental (Robin et al. 2009) pericytes that apparently are heterogeneous [as in others (Prazeres et al. 2017)], and both participate in the angiogenesis processes (Bergers and Song 2005).

The vascular endothelial growth factor (VEGF) is expressed in the placental pericytes and induces angiogenesis (Aronoff et al. 2017). A hypoxia-induced proliferation could be influenced by autocrine VEGF signaling (Nomura et al. 1995; Yamagishi et al. 1999a) to promote angiogenesis (Yonekura et al. 1999). It can also influence pericyte proliferation and migration (Yamagishi et al. 1999b). In the placenta, placental growth factor (PGF) also participates in this process (Yonekura et al. 1999).

Together with VEGF, the platelet-derived growth factor B (PDGF-B), a potent mitogenic agent for connective tissue cells, is involved in the angiogenesis (Moreau et al. 2014). PDGF-B can be produced by several cells, like megakaryocytes/

platelets, monocytes/macrophages, and villous cytotrophoblast cells, and can control pericyte recruitment (Holmgren et al. 1991; Bjarnegard 2004). In addition, non-endothelial source of PDGF-B may have long-range effects on pericyte growth and recruitment (Bjarnegard 2004). The PDGF-B expression could be controlled by several factors, such as hypoxia (Herrmann et al. 2016), similar to VEGF.

The PDGF signaling is important to attract pericytes by PDGF receptor β (PDGFRβ) and control pericyte recruitment, proliferation, and differentiation (Moreau et al. 2014). Apparently, PDGFRβ is also expressed in mice sinusoidal trophoblast giant cells (Moreau et al. 2014). The PDGF-B/PDGFRβ signaling between endothelium and pericytes controls the pericyte recruitment. The lack of this signaling, by knockout of those genes or in some placental diseases (that will be discussed later in this chapter), could decrease pericyte number and coverage and increase capillary diameter and permeability, and it also interferes in trophoblast development. Likewise, Cited2 is involved in pericyte recruitment, proliferation, and differentiation, acting directly on capillary formation (Moreau et al. 2014); however, the mechanism is still unknown.

In swine placenta, an epitheliochorial model, the PDGF receptors in the endothelial/perivascular areas of the subepithelial layer from day 24 to 40 of swine pregnancy show stronger expression compared to earlier stages (Rodriguez-Martinez et al. 1992; Persson and Rodriguez-Martinez 1997) suggesting that PDGF plays also a role in stroma reorganization, particularly during maternal angiogenesis.

After maturation of the vessels, the maternal-fetal interface is composed by the capillary system at the fetal side and another capillary system of blood space at the maternal side. Then, the arterioles before maternal-fetal interface are surrounded by only one or two smooth muscle cell layers; however, some venules after maternal-fetal interface are surrounded only by pericytes (Rhodin 1968; Kaufmann et al. 1988). Moreover, maternal precapillary sphincters, before blood spaces in hemochorial placentas, due to the absence neural innervation to control vasoconstriction or vasodilatation (Lachenmayer 1971a; Nikolov and Schiebler 1973; Myatt 1992) could be controlled by pericytes, as in neural capillary beds, by alpha smooth muscle actin (α-SMA) and some cholinergic and adrenergic receptors (Rucker et al. 2000).

8.5.2 Pericytes in Knockout-/Null-Induced Placental Abnormalities or Diseases

It is a consensus that pericytes participate in capillary wall formation also in the placenta (Challier et al. 1999). For decades, in some hemochorial placenta types, pericytes were described to form an incomplete layer adjacent to the endothelial cells (Heinrich et al. 1988). However, in late human pregnancy, during angiogenesis, the capillaries sprouted by the proliferation of endothelial cells and pericytes are evolved by thick layers of basement membrane (Demir et al. 1989). Nevertheless, knowledge about the distribution and function of placental pericytes still needs more investigation. Something is known about the pericyte ultrastructure (Challier et al. 1997; Ohlsson et al. 1999; Bjarnegard 2004; Kučera et al. 2010; Jones and Desoye 2011; Maier and Pober 2011; Deveci et al. 2013; Kim et al. 2015), their perivascular localization in the capillaries, and the presence of vesicles nearby (Jones and Desoye 2011). Those vesicles are probably delivery by pericytes in the direction of the trophoblast basement membrane and some macrophages (Jones and Desoye 2011).

Even though the placental pericytes are poorly characterized, it is known that they express α-SMA and NG2 (Ohlsson et al. 1999; Sims 2000; Looman et al. 2007), and pericyte dysfunction, coverage, and distribution may be modified in relation to pregnancy abnormalities (Maier et al. 2010). In high-altitude pregnancies that have higher blood pressure, the placental capillaries have lower pericyte coverage and increased diameter than in lowlands (Zhang et al. 2002). In pregnancies complicated by type I diabetes mellitus, no differences in pericyte coverage in placental vessels were observed (Kučera et al. 2010). In complete hydatidiform mole pregnancies, a lack in pericyte recruitment is evident, with no association of pericytes and the villi stroma vessels, spaces in between endothelial cells, and no basal lamina that results in an immature vasculature (Kim et al. 2015). Similarly, lesions of villous capillary (such as chorangioma, chorangiomatosis, and chorangiosis) also decrease pericyte number in the capillaries; however, they form a more continuous and preserved layer (Ogino and Redline 2000).

Pericytes are the most permissive cell for congenital human cytomegalovirus (HCMV) infection in the villi, such as in the blood-brain barrier and blood-retinal barrier. This results in the decrease of pericytes due to infection and could interfere in the placental vascular permeability and increase placental inflammation, angiogenesis, and microvascular abnormalities (Aronoff et al. 2017).

On the other side, in the mice model of pregnancy-associated hypertension, the small vessels have abnormal diameter and number and also were not lined by a basement membrane and poorly covered by pericytes, majorly in the labyrinth, showing immature vascular network (Furuya et al. 2008). A similar phenotype is observed in Cited2 null mice, where pericytes around embryonic vessels were disorganized (Moreau et al. 2014).

In turn, the PDGF-B, a mitogenic agent that can be produced by several cells (Holmgren et al. 1991), controls the proliferative potential of extravillous trophoblast (Holmgren et al. 1991, 1992). A deficiency in PDGF-B, or its receptor, could lead to a perinatal death in mice with several malformations due to lack of microvascular pericytes and rupture of microvasculature (Ohlsson et al. 1999). On the other hand, it induces dilated blood vessels; and it reduces the number of pericytes, the trophoblast layer, and the surface area of maternal blood spaces in mice labyrinth; and it increases hemorrhage due to microvasculature rupture (Ohlsson et al. 1999).

8.5.3 Pericytes in Placental Immune Cell Regulation

It is known that the placental mesenchymal stem cells from the maternal origin are more responsible to immunomodulatory stimulus than fetal ones. Those maternal cells regulate the induction of tolerogenic dendritic cells and regulatory T cells (Zhu et al. 2014), having a better immunomodulatory potential than cells from the umbilical funiculus (Talwadekar et al. 2015). Such stem cells too can increase the survival of skin grafts (Zhu et al. 2014). Mesenchymal stem cells are stromal cells, and they can be localized perivascularly in placental tissues [reviewed by Lobo et al. (2016) and Favaron and Miglino (2017)], and they share some markers with pericytes (Crisan et al. 2008; Muñoz-Fernández et al. 2018). In parallel, human decidual stromal cells are also perivascular cells distributed in the endometrium and decidua (Wynn 1974) and recently are described as decidual pericytes, due to the expression of pericyte-associated antigens (such as α-SMA, nestin, PDGF-B, and PDGFRβ), expression of angiogenic factors, cytokine-induced contraction, and attraction of peripheral blood NK cells (Muñoz-Fernández et al. 2018). Then those cells may be pericytes specialized and involved in immune responses during the pregnancy.

Placental pericytes, like other stromal or mesenchymal cells, decrease T cell proliferation by cytokine signaling. Besides that, placental pericytes have low stimulation of T cell; however, MHC-II molecules from pericytes are recognized by and desensitize T cell through induction of T cell anergy by TCR-dependent mechanism after IFN-γ stimulation (Maier and Pober 2011). Placenta-derived pericytes, such as brain pericytes, can induce the formation of putative T_{reg} cells ($CD4^{+}CD25^{high}FOXP3^{+}$) with suppressive activity and reduce effector T cells, and this pericyte phenotype was repressed using TGF-β inhibitor (Domev et al. 2014).

Abnormal interaction between extravillous trophoblast and decidual NK cells could be observed in impaired pseudovasculogenesis in early human preeclamptic placenta (Furuya et al. 2008). And pericytes from microvasculature are important in guiding leukocytes to inflammation site (Stark et al. 2013).

The extravillous trophoblast (EVT) downregulates the expression of MHC class Ia (MHC1a) (Helige et al. 2008; Jiang et al. 2015) to inhibit maternal immune rejection by inhibition of uNK, macrophages, and other immune cell receptors (Caumartin et al. 2007; Li et al. 2009). However, EVT overexpresses MHC1b that active uNK cells against fetal tissues, altogether works to activate T_{reg} cells ($CD4^{+}CD25^{+}FOXP3^{+}$) that inhibits effector T cells (Djurisic et al. 2014; Tilburgs et al. 2015). The T_{reg} cells have a bidirectional signaling with dendritic cells (DC) (Guerin et al. 2009). When the CTLA4 in the T_{reg} surface stimulates the CD80/CD86 in DC surface (Aluvihare et al. 2004), the DC produces indoleamine 2,3-dioxygenase (IDO) (Sato et al. 2003) that is metabolized and induces FOXP3 production in CD4 + CD25+ T cell, inducing the generation of more T_{reg} cells (Mezrich et al. 2010). As pericytes could induce dendritic and T_{reg} cells (Zhu et al. 2014), probably by a unknown signaling to DC and in consequence increase T_{reg} cells. Also, decidual pericytes could be related to the attraction of peripheral blood NK cells to the decidua (Muñoz-Fernández et al. 2018). As the decrease of uNK ($CD56^{bright}CD16^{-}$) could reduce EVT migration

(Lash et al. 2010) and cause defects in maternal spiral arteries (Leonard et al. 2006) resulting in miscarriage increase (Rizzo et al. 2015). Then, indirectly, pericytes could be related to EVT recognition by the maternal immune system and spiral arteries formation, as hypothesized in Fig. 8.4.

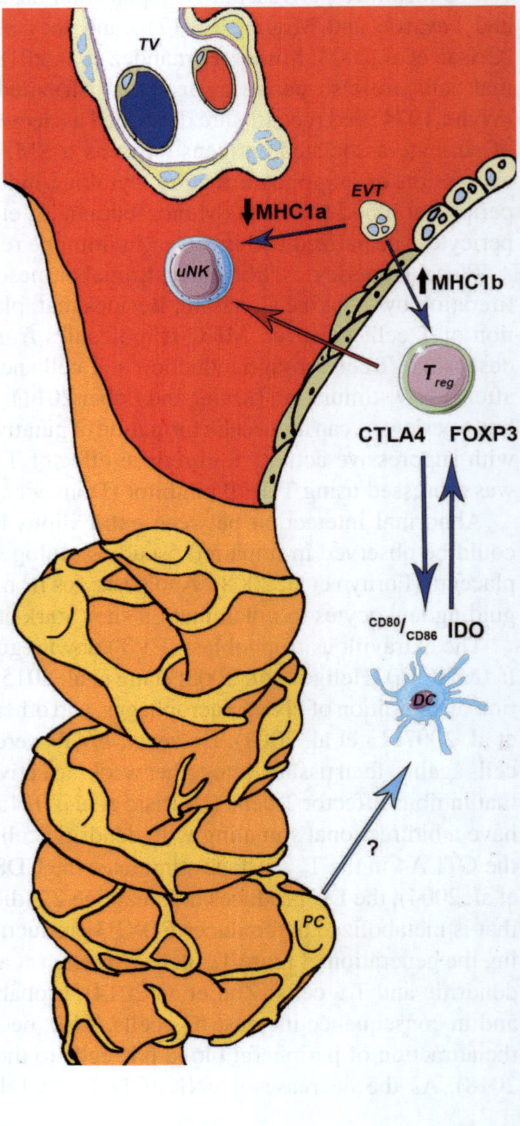

Fig. 8.4 Schema of human spiral arteriole. During spiral arteriole formation, there is interaction between trophoblast cells and some immune cells. In the terminal villi (TV), some cells with migratory profile, the extravillous trophoblast (EVT), migrate to spiral artery replacing endothelial cells. EVT have downregulation of MCH1a that could activate uterine natural killer (uNK) cells. However, EVT upregulates MHC1b that activates regulatory T (T_{reg}) cells. T_{reg} and dendritic cells (DC) activate each other, and this inhibits the attack of EVT by uNK. Also, the pericytes (PC) from spiral arteriole wall can interact with some DC (by unknown mechanism) and help in immune cell activation and trophoblast cell migration. Dark blue arrow (activation), red arrow (inactivation), and light blue arrow (activation by unknown mechanism)

8.5.4 *Pericytes in Placental Cell Differentiation*

The pericyte function and plasticity potential are tissue dependent, i.e., skeletal muscle pericytes have better ability to differentiate to myogenic lineage than myocardial-derived ones (Chen et al. 2015). Besides that, pericyte heterogeneity could be present even inside the same tissue (Prazeres et al. 2017). However, little is known about specific pericyte differentiation in the placenta.

In general, mesenchymal stem cells can be localized perivascularly [reviewed by Lobo et al. (2016) and Favaron and Miglino (2017)] such as in placental tissues. Besides mesenchymal stem cells, placental tissues are a niche of hematopoietic stem cells (Lee et al. 2010; Dzierzak and Robin 2010), since their early differentiation from hemangioblast during yolk sac formation (Ueno and Weissman 2010; Golub and Cumano 2013), in the labyrinth development in mice placenta [reviewed by Ottersbach and Dzierzak (2010)], or in the major vessels of the chorionic plate (Gekas et al. 2010). Therefore, the placenta is a niche of hematopoietic stem cells acting in the generation, renewal, and maintenance of undifferentiated status (Gekas et al. 2010), with stem cell factor (a pericyte-related marker) as a central molecule to maintain this function (Khodadi et al. 2016).

Endometrial stromal cells have pericyte phenotype with immunogenic profile (Muñoz-Fernández et al. 2018), as cited before. Also, pericytes that support hematopoiesis are present in small number in maternal tissue as early as week 3, but it increases considerably at the fetal side in week 6 (Robin et al. 2009), suggesting that maternal pericytes support initial hematopoiesis, and just after fetal pericytes assume this function in the growing and development placenta. In addition, one should consider the existence of different populations of pericytes, evidenced by the distinct functions performed by these cells within the same tissue (Prazeres et al. 2017). A similar event happens in mice, where at E11 low pericyte amount is recovered; however, between E12 and E13, they reach the highest number (Gekas et al. 2010; Ottersbach and Dzierzak 2010), on the same period of higher influx of uNK and trophoblast hyperplasia (Wooding and Burton 2008b).

The ERK/MAPK pathway in pericytes must be required for self-function maintenance and allow differentiation and formation of syncytiotrophoblast layers. Then a deficiency in this pathway affects syncytiotrophoblast layer formation that implicates in reduced maternal labyrinthic vascularization due to a decrease in interaction with neighbor trophoblast, but without interfering in pericyte migration (Nadeau and Charron 2014).

The endothelial cell hyperplasia, hypervariable diameter, abundant microaneurysm, abnormal endothelial ultrastructure, and altered permeability could be observed in the microvasculature of PDGF-B or PDGFRβ null mice. Probably by reduction of pericytes recruitment in the placenta. Withal, reduction of trophoblast development and proliferation was observed in these mice null model and in endothelial PDGF-B-deficient mice (Bjarnegard 2004).

8.6 Final Remarks

Despite the variability of placental morphology among species and even more in the maternal-fetal interface, the gestational goal is acquired—the production of viable offspring. Also, the placental microenvironment is unique, due to the interaction between maternal and fetal tissues, constant cell differentiation, peripheral maternal immune cell influx, and maintenance of stem cell niche. In parallel, pericytes have a plethora of functions that meet placental homeostasis. Pericytes have been associated with some placental function, such as vascular control, immune cell activation, and trophoblast differentiation; however, the detailed mechanisms are still unknown.

Acknowledgments Alexander Birbrair is supported by a grant from Instituto Serrapilheira/Serra-1708-15285, a grant from Pró-reitoria de Pesquisa/Universidade Federal de Minas Gerais (PRPq/UFMG) (Edital 05/2016), a grant from the National Institute of Science and Technology in Theranostics and Nanobiotechnology (CNPq/CAPES/FAPEMIG, Process No. 465669/2014-0), a grant from FAPEMIG [Rede Mineira de Engenharia de Tecidos e Terapia Celular (REMETTEC, RED-00570-16)], and a grant from FAPEMIG [Rede De Pesquisa Em Doenças Infecciosas Humanas E Animais Do Estado De Minas Gerais (RED-00313-16)].

References

(1955) The normal anatomy of the uterine artery. Acta Radiol. 43:21–36. doi: 10.3109/00016925509170755, http://dx.doi.org/10.3109/00016925509170755

Abrahamsohn PA, Zorn TMT (1993) Implantation and decidualization in rodents. J Exp Zool 266:603–628. https://doi.org/10.1002/jez.1402660610

Adamson SL, Lu Y, Whiteley KJ, Holmyard D, Hemberger M, Pfarrer C, Cross JC (2002) Interactions between trophoblast cells and the maternal and fetal circulation in the mouse placenta. Dev Biol 250:358–373

Al Ahmad A, Taboada CB, Gassmann M, Ogunshola OO (2011) Astrocytes and Pericytes differentially modulate blood—brain barrier characteristics during development and hypoxic insult. J Cereb Blood Flow Metab 31:693–705. https://doi.org/10.1038/jcbfm.2010.148

Aluvihare VR, Kallikourdis M, Betz AG (2004) Regulatory T cells mediate maternal tolerance to the fetus. Nat Immunol 5:266–271. https://doi.org/10.1038/ni1037

Antoniadou E, David AL (2016) Placental stem cells. Best Pract Res Clin Obstet Gynaecol 31:13–29. https://doi.org/10.1016/J.BPOBGYN.2015.08.014

Arck PC, Hecher K (2013) Fetomaternal immune cross-talk and its consequences for maternal and offspring's health. Nat Med 19:548–556. https://doi.org/10.1038/nm.3160

Armulik A, Genové G, Mäe M, Nisancioglu MH, Wallgard E, Niaudet C, He L, Norlin J, Lindblom P, Strittmatter K, Johansson BR, Betsholtz C (2010) Pericytes regulate the blood–brain barrier. Nature 468:557–561. https://doi.org/10.1038/nature09522

Armulik A, Genové G, Betsholtz C (2011) Pericytes: developmental, physiological, and pathological perspectives, problems, and promises. Dev Cell 21:193–215. https://doi.org/10.1016/j.devcel.2011.07.001

Aronoff DM, Correa H, Rogers LM, Arav-Boger R, Alcendor DJ (2017) Placental pericytes and cytomegalovirus infectivity: implications for HCMV placental pathology and congenital disease. Am J Reprod Immunol 78. https://doi.org/10.1111/aji.12728

Arts NF (1961) Investigations on the vascular system of the placenta. I. General introduction and the fetal vascular system. Am J Obstet Gynecol 82:147–158

Balabanov R, Washington R, Wagnerova J, Dore-Duffy P (1996) CNS microvascular pericytes express macrophage-like function, cell surface integrin alpha M, and macrophage marker ED-2. Microvasc Res 52:127–142. https://doi.org/10.1006/mvre.1996.0049

Balabanov R, Beaumont T, Dore-Duffy P (1999) Role of central nervous system microvascular pericytes in activation of antigen-primed splenic T-lymphocytes. J Neurosci Res 55:578–587. https://doi.org/10.1002/(SICI)1097-4547(19990301)55:5<578::AID-JNR5>3.0.CO;2-E

Bandeira DS, Casamitjana J, Crisan M (2017) Pharmacology & Therapeutics Pericytes , integral components of adult hematopoietic stem cell niches. Pharmacol Ther 171:104–113. https://doi.org/10.1016/j.pharmthera.2016.11.006

Barreto RSN, Bressan FF, Oliveira LJ, Pereira FTV, Perecin F, Ambrósio CE, Meirelles FV, Miglino MA (2011) Gene expression in placentation of farm animals: an overview of gene function during development. Theriogenology 76:589–597. https://doi.org/10.1016/j.theriogenology.2011.03.001

Baur R (1977) Morphometry of the placental exchange area. Adv Anat Embryol Cell Biol 53:3–65

Bell RD, Winkler EA, Sagare AP, Singh I, LaRue B, Deane R, Zlokovic BV (2010) Pericytes control key neurovascular functions and neuronal phenotype in the adult brain and during brain aging. Neuron 68:409–427. https://doi.org/10.1016/j.neuron.2010.09.043

Bergers G, Song S (2005) The role of pericytes in blood-vessel formation and maintenance. Neuro-Oncology 7:452–464. https://doi.org/10.1215/S1152851705000232

Bjarnegard M (2004) Endothelium-specific ablation of PDGFB leads to pericyte loss and glomerular, cardiac and placental abnormalities. Development 131:1847–1857. https://doi.org/10.1242/dev.01080

Bouchard BA, Shatos MA, Tracy PB (1997) Human brain pericytes differentially regulate expression of procoagulant enzyme complexes comprising the extrinsic pathway of blood coagulation. Arterioscler Thromb Vasc Biol 17:1–9

Burton GJ, Charnock-Jones DS, Jauniaux E (2009) Regulation of vascular growth and function in the human placenta. Reproduction 138:895–902. https://doi.org/10.1530/REP-09-0092

Carter AM, Mess A (2007) Evolution of the placenta in eutherian mammals. Placenta 28:259–262. https://doi.org/10.1016/j.placenta.2006.04.010

Castejón OJ (2011) Ultrastructural pathology of cortical capillary pericytes in human traumatic brain oedema. Folia Neuropathol 49:162–173

Caumartin J, Favier B, Daouya M, Guillard C, Moreau P, Carosella ED, LeMaoult J (2007) Trogocytosis-based generation of suppressive NK cells. EMBO J 26:1423–1433. https://doi.org/10.1038/sj.emboj.7601570

Cecati M, Giannubilo SR, Emanuelli M, Tranquilli AL, Saccucci F (2011) HLA-G and pregnancy adverse outcomes. Med Hypotheses 76:782–784. https://doi.org/10.1016/j.mehy.2011.02.017

Challier JC, Kacemi A, Galtier M, Boucher M, Tangapregassom MJ, Vervelle C, Bintein T, Espié MJ, Olive G (1997) Phenotype of cultured fetal perivascular cells from human placenta studied by scanning electron microscopy. Anat Embryol (Berl) 195:79–86. https://doi.org/10.1007/s004290050027

Challier JC, Galtier M, Kacemi A, Guillaumin D (1999) Pericytes of term human foeto-placental microvessels: ultrastructure and visualization. Cell Mol Biol (Noisy-le-Grand) 45:89–100

Chen S, Lechleider RJ (2004) Transforming growth factor—induced differentiation of smooth muscle from a neural crest stem cell line. Circ Res 94:1195–1202. https://doi.org/10.1161/01.RES.0000126897.41658.81

Chen C-W, Okada M, Proto JD, Gao X, Sekiya N, Beckman SA, Corselli M, Crisan M, Saparov A, Tobita K, Péault B, Huard J (2013) Human pericytes for ischemic heart repair. Stem Cells 31:305–316. https://doi.org/10.1002/stem.1285

Chen WCW, Baily JE, Corselli M, Díaz ME, Sun B, Xiang G, Gray GA, Huard J, Péault B (2015) Human myocardial pericytes: multipotent mesodermal precursors exhibiting cardiac specificity. Stem Cells 33:557–573. https://doi.org/10.1002/stem.1868

Corselli M, Chin CJ, Parekh C, Sahaghian A, Wang W, Ge S, Evseenko D, Wang X, Montelatici E, Lazzari L, Crooks GM, Péault B (2013) Perivascular support of human hematopoietic stem/progenitor cells. Blood 121:2891–2901. https://doi.org/10.1182/blood-2012-08-451864

Crisan M, Yap S, Casteilla L, Chen C, Corselli M, Park TS, Andriolo G, Sun B, Zheng B, Zhang L, Norotte C, Teng P, Traas J, Schugar R, Deasy BM, Badylak S, Lazzari L, Huard J, Pe B (2008) A perivascular origin for mesenchymal stem cells in multiple human organs. Cell Stem Cell 3:301–313. https://doi.org/10.1016/j.stem.2008.07.003

Croy BA, Yamada AT, DeMayo FJ, Adamson SL (2015) The guide to investigation of mouse pregnancy. Elsevier, San Diego, CA

Cuevas P, Gutierrez-Diaz JA, Reimers D, Dujovny M, Diaz FG, Ausman JI (1984) Pericyte endothelial gap junctions in human cerebral capillaries. Anat Embryol (Berl) 170:155–159

Daneman R, Zhou L, Kebede AA, Barres BA (2010) Pericytes are required for blood-brain barrier integrity during embryogenesis. Nature 468:562–566. https://doi.org/10.1038/nature09513

Dantzer V (2002) Endometrium of epitheliochorial and endotheliochorial placentae. In: Glasser SR, Aplin JD, Giudice LC, Tabibzadeh S (eds) The endometrium. Taylor & Francis, New York, pp 352–368

Dellavalle A, Maroli G, Covarello D, Azzoni E, Innocenzi A, Perani L, Antonini S, Sambasivan R, Brunelli S, Tajbakhsh S, Cossu G (2011) Pericytes resident in postnatal skeletal muscle differentiate into muscle fibres and generate satellite cells. Nat Commun 2:499. https://doi.org/10.1038/ncomms1508

Demir R, Kaufmann P, Castelluce M, Erbeng T, Kotowsk A (1989) Fetal vasculogenesis and angiogenesis in human placental villi. Cells Tissues Organs 136:190–203. https://doi.org/10.1159/000146886

Deveci E, Soker S, Turgut A, Aktas A, Ayaz E, Sak S, Seker U (2013) The immunohistochemical and ultrastructural evaluation of pericytes in human full term placentas of gestasyonal diabetes mellitus. Acta Medica Mediterr 29:697–700

Ding R, Darland DC, Parmacek MS, D'amore PA (2004) Endothelial–mesenchymal interactions in vitro reveal molecular mechanisms of smooth muscle/Pericyte differentiation. Stem Cells Dev 13:509–520. https://doi.org/10.1089/scd.2004.13.509

Ding L, Saunders TL, Enikolopov G, Morrison SJ (2012) Endothelial and perivascular cells maintain haematopoietic stem cells. Nature 481:457–462. https://doi.org/10.1038/nature10783

Djurisic S, Teiblum S, Tolstrup CK, Christiansen OB, Hviid TVF (2014) Allelic imbalance modulates surface expression of the tolerance-inducing HLA-G molecule on primary trophoblast cells. Mol Hum Reprod 21:281–295. https://doi.org/10.1093/molehr/gau108

Domev H, Milkov I, Itskovitz-Eldor J, Dar A (2014) Immunoevasive Pericytes from human pluripotent stem cells preferentially modulate induction of allogeneic regulatory T cells. Stem Cells Transl Med 3:1169–1181. https://doi.org/10.5966/sctm.2014-0097

Dubova EA, Buranova FB, Fyodorova TA, Shchyogolev AI, Sukhikh GT (2013) Morphological characteristics of the terminal villi in placental failure. Bull Exp Biol Med 155:507–511

Duffield JS, Lupher M, Thannickal VJ, Wynn TA (2013) Host responses in tissue repair and fibrosis. Annu Rev Pathol 8:241–276. https://doi.org/10.1146/annurev-pathol-020712-163930

Dulauroy S, Di Carlo SE, Langa F, Eberl G, Peduto L (2012) Lineage tracing and genetic ablation of ADAM12+ perivascular cells identify a major source of profibrotic cells during acute tissue injury. Nat Med 18:1262–1270. https://doi.org/10.1038/nm.2848

Dzierzak E, Robin C (2010) Placenta as a source of hematopoietic stem cells. Trends Mol Med 16:361–367. https://doi.org/10.1016/j.molmed.2010.05.005

Enders AC, Blankenship TN, Lantz KC, Enders SS (1998) Morphological variation in the interhemal areas of chorioallantoic placentae: a review. Placenta 19:1–19. https://doi.org/10.1016/S0143-4004(98)80030-1

Enge M, Bjarnegård M, Gerhardt H, Gustafsson E, Kalén M, Asker N, Hammes H-P, Shani M, Fässler R, Betsholtz C (2002) Endothelium-specific platelet-derived growth factor-B ablation mimics diabetic retinopathy. EMBO J 21:4307–4316

Fabry Z, Fitzsimmons KM, Herlein JA, Moninger TO, Dobbs MB, Hart MN (1993) Production of the cytokines interleukin 1 and 6 by murine brain microvessel endothelium and smooth muscle pericytes. J Neuroimmunol 47:23–34

Farrington-Rock C, Crofts NJ, Doherty MJ, Ashton BA, Griffin-Jones C, Canfield AE (2004) Chondrogenic and adipogenic potential of microvascular pericytes. Circulation 110:2226–2232. https://doi.org/10.1161/01.CIR.0000144457.55518.E5

Favaron PO, Miglino MA (2017) Fetal membranes-derived stem cells microenvironment. Adv Exp Med Biol 1041:235–244

Feng J, Mantesso A, De Bari C, Nishiyama A, Sharpe PT (2011) Dual origin of mesenchymal stem cells contributing to organ growth and repair. Proc Natl Acad Sci U S A 108:6503–6508. https://doi.org/10.1073/pnas.1015449108

Firth JA, Leach L (1996) Not trophoblast alone: a review of the contribution of the fetal microvasculature to transplacental exchange. Placenta 17:89–96. https://www.ncbi.nlm.nih.gov/pubmed/8730878

Fisher M (2009) Pericyte Signaling in the neurovascular unit. Stroke 40:S13–S15. https://doi.org/10.1161/STROKEAHA.108.533117

Furuya M, Ishida J, Inaba S, Kasuya Y, Kimura S, Nemori R, Fukamizu A (2008) Impaired placental neovascularization in mice with pregnancy-associated hypertension. Lab Investig 88:416–429. https://doi.org/10.1038/labinvest.2008.7

Gekas C, Rhodes KE, Vanhandel B, Chhabra A, Ueno M, Mikkola HKA (2010) Hematopoietic stem cell development in the placenta. Int J Dev Biol 54:1089–1098. https://doi.org/10.1387/ijdb.103070cg

Golub R, Cumano A (2013) Embryonic hematopoiesis. Blood Cells Mol Dis 51:226–231. https://doi.org/10.1016/j.bcmd.2013.08.004

Guerin LR, Prins JR, Robertson SA (2009) Regulatory T-cells and immune tolerance in pregnancy: a new target for infertility treatment? Hum Reprod Update 15:517–535. https://doi.org/10.1093/humupd/dmp004

Hasan M, Glees P (1990) The fine structure of human cerebral perivascular pericytes and juxtavascular phagocytes: their possible role in hydrocephalic edema resolution. J Hirnforsch 31:237–249

He N, van Iperen L, de Jong D, Szuhai K, Helmerhorst FM, van der Westerlaken LAJ, Chuva de Sousa Lopes SM (2017) Human extravillous trophoblasts penetrate decidual veins and lymphatics before remodeling spiral arteries during early pregnancy. PLoS One 12:e0169849. https://doi.org/10.1371/journal.pone.0169849

Heinrich D, Aoki A, Metz J (1988) Fetal capillary organization in different types of placenta. Trophobl Res 3:149–162. https://doi.org/10.1007/978-1-4615-8109-3_11

Helige C, Ahammer H, Hammer A, Huppertz B, Frank H-G, Dohr G (2008) Trophoblastic invasion in vitro and in vivo: similarities and differences. Hum Reprod 23:2282–2291. https://doi.org/10.1093/humrep/den198

Hellerbrand C (2013) Hepatic stellate cells—the pericytes in the liver. Pflügers Arch Eur J Physiol 465:775–778. https://doi.org/10.1007/s00424-012-1209-5

Hellström M, Gerhardt H, Kalén M, Li X, Eriksson U, Wolburg H, Betsholtz C (2001) Lack of Pericytes leads to endothelial hyperplasia and abnormal vascular morphogenesis. J Cell Biol 153:543–553

Herrmann M, Bara JJ, Sprecher CM, Menzel U, Jalowiec JM, Osinga R, Scherberich A, Alini M, Verrier S (2016) Pericyte plasticity—comparative investigation of the angiogenic and multilineage potential of pericytes from different human tissues. Eur Cells Mater 31:236–249. https://doi.org/10.22203/eCM.v031a16

Holmgren L, Glaser A, Pfeifer-Ohlsson S, Ohlsson R (1991) Angiogenesis during human extraembryonic development involves the spatiotemporal control of PDGF ligand and receptor gene expression. Development 113:749–754

Holmgren L, Claesson-welsh L, Heldin C-H, Ohlsson R (1992) The expression of PDGF α- and β-receptors in subpopulations of PDGF-producing cells implicates autocrine stimulatory loops in the control of proliferation in Cytotrophoblasts that have invaded the maternal endometrium. Growth Factors 6:219–231. https://doi.org/10.3109/08977199209026929

Hurtado-Alvarado G, Cabañas-Morales AM, Gómez-Gónzalez B (2014) Pericytes: brain-immune interface modulators. Front Integr Neurosci 7:80. https://doi.org/10.3389/fnint.2013.00080

James AW, Zara JN, Zhang X, Askarinam A, Goyal R, Chiang M, Yuan W, Chang L, Corselli M, Shen J, Pang S, Stoker D, Wu B, Ting K, Péault B, Soo C (2012) Perivascular stem cells: a prospectively purified mesenchymal stem cell population for bone tissue engineering. Stem Cells Transl Med 1:510–519. https://doi.org/10.5966/sctm.2012-0002

Jeynes B (1985) Reactions of granular pericytes in a rabbit cerebrovascular ischemia model. Stroke 16:121–125

Jiang F, Zhao H, Wang L, Guo X, Wang X, Yin G, Hu Y, Li Y, Yao Y (2015) Role of HLA-G1 in trophoblast cell proliferation, adhesion and invasion. Biochem Biophys Res Commun 458:154–160. https://doi.org/10.1016/j.bbrc.2015.01.085

Jones CJP, Desoye G (2011) A new possible function for placental pericytes. Cells Tissues Organs 194:76–84. https://doi.org/10.1159/000322394

Kamouchi M, Ago T, Kitazono T (2011) Brain Pericytes: emerging concepts and functional roles in brain homeostasis. Cell Mol Neurobiol 31(2):175–193. https://doi.org/10.1007/s10571-010-9605-x

Katare R, Riu F, Mitchell K, Gubernator M, Campagnolo P, Cui Y, Fortunato O, Avolio E, Cesselli D, Beltrami AP, Angelini G, Emanueli C, Madeddu P (2011) Transplantation of human pericyte progenitor cells improves the repair of infarcted heart through activation of an angiogenic program involving micro-RNA-132. Circ Res 109:894–906. https://doi.org/10.1161/CIRCRESAHA.111.251546

Kaufmann P, Luckhardt M, Leiser R (1988) Three-dimensional representation of the Fetal vessel system in the human placenta. Trophobl Res 3:113–137. https://doi.org/10.1007/978-1-4615-8109-3_9

Kaufmann P, Mayhew TM, Charnock-Jones DS (2004) Aspects of human Fetoplacental Vasculogenesis and angiogenesis. II. Changes during Normal pregnancy. Placenta 25:114–126. https://doi.org/10.1016/j.placenta.2003.10.009

Khodadi E, Shahrabi S, Shahjahani M, Azandeh S, Saki N (2016) Role of stem cell factor in the placental niche. Cell Tissue Res 366:523–531. https://doi.org/10.1007/s00441-016-2429-3

Kim JA, Tran ND, Li Z, Yang F, Zhou W, Fisher MJ (2006) Brain endothelial Hemostasis regulation by Pericytes. J Cereb Blood Flow Metab 26:209–217. https://doi.org/10.1038/sj.jcbfm.9600181

Kim KR, Sung CO, Kwon TJ, Lee JB, Robboy SJ (2015) Defective pericyte recruitment of villous stromal vessels as the possible etiologic cause of hydropic change in complete hydatidiform mole. PLoS One 10:1–13. https://doi.org/10.1371/journal.pone.0122266

Knox K, Baker JC (2008) Genomic evolution of the placenta using co-option and duplication and divergence. Genome Res 18:695–705. https://doi.org/10.1101/gr.071407.107

Krautler NJ, Kana V, Kranich J, Tian Y, Perera D, Lemm D, Schwarz P, Armulik A, Browning JL, Tallquist M, Buch T, Oliveira-Martins JB, Zhu C, Hermann M, Wagner U, Brink R, Heikenwalder M, Aguzzi A (2012) Follicular dendritic cells emerge from ubiquitous perivascular precursors. Cell 150:194–206. https://doi.org/10.1016/j.cell.2012.05.032

Krueger M, Bechmann I (2010) CNS pericytes: concepts, misconceptions, and a way out. Glia 58:1–10. https://doi.org/10.1002/glia.20898

Kučera T, Vyletěl I, Moravcová M, Krejčí V, Žižka Z, Jirkovská M (2010) Pericyte coverage of fetoplacental vessels in pregnancies complicated by type 1 diabetes mellitus. Placenta 31:1120–1122. https://doi.org/10.1016/j.placenta.2010.09.014

Kunisaki Y, Bruns I, Scheiermann C, Ahmed J, Pinho S, Zhang D, Mizoguchi T, Wei Q, Lucas D, Ito K, Mar JC, Bergman A, Frenette PS (2013) Arteriolar niches maintain haematopoietic stem cell quiescence. Nature 502:637–643. https://doi.org/10.1038/nature12612

Lachenmayer L (1971a) Adrenergic innervation of the umbilical vessels. Light- and fluorescence microscopic studies. Z Zellforsch Mikrosk Anat 120:120–136

Lachenmayer L (1971b) Adrenergic innervation of the umbilical vessels. Zeitschrift für Zellforsch und Mikroskopische Anat 120:120–136. https://doi.org/10.1007/BF00331246

Lash GE, Otun HA, Innes BA, Percival K, Searle RF, Robson SC, Bulmer JN (2010) Regulation of extravillous trophoblast invasion by uterine natural killer cells is dependent on gestational age. Hum Reprod 25:1137–1145. https://doi.org/10.1093/humrep/deq050

Lee LK, Ueno M, Van Handel B, Mikkola HK (2010) Placenta as a newly identified source of hematopoietic stem cells. Curr Opin Hematol 17:313–318. https://doi.org/10.1097/MOH.0b013e328339f295
Leiser R, Kaufmann P (1994) Placental structure: in a comparative aspect. Exp Clin Endocrinol 102:122–134
Leiser R, Kohler T (1983) The blood vessels of the cat girdle placenta. Observations on corrosion casts, scanning electron microscopical and histological studies—I. Maternal vasculature. Anat Embryol (Berl) 167:85–93. https://doi.org/10.1007/BF00304602
Leiser R, Kohler T (1984) The blood vessels of the cat girdle placenta. Observations on corrosion casts, scanning electron microscopical and histological studies—II. Fetal vasculature. Anat Embryol (Berl) 170:209–216. https://doi.org/10.1007/BF00319006
Leiser R, Krebs C, Ebert B, Dantzer V (1997a) Placental vascular corrosion cast studies: a comparison between ruminants and humans. Microsc Res Tech 38:76–87. https://doi.org/10.1002/(SICI)1097-0029(19970701/15)38:1/2<76::AID-JEMT9>3.0.CO;2-S
Leiser R, Krebs C, Klisch K, Ebert B, Dantzer V, Schuler G, Hoffmann B (1997b) Fetal villosity and microvasculature of the bovine placentome in the second half of gestation. J Anat 191:517–527. https://doi.org/10.1017/S0021878297002690
Leiser R, Pfarrer C, Abd-Elnaeim M, Dantzer V (1998) Feto-maternal anchorage in epitheliochorial and endotheliochorial placental types studied by histology and microvascular corrosion casts. Placenta 19:21–39. https://doi.org/10.1016/S0143-4004(98)80031-3
Leno-Durán E, Hatta K, Bianco J, Yamada ÁT, Ruiz-Ruiz C, Olivares EG, Croy BA (2010) Fetal-placental hypoxia does not result from failure of spiral arterial modification in mice. Placenta 31:731–737. https://doi.org/10.1016/j.placenta.2010.06.002
Leonard S, Murrant C, Tayade C, Vandenheuvel M, Watering R, Croy B (2006) Mechanisms regulating immune cell contributions to spiral artery modification—facts and hypotheses—a review. Placenta 27:40–46. https://doi.org/10.1016/j.placenta.2005.11.007
Levéen P, Pekny M, Gebre-Medhin S, Swolin B, Larsson E, Betsholtz C (1994) Mice deficient for PDGF B show renal, cardiovascular, and hematological abnormalities. Genes Dev 8:1875–1887
Li C, Houser BL, Nicotra ML, Strominger JL (2009) HLA-G homodimer-induced cytokine secretion through HLA-G receptors on human decidual macrophages and natural killer cells. Proc Natl Acad Sci U S A 106:5767–5772. https://doi.org/10.1073/pnas.0901173106
Lindahl P, Johansson BR, Levéen P, Betsholtz C (1997) Pericyte loss and microaneurysm formation in PDGF-B-deficient mice. Science 277:242–245
Lobo SE, Leonel LCPC, Miranda CMFC, Coelho TM, Ferreira GAS, Mess A, Abrão MS, Miglino MA (2016) The placenta as an organ and a source of stem cells and extracellular matrix: a review. Cells Tissues Organs 201:239–252. https://doi.org/10.1159/000443636
Looman C, Sun T, Yu Y, Zieba A, Ahgren A, Feinstein R, Forsberg H, Hellberg C, Heldin CH, Zhang XQ, Forsberg-Nilsson K, Khoo N, Fundele R, Heuchel R (2007) An activating mutation in the PDGF receptor-beta causes abnormal morphology in the mouse placenta. Int J Dev Biol 51:361–370. https://doi.org/10.1387/ijdb.072301cl
Macara L, Kingdom JC, Kaufmann P, Kohnen G, Hair J, More IA, Lyall F, Greer IA (1996) Structural analysis of placental terminal villi from growth-restricted pregnancies with abnormal umbilical artery Doppler waveforms. Placenta 17:37–48
Maier CL, Pober JS (2011) Human placental pericytes poorly stimulate and actively regulate allogeneic CD4 T cell responses. Arterioscler Thromb Vasc Biol 31:183–189. https://doi.org/10.1161/ATVBAHA.110.217117
Maier CL, Shepherd BR, Yi T, Pober JS (2010) Explant outgrowth, propagation and characterization of human pericytes. Microcirculation 17:367–380. https://doi.org/10.1111/j.1549-8719.2010.00038.x
Méndez-Ferrer S, Michurina TV, Ferraro F, Mazloom AR, MacArthur BD, Lira SA, Scadden DT, Ma'ayan A, Enikolopov GN, Frenette PS (2010) Mesenchymal and haematopoietic stem cells form a unique bone marrow niche. Nature 466:829–834. https://doi.org/10.1038/nature09262
Mess A, Carter AM (2007) Evolution of the placenta during the early radiation of placental mammals. Comp Biochem Physiol A Mol Integr Physiol 148(4):769–779

Mezrich JD, Fechner JH, Zhang X, Johnson BP, Burlingham WJ, Bradfield CA (2010) An interaction between kynurenine and the aryl hydrocarbon receptor can generate regulatory T cells. J Immunol 185:3190–3198. https://doi.org/10.4049/jimmunol.0903670

Moffett A, Loke C (2006) Immunology of placentation in eutherian mammals. Nat Rev Immunol 6:584–594. https://doi.org/10.1038/nri1897

Mónica Brauer M, Smith PG (2015) Estrogen and female reproductive tract innervation: cellular and molecular mechanisms of autonomic neuroplasticity. Auton Neurosci 187:1–17

Moreau JLM, Artap ST, Shi H, Chapman G, Leone G, Sparrow DB, Dunwoodie SL (2014) Cited2 is required in trophoblasts for correct placental capillary patterning. Dev Biol 392:62–79. https://doi.org/10.1016/j.ydbio.2014.04.023

Moser G, Weiss G, Sundl M, Gauster M, Siwetz M, Lang-Olip I, Huppertz B (2017) Extravillous trophoblasts invade more than uterine arteries: evidence for the invasion of uterine veins. Histochem Cell Biol 147:353–366. https://doi.org/10.1007/s00418-016-1509-5

Mossman HW (1987) Vertebrate fetal membranes. Rutgers University Press, New Brunswisk

Muñoz-Fernández R, de la Mata C, Prados A, Perea A, Ruiz-Magaña MJ, Llorca T, Fernández-Rubio P, Blanco O, Abadía-Molina AC, Olivares EG (2018) Human predecidual stromal cells have distinctive characteristics of pericytes: cell contractility, chemotactic activity, and expression of pericyte markers and angiogenic factors. Placenta 61:39–47. https://doi.org/10.1016/j.placenta.2017.11.010

Murphy WJ, Eizirik E, Johnson WE, Zhang YP, Ryder OA, O'Brien SJ (2001) Molecular phylogenetics and the origins of placental mammals. Nature 409:614–618. https://doi.org/10.1038/35054550

Murrant CL (2014) Intravital imaging of Vasoactivity in the uterine arterial vasculature tree during pregnancy. In: Croy BA, Yamada A, DeMayo F, Lee Adamson S (eds) The guide to investigation of mouse pregnancy. Elseiver, San Diego, CA

Myatt L (1992) Control of vascular resistance in the human placenta. Placenta 13:329–341. https://doi.org/10.1016/0143-4004(92)90057-Z

Nadeau V, Charron J (2014) Essential role of the ERK/MAPK pathway in blood-placental barrier formation. Development 141:2825–2837. https://doi.org/10.1242/dev.107409

Nakagawa S, Deli MA, Nakao S, Honda M, Hayashi K, Nakaoke R, Kataoka Y, Niwa M (2007) Pericytes from brain microvessels strengthen the barrier integrity in primary cultures of rat brain endothelial cells. Cell Mol Neurobiol 27:687–694. https://doi.org/10.1007/s10571-007-9195-4

Nakamura K, Kamouchi M, Kitazono T, Kuroda J, Matsuo R, Hagiwara N, Ishikawa E, Ooboshi H, Ibayashi S, Iida M (2008) Role of NHE1 in calcium signaling and cell proliferation in human CNS pericytes. AJP Hear Circ Physiol 294:H1700–H1707. https://doi.org/10.1152/ajpheart.01203.2007

Nees S, Weiss DR, Juchem G (2013) Focus on cardiac pericytes. Pflügers Arch - Eur J Physiol 465:779–787. https://doi.org/10.1007/s00424-013-1240-1

Nikolov SD, Schiebler TH (1973) Über das fetale Gefäßsystem der reifen menschlichen Placenta. Zeitschrift für Zellforsch und Mikroskopische Anat 139:333–350. https://doi.org/10.1007/BF00306590

Nomura M, Yamagishi SI, Harada SI, Hayashi Y, Yamashima T, Yamashita J, Yamamoto H (1995) Possible participation of autocrine and paracrine vascular endothelial growth factors in hypoxia-induced proliferation of endothelial cells and pericytes. J Biol Chem 270:28316–28324. https://doi.org/10.1074/jbc.270.47.28316

Ogino S, Redline RW (2000) Villous capillary lesions of the placenta: distinctions between chorangioma, chorangiomatosis, and chorangiosis. Hum Pathol 31:945–954. https://doi.org/10.1053/hupa.2000.9036

Ohlsson R, Falck P, Hellstrom M, Lindahl P, Bostrom H, Franklin G, Ährlund-Richter L, Pollard J, Soriano P, Betsholtz C (1999) PDGFB regulates the development of the labyrinthine layer of the mouse fetal placenta. Dev Biol 212:124–136. https://doi.org/10.1006/dbio.1999.9306

Oliveira MS, Barreto-Filho JB (2015) Placental-derived stem cells: culture, differentiation and challenges. World J Stem Cells 7:769–775. https://doi.org/10.4252/wjsc.v7.i4.769

Ottersbach K, Dzierzak E (2010) The placenta as a haematopoietic organ. Int J Dev Biol 54:1099–1106. https://doi.org/10.1387/ijdb.093057ko

Pallone TL, Silldorff EP (2001) Pericyte regulation of renal medullary blood flow. Exp Nephrol 9:165–170. doi: 52608

Pallone TL, Silldorff EP, Turner MR (1998) Intrarenal blood flow: microvascular anatomy and the regulation of medullary perfusion. Clin Exp Pharmacol Physiol 25:383–392

Pallone TL, Zhang Z, Rhinehart K (2003) Physiology of the renal medullary microcirculation. Am J Physiol Ren Physiol 284:F253–F266. https://doi.org/10.1152/ajprenal.00304.2002

Pan S-Y, Chang Y-T, Lin S-L (2014) Microvascular pericytes in healthy and diseased kidneys. Int J Nephrol Renovasc Dis 7:39–48. https://doi.org/10.2147/IJNRD.S37892

Pereira FTV, Oliveira LJ, Barreto RSN, Mess A, Perecin F, Bressan FF, Mesquita LG, Miglino MA, Pimentel JR, Neto PF, Meirelles FV (2013) Fetal-maternal interactions in the Synepitheliochorial placenta using the eGFP cloned cattle model. PLoS One 8:e64399. https://doi.org/10.1371/journal.pone.0064399

Persson E, Rodriguez-Martinez H (1997) Immunocytochemical localization of growth factors and intermediate filaments during the establishment of the porcine placenta. Microsc Res Tech 38:165–175. https://doi.org/10.1002/(SICI)1097-0029(19970701/15)38:1/2<165::AID-JEMT17>3.0.CO;2-N

Pfister F, Przybyt E, Harmsen MC, Hammes H-P (2013) Pericytes in the eye. Pflügers Arch Eur J Physiol 465:789–796. https://doi.org/10.1007/s00424-013-1272-6

Prazeres PHDM, Sena IFG, IDT B, de Azevedo PO, Andreotti JP, de Paiva AE, de Almeida VM, de Paula Guerra DA, Pinheiro Dos Santos GS, Mintz A, Delbono O, Birbrair A (2017) Pericytes are heterogeneous in their origin within the same tissue. Dev Biol 427:6–11. https://doi.org/10.1016/j.ydbio.2017.05.001

Ramsey EM, Martin CB, Donner MW (1967) Fetal and maternal placental circulations. Simultaneous visualization in monkeys by radiography. Am J Obstet Gynecol 98:419–423

Resta L, Capobianco C, Marzullo A, Piscitelli D, Sanguedolce F, Schena FP, Gesualdo L (2006) Confocal laser scanning microscope study of terminal villi vessels in Normal term and pre-eclamptic placentas. Placenta 27:735–739. https://doi.org/10.1016/j.placenta.2005.07.006

Rhodin J (1968) Ultrastructure of mammalian venous capillaries, venules, and small collecting veins. J Ultrastruct Res 25:452–500. https://doi.org/10.1016/S0022-5320(68)80098-X

Rizzo R, Lo Monte G, Bortolotti D, Graziano A, Gentili V, Di Luca D, Marci R (2015) Impact of soluble HLA-G levels and endometrial NK cells in uterine flushing samples from primary and secondary unexplained infertile women. Int J Mol Sci 16:5510–5516. https://doi.org/10.3390/ijms16035510

Robin C, Bollerot K, Mendes S, Haak E, Crisan M, Cerisoli F, Lauw I, Kaimakis P, Jorna R, Vermeulen M, Kayser M, van der Linden R, Imanirad P, Verstegen M, Nawaz-Yousaf H, Papazian N, Steegers E, Cupedo T, Dzierzak E (2009) Human placenta is a potent hematopoietic niche containing hematopoietic stem and progenitor cells throughout development. Cell Stem Cell 5(4):385–395. https://doi.org/10.1016/j.stem.2009.08.020

Rodriguez-Martinez H, Persson E, Hurst M, Stanchev P (1992) Immunohistochemical localization of platelet-derived growth factor receptors in the porcine uterus during the oestrous cycle and pregnancy. J Vet Med Ser A 39:1–10. https://doi.org/10.1111/j.1439-0442.1992.tb00151.x

Rouget C-MB (1873) Mémoire sur le développement, la structure et les proprietés physiologiques des capillaires sanguins et lymphatiques. Arch Psys 5:603–610

Rucker HK, Wynder HJ, Thomas WE (2000) Cellular mechanisms of CNS pericytes. Brain Res Bull 51:363–369

Sato K, Yamashita N, Baba M, Matsuyama T (2003) Modified myeloid dendritic cells act as regulatory dendritic cells to induce anergic and regulatory T cells. Blood 101:3581–3589. https://doi.org/10.1182/blood-2002-09-2712

Shimizu F, Sano Y, Maeda T, Abe M, Nakayama H, Takahashi R, Ueda M, Ohtsuki S, Terasaki T, Obinata M, Kanda T (2008) Peripheral nerve pericytes originating from the blood-nerve barrier expresses tight junctional molecules and transporters as barrier-forming cells. J Cell Physiol 217:367–380. https://doi.org/10.1002/jcp.21508

Sims DE (2000) Diversity within Pericytes. Clin Exp Pharmacol Physiol 27:842–846. https://doi.org/10.1046/j.1440-1681.2000.03343.x
Soriano P (1994) Abnormal kidney development and hematological disorders in PDGF beta-receptor mutant mice. Genes Dev 8:1888–1896
Stark K, Eckart A, Haidari S, Tirniceriu A, Lorenz M, Von Brühl ML, Gärtner F, Khandoga AG, Legate KR, Pless R, Hepper I, Lauber K, Walzog B, Massberg S (2013) Capillary and arteriolar pericytes attract innate leukocytes exiting through venules and "instruct" them with pattern-recognition and motility programs. Nat Immunol 14(1):41–51. https://doi.org/10.1038/ni.2477
Talwadekar MD, Kale VP, Limaye LS (2015) Placenta-derived mesenchymal stem cells possess better immunoregulatory properties compared to their cord-derived counterparts–a paired sample study. Sci Rep 5:15784. https://doi.org/10.1038/srep15784
Thanabalasundaram G, Schneidewind J, Pieper C, Galla H-J (2011) The impact of pericytes on the blood-brain barrier integrity depends critically on the pericyte differentiation stage. Int J Biochem Cell Biol 43(9):1284–1293. https://doi.org/10.1016/j.biocel.2011.05.002
Thomas WE (1999) Brain macrophages: on the role of pericytes and perivascular cells. Brain Res Brain Res Rev 31:42–57
Thornburg KL, Bagby SP, Giraud GD (2006) Maternal adaptation to pregnancy. In: Neill JD (ed) Knobil and Neill's physiology of reproduction. Elseiver, Amsterdam
Tilburgs T, Crespo ÂC, van der Zwan A, Rybalov B, Raj T, Stranger B, Gardner L, Moffett A, Strominger JL (2015) Human HLA-G+ extravillous trophoblasts: immune-activating cells that interact with decidual leukocytes. Proc Natl Acad Sci 112:7219–7224. https://doi.org/10.1073/pnas.1507977112
Tu Z, Li Y, Smith DS, Sheibani N, Huang S, Kern T, Lin F (2011) Retinal pericytes inhibit activated T cell proliferation. Invest Ophthalmol Vis Sci 52:9005–9010. https://doi.org/10.1167/iovs.11-8008
Ueno H, Weissman IL (2010) The origin and fate of yolk sac hematopoiesis: application of chimera analyses to developmental studies. Int J Dev Biol 54:1019–1031. https://doi.org/10.1387/ijdb.093039hu
van der Heijden GW, Dieker JW, Derijck AAHA, Muller S, Berden JHM, Braat DDM, van der Vlag J, de Boer P (2005) Asymmetry in histone H3 variants and lysine methylation between paternal and maternal chromatin of the early mouse zygote. Mech Dev 122:1008–1022. https://doi.org/10.1016/j.mod.2005.04.009
Verbeek MM, Westphal JR, Ruiter DJ, de Waal RM (1995) T lymphocyte adhesion to human brain pericytes is mediated via very late antigen-4/vascular cell adhesion molecule-1 interactions. J Immunol 154:5876–5884
Vogel P (2005) The current molecular phylogeny of eutherian mammals challenges previous interpretations of placental evolution. Placenta 26:591–596. https://doi.org/10.1016/j.placenta.2004.11.005
Wehrenberg WB, Chaichareon DP, Dierschke DJ, Rankin JH, Ginther OJ (1977) Vascular dynamics of the reproductive tract in the female rhesus monkey: relative contributions of ovarian and uterine arteries. Biol Reprod 17:148–153
Wooding FBP, Burton GJ (2008a) Comparative placentation: structures, functions and evolution. Springer, New York
Wooding P, Burton GJ (2008b) Haemochorial placentation: mouse, rabbit, man, apes, monkeys. In: Comparative placentation. Springer, Berlin, pp 185–230
Wooding FBP, Flint APF (1994) Placentation. In: Lamming GE (ed) Marshall's physiology of reproduction, 4th edn. Springer-Science, London, pp 233–429
Wynn RM (1974) Ultrastructural development of the human decidua. Am J Obstet Gynecol 118(5):652–670. https://doi.org/10.1016/S0002-9378(16)33740-1
Yamada AT, Bianco JR, Lippe EMO, Degaki KY, Dalmorin AF, Edwards AK, Lima PDA, Paffaro VA (2014) Unique features of endometrial dynamics during pregnancy. In: Anne Croy B, Yamada A, DeMayo F, Lee Adamson S (eds) The guide to investigation of mouse pregnancy. Elseiver, San Diego, CA

Yamagishi S, Kawakamia T, Fujimoria H, Yonekura H, Tanaka N, Yamamoto Y, Urayama H, Watanabe Y, Yamamoto H (1999a) Insulin stimulates the growth and tube formation of human microvascular endothelial cells through autocrine vascular endothelial growth factor. Microvasc Res 57:329–339

Yamagishi S, Yonekura H, Yamamoto Y, Fujimori H, Sakurai S, Tanaka N, Yamamoto H (1999b) Vascular endothelial growth factor acts as a pericyte mitogen under hypoxic conditions. Lab Investig 79(4):501–509

Yonekura H, Sakurai S, Liu X, Migita H, Wang H, Yamagishi S, Nomura M, Abedin MJ, Unoki H, Yamamoto Y, Yamamoto H (1999) Placenta growth factor and vascular endothelial growth factor B and C expression in microvascular endothelial cells and Pericytes. J Biol Chem 274:35172–35178

Zebardast N, Lickorish D, Davies JE (2010) Human umbilical cord perivascular cells (HUCPVC): a mesenchymal cell source for dermal wound healing. Organogenesis 6:197–203. https://doi.org/10.4161/org.6.4.12393

Zhang EG, Burton GJ, Smith SK, Charnock-Jones DS (2002) Placental vessel adaptation during gestation and to high altitude: changes in diameter and perivascular cell coverage. Placenta 23(10):751–762. https://doi.org/10.1016/S0143-4004(02)90856-8

Zhao H, Feng J, Seidel K, Shi S, Klein O, Sharpe P, Chai Y (2014) Secretion of shh by a neurovascular bundle niche supports mesenchymal stem cell homeostasis in the adult mouse incisor. Cell Stem Cell 14:160–173. https://doi.org/10.1016/j.stem.2013.12.013

Zhu Y, Yang Y, Zhang Y, Hao G, Liu T, Wang L, Yang T, Wang Q, Zhang G, Wei J, Li Y (2014) Placental mesenchymal stem cells of fetal and maternal origins demonstrate different therapeutic potentials. Stem Cell Res Ther 5:48. https://doi.org/10.1186/scrt436

Zygmunt M, Herr F, Münstedt K, Lang U, Liang OD (2003) Angiogenesis and vasculogenesis in pregnancy. Eur J Obstet Gynecol Reprod Biol 110(Suppl 1):S10–S18

Chapter 9
Pericytes in the Liver

Enis Kostallari and Vijay H. Shah

Abstract Liver pericytes, commonly named hepatic stellate cells (HSCs), reside in the space between liver sinusoidal endothelial cells (LSECs) and hepatocytes. They display important roles in health and disease. HSCs ensure the storage of the majority of vitamin A in a healthy body, and they represent the major source of fibrotic tissue in liver disease. Surrounding cells, such as LSECs, hepatocytes, and Kupffer cells, present a significant role in modulating HSC behavior. Therapeutic strategies against liver disease are being currently developed, where HSCs represent an ideal target. In this chapter, we will discuss HSC quiescence and activation in the context of healthy liver and diseases, such as fibrosis, steatohepatitis, and hepatocellular carcinoma.

Keywords Liver · Hepatic stellate cells · Pericytes · Healthy liver · Regeneration · Fibrosis · NASH · NAFLD · Hepatocellular carcinoma · Hepatocytes · Sinusoidal endothelial cells · Kupffer cells

9.1 Introduction

The liver-specific pericytes are called hepatic stellate cells (HSCs). They reside in the subendothelial space of Disse, defined as the space between the liver sinusoidal endothelial cells and the parenchymal cells. HSCs were first described in 1876 by the German anatomist Carl von Kupffer who visualized them using gold chloride preparations of human liver. He called these cells "sternzellen" meaning stellate cells due to their shape (Kupffer 1876). In 1952, Toshito Ito described perisinusoidal cells as fat-storing cells based on their capacity to accumulate lipid droplets (Ito and Nemoto 1952), also called later lipocytes or Ito cells. The term HCS was reused in 1971 by the Japanese Kenjiro Wake (Wake 1971), who also described the storage of vitamin A in rat quiescent HSCs (Wake 1974) and who opened the era of their

E. Kostallari · V. H. Shah (✉)
Division of Gastroenterology and Hepatology, Mayo Clinic, Rochester, MN, USA
e-mail: shah.vijay@mayo.edu

A. Birbrair (ed.), *Pericyte Biology in Different Organs*,
Advances in Experimental Medicine and Biology 1122,
https://doi.org/10.1007/978-3-030-11093-2_9

characterization. The existence of many names describing the same cell, such as Ito cell, fat-storing cell, lipocyte, and perisinusoidal or parasinusoidal cell, prompted the investigators of the field in 1996 to standardize the term HSC when referring to these cells (Hepatology 1996;23:193). In this chapter, we will discuss the origin of HSCs, their role in the adult healthy liver, and finally their implication in liver diseases where HSCs are designated as the main mediators of fibrosis.

9.2 The Origin of HSCs During Development

The developmental origin of HSCs has been debated since they express both neuronal and mesenchymal cell markers, such as nestin, glial fibrillary acidic protein (GFAP), p75 neurotrophin receptor ($p75^{NTR}$), desmin, type 1 collagen, and vimentin. For this reason, it has been thought that HSCs derive from neural crest and from the mesenchyme (Asahina 2012). To investigate the neural crest origin of HSCs, yellow fluorescent protein $(YFP)^{+}$ neural crest descendants were obtained by crossing wingless-type MMTV integration site family member 1 $(Wnt1)^{Cre}$ mice with $Rosa26^{YFP}$ reporter mice (Cassiman et al. 2006). In this model, HSCs failed to express the YFP leading to the conclusion that they do not derive from the neural crest (Cassiman et al. 2006). Nevertheless, a more recent study shows that in zebrafish, HSCs express heart and neural crest derivatives expressed 2 (hand2), a mesoderm and neural crest marker suggesting a link between HSCs and neural crest (Yin et al. 2012). This controversy may be explained by interspecies differences and needs further investigation.

The most accepted developmental origin for HSCs is a sheet of mesoderm that develops along the foregut endoderm near the cardiac mesoderm, called the septum transversum (Asahina et al. 2009, 2011; Loo and Wu 2008; Pérez-Pomares et al. 2004; Toi et al. 2018). The septum transversum is also important for the prenatal development of the liver by secreting bone morphogenetic proteins (BMPs) and fibroblast growth factors (FGFs) (Zaret 2002). To test whether septum transversum contributes to HSC population in mice, Wilm's tumor 1 (WT1) was used as a marker. Before liver development, WT1 is only expressed in the septum transversum, while in later stages of liver morphogenesis, WT1 is expressed in mesothelial and submesothelial cells. These cells give rise to $WT1^{-}/\beta\text{-galactosidase}^{+}$ HSCs (Asahina et al. 2011). The septum transversum origin of HSCs is also demonstrated in rats and in developing human liver (Toi et al. 2018; Loo and Wu 2008). Moreover, in human fetuses a cell population expressing hematopoietic stem cell markers (CD34 and cytokeratin 7/8) and HSC markers [desmin and α-smooth muscle actin (αSMA)] was identified suggesting a link between hematopoietic and hepatic systems (Suskind and Muench 2004). In addition to these studies, further investigation is necessary to understand the developmental origin of HSCs in different species and whether there exist several subpopulations of HSCs deriving from different precursors.

9.3 Role of HSCs in Healthy Liver

In a healthy adult liver, quiescent HSCs represent ~10% of total resident cells (Tsuchida and Friedman 2017). They are mainly recognized for their role in storage and controlled release of vitamin A. Storage of retinols or vitamin A in cytoplasmic lipid droplets is the most distinctive feature of quiescent HSCs. These lipid droplets are routinely used for HSC isolation and are autofluorescent when excited with the light of ~328 nm (Popper 1944) (Fig. 9.1). In a healthy liver, 50–80% of total body retinol is stocked in the liver (Blomhoff et al. 1990), of which 80–90% is stored in HSCs (Hendriks et al. 1985). While it remains unknown how retinol is transferred to HSCs, more studies exist regarding retinol metabolism in HSCs. Upon entering in HSCs, retinol is esterified and mostly stored as retinyl esters by lecithin retinol acyltransferase (LRAT) (Friedman 2008). Retinyl esters constitute 30–50% of the

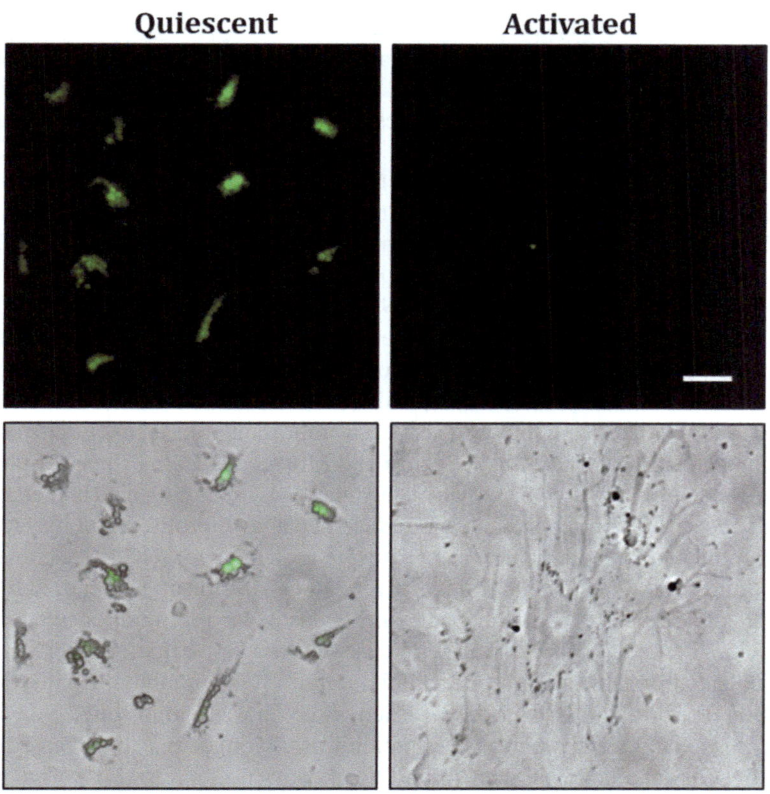

Fig. 9.1 (Left) Primary quiescent HSCs isolated from mouse liver presenting autofluorescent lipid droplets (day 1 after isolation). (Right) Primary activated HSC isolated from mouse liver (day 10 after isolation). It presents several processes, and lipid droplets are absent. (Upper panels) Autofluorescent lipid droplets. (Lower panels) Bright-field capture of the cells merged with the autofluorescent lipid droplets. Scale bar: 10 μm

lipids of HSC lipid droplets (Senoo 2004). Vitamin A is important for maintaining HSCs in quiescence since treatment of cultured HSCs with vitamin A induced its accumulation in cytoplasmic lipid droplets and decreased the expression of HSC activation markers (Yoneda et al. 2016). Moreover, all-trans-retinoic acid, a derivate of vitamin A, was demonstrated to reduce HSC activation (Shimizu et al. 2018). However, the absence of lipid droplets in LRAT-deficient mice was not associated with liver fibrosis (Kluwe et al. 2011). This controversy raises the question whether vitamin A loss is the cause or the consequence of HSC activation, leading to the need of further studies to elucidate this question.

HSCs also secrete molecules that act in an autocrine or paracrine manner, part of which are lipoproteins, growth factors, and cytokines (Friedman 2008). As other quiescent perivascular cells, HSCs secrete apolipoprotein E (ApoE) (Ramadori et al. 1989). At basal level, ApoE-deficient mice showed increased transforming growth factor beta (TGFβ), monocyte chemoattractant protein-1 (MCP-1), and tissue inhibitor of metalloproteinase-1 (TIMP-1) expression in the liver when compared to wild-type (WT) mice (Ferré et al. 2009), suggesting a role for ApoE in maintaining liver homeostasis. It would have been interesting to investigate whether HSC-specific ApoE deletion would have an incidence on liver injury. Follistatin and interleukin 10 (IL-10) are other molecules secreted by HSCs that have shown anti-fibrotic effects (Friedman 2008). Indeed, treatment of carbon tetrachloride (CCl_4)-exposed rats with follistatin reduced liver fibrosis and constrained HSC proliferation (Patella et al. 2006). Nevertheless, another study showed that follistatin is expressed only by activated HSCs and could be a marker of liver fibrosis (Boers et al. 2006). HSCs can also secrete IL-10, a hepatoprotective cytokine recognized for its role in reducing liver injury (Thompson et al. 1998a, b; Byun et al. 2013). Indeed, IL-10-deficient mice develop more severe liver fibrosis following CCl_4 administration (Thompson et al. 1998a, b). The anti-fibrotic molecules produced by HSCs deserve more elucidation as they have a therapeutic potential.

HSCs have also been shown to have an important role in liver regeneration following a partial hepatectomy (Preziosi and Monga 2017). In the earlier phase of liver regeneration, they regulate matrix degradation by secreting matrix metalloproteinase and several proteoglycans and induce hepatocyte proliferation by secreting pleiotrophin and the potent mitogen hepatocyte growth factor (HGF) (Asahina et al. 2002; Taub 2004; Preziosi and Monga 2017). When HSC death was caused by gliotoxin 24 hours before partial hepatectomy, hepatocyte proliferation was significantly impaired due to a lack of HGF (Nejak-Bowen et al. 2013). At later phases of liver regeneration, HSCs secrete the mitoinhibitory TGFβ to inhibit hepatocyte proliferation (Preziosi and Monga 2017). When gliotoxin was administered 5 days after partial hepatectomy, hepatocyte proliferation was prolonged due to decreased HSC-derived TGFβ and collagen deposition (Nejak-Bowen et al. 2013). An increasing interest has been shown on the aspect of HSCs as progenitor cells. Several groups have demonstrated that HSCs can differentiate in hepatocytes during liver regeneration (Yang et al. 2008b; Kordes et al. 2014), suggesting a mesenchymal stem cell role. All these studies show that HSC presence is essential for normal liver function and regeneration. Nevertheless, further investigation is needed to elucidate the existing controversy on the role of HSC-derived signaling in healthy liver.

9.4 Role of HSCs in Liver Disease

HSCs have mainly been studied in liver diseases, due to their capacity to transdifferentiate in myofibroblasts, the major source of collagen deposition during fibrogenesis. In this sub-chapter, the HSC behavior will be discussed in the context of liver fibrosis, steatohepatitis, and hepatocellular carcinoma.

9.4.1 HSCs and Liver Fibrosis

Liver fibrosis is the deposition of excessive scar tissue in response to a chronic injury. The identification of HSC activation in hepatic fibrosis has been a significant discovery in the understanding of liver's response to injury. HSC activation refers to the passage from a quiescent vitamin-rich cell to a proliferative, contractile, migratory, and fibrogenic cell (Friedman 2008) (Fig. 9.1). Some of the most studied signals that activate HSCs are platelet growth factor (PDGF), TGFβ, Fas cell surface death receptor (Fas), apoptotic bodies, extracellular vesicles, and stiffness (Friedman 2008; Kostallari and Shah 2016; Dou et al. 2018). These signals derive from several cell types, such as sinusoidal endothelial cells (SECs) (Kostallari and Shah 2016), Kupffer cells (Bilzer et al. 2006), injured hepatocytes (Canbay et al. 2002, 2003), and HSCs (Kostallari et al. 2018) (Fig. 9.2).

In liver injury, SECs become activated or capillarized by losing their fenestrae (Brunt et al. 2014). Defenestrated SECs increase endothelin-1 (ET-1) secretion and decrease nitric oxide (NO) release, which contribute to HSC contraction (Kostallari and Shah 2016; Tsuchida and Friedman 2017). ET-1 is found to also be secreted by HSCs and to act in an autocrine manner (Pinzani et al. 1996). SECs stimulate HSC migration by producing stromal-derived factor-1 (SDF-1)/C-X-C motif chemokine ligand 12 (CXCL12) (Kordes and Häussinger 2013) and releasing sphingosine kinase (SK)-1-containing EVs (Wang et al. 2015). Moreover, in liver fibrogenesis, SECs upregulate fibroblast growth factor receptor 1 (FGFR1) and C-X-C motif chemokine receptor 4 (CXCR4) expressions, enforcing the fibrogenic phenotype of HSCs (Ding et al. 2014). Another cell type influencing HSC behavior is the Kupffer cell. Kupffer cells are the liver resident macrophages lying in the sinusoids and penetrating between the hepatocytes during liver injury. They have an important role in HSC migration through PDGF secretion and HSC differentiation into myofibroblasts through TGFβ and reactive oxygen species (ROS) release (Tsuchida and Friedman 2017; Kiagiadaki et al. 2018). Moreover, Kupffer cells secrete MCP-1 in a SK-1-dependent manner (Lan et al. 2018). MCP-1 binds to C-C motif chemokine receptor 2 (CCR2), which is expressed by both HSCs and Kupffer cells, leading to their respective migration (Marra et al. 1999; Lan et al. 2018) and ultimately to liver fibrosis (Seki et al. 2009). Treatment of mice with a specific inhibitor of SK-1 prevented liver fibrosis in CCl_4- and bile duct ligation (BDL)-induced liver injuries (Lan et al. 2018). Hepatocytes, which constitute 70–85% of the liver mass, also

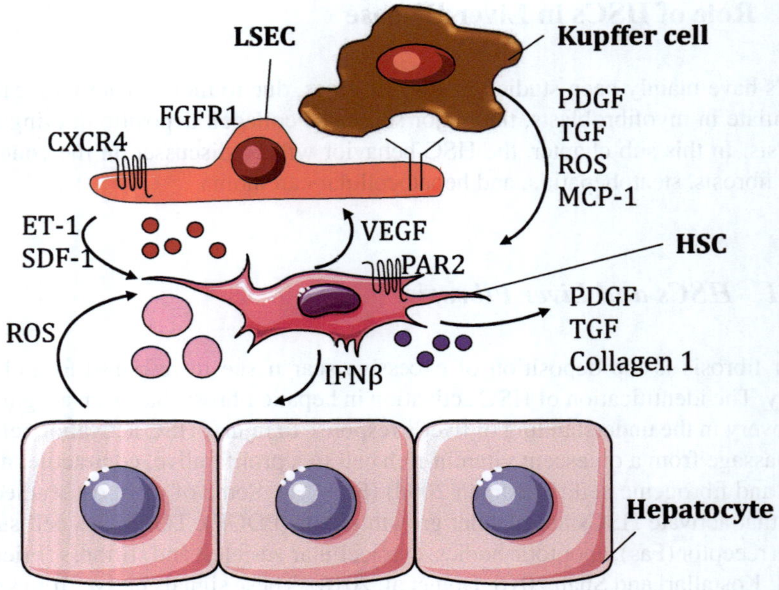

Fig. 9.2 Cross-talking between HSCs and the surrounding cells in the context of liver fibrosis. Defenestrated LSEC upregulates the expression of CXCR4 and FGFR1 and secretion of ET-1, SDF-1, and SK-1-containing EVs (small red circles), which induce HSC activation. Kupffer cells participate in this process by secreting PDGF, TGF, ROS, and MCP-1. ROS are also released by injured hepatocytes. HSCs secrete PDGF, TGF collagen 1, and PDGFRα-enriched EVs (small purple circles) which act in an autocrine manner. In addition, activated HSCs release IFNβ that induces hepatocyte apoptosis and hepatocyte-derived apoptotic bodies (pink circles) implicated in HSC further activation

have a role in HSC activation. In a healthy liver, hepatocytes are polarized cells presenting microvilli, which contribute to the absorption of proteins and other molecules coming from the blood. Injured hepatocytes lose their microvilli and can participate to HSC activation (Canbay et al. 2003; Tsuchida and Friedman 2017). Indeed, injured hepatocytes underwent apoptosis through a Fas-dependent mechanism and released apoptotic bodies, which were engulfed by HSCs (Canbay et al. 2002, 2003). This induced HSC fibrogenic activity by increasing collagen α1, TGFβ1, and αSMA expression (Canbay et al. 2003). Moreover, damaged hepatocytes release ROS, which activate HSCs and lead to liver fibrogenesis (Jiang et al. 2010; Tsuchida and Friedman 2017).

Besides SECs, Kupffer cells, and hepatocytes, activated HSCs can also release signals in a paracrine manner, which contribute to fibrosis. Activated HSCs are well recognized to secrete the canonical pro-fibrogenic molecules such as collagen 1α1, TGFβ, and PDGF-BB (Drinane et al. 2017; Friedman 2008). Moreover, a decrease of retinoic acid signaling in HSCs is associated with induction of TGFβ1 expression and activation of HSCs, suggesting a role for retinoic acid in HSCs quiescence (Ohata et al. 1997). The presence of bacterial lipopolysaccharide (LPS), due to increased gut permeability in liver disease, has been shown to have a significant role in retinoic acid signaling decrease by inducing the autophagy of HSC lipid droplets and leading to HSC activation (Chen et al. 2017). LPS-dependent activation of HSCs induces interferon beta (IFNβ) release, which promotes hepatocyte apoptosis (Dangi et al. 2016). In turn, apoptotic hepatocytes participate to further activation of HSCs (Canbay et al. 2002, 2003). Moreover, activated HSCs and myofibroblasts secrete vascular endothelial growth factor (VEGF), a potent angiogenic molecule that binds to VEGF receptor (VEGFR) expressed by SECs, and induce fibrosis-associated angiogenesis (Das et al. 2010). In a recent study, activated HSCs also secrete PDGFRα-enriched EVs, which have a paracrine effect in promoting in vitro cell migration and in vivo fibrogenesis (Kostallari et al. 2018). The PDGFRα enrichment in HSC-derived EVs is SHP2 dependent, and treatment of mice with SHP2 inhibitor, SHP099, reduced significantly liver fibrosis (Kostallari et al. 2018). Interestingly, G protein-coupled receptors (GPCRs) are emerging as new targets in liver fibrosis. Indeed, protease-activated receptor-2 (PAR2) is expressed in activated HSCs. Inhibiting PAR2 by PZ-235 decreased HSC proliferation and collagen deposition (Shearer et al. 2016).

Another interesting stimulator of HSC activation is liver stiffness. Indeed, when plated on a soft substrate, freshly isolated HSCs maintain their lipid droplets and high levels of PPARγ. However, when plated on a stiff substrate, HSCs lose their lipid droplets and upregulate αSMA and collagen 1 expression (Guvendiren et al. 2014). Moreover, in a recent study, it has been demonstrated that substrate stiffness induces HSC activation through translocation into the nucleus of histone acetyltransferase p300, which upregulated transcription of HSC activation genes (Dou et al. 2018). In line with these studies and in a therapeutic perspective, it could be of interest to further investigate the cause of the stiffness in order to modulate it in vivo and to explore the role of other GPCRs in liver fibrosis.

9.4.2 HSCs and Steatohepatitis

Steatohepatitis is one of the leading causes of liver diseases in the United States. It can occur due to a high consumption of alcohol leading to alcoholic steatohepatitis (ASH) or a high-fat diet leading to nonalcoholic steatohepatitis (NASH). Despite the different etiologies, these two entities have similar pathogenic mechanisms. They usually start with steatosis, which is the fat accumulation within the hepatocytes (ballooned hepatocytes), followed by hepatocyte cell death response, inflammation, and ultimately fibrogenesis (Greuter et al. 2017; Ibrahim et al. 2018). HSCs have a significant role in driving inflammation and fibrogenesis in ASH and NASH by receiving and emitting signals. Lipotoxicity-induced hepatocyte-derived extracellular vesicles promote HSC activation, proliferation, and migration by inhibiting the quiescence marker peroxisome proliferator-activated receptor gamma (PPARγ) through miR-128-3p (Povero et al. 2015). Lipotoxicity-induced hepatocyte can also secrete sonic hedgehog (Shh), which is an activator of HSCs (Yang et al. 2008a;

Rangwala et al. 2011). In turn, activated HSCs can increase the inflammatory reaction by secreting pro-inflammatory cytokines such as MCP1, IL6, TGFβ, and neural cell adhesion molecules (Friedman 2008; Lee and Jeong 2012).

In alcoholic and nonalcoholic liver disease, intestinal epithelial permeability is increased, which can be detected by a high level of bacterial LPS in the liver. LPS can stimulate HSCs through Toll-like receptors (TLRs) (Tsuchida and Friedman 2017). HSCs express several TLRs, such as TLR2, TLR3, TLR4, TLR7, and TLR9 (Seki et al. 2007; Gäbele et al. 2008; Chou et al. 2012; Miura et al. 2013; Byun et al. 2013). In some mouse models of steatohepatitis, activation of TLR4 induces Kupffer cell chemotaxis and inflammation (Seki et al. 2007; Guo et al. 2009). Moreover TLR activation in HSCs induces liver fibrosis (Huang et al. 2007; Seki et al. 2007). HSC nuclear receptors, such as farnesoid X receptor (FXR), peroxisome proliferator-activated receptor (PPAR), or vitamin D3 receptor (VDR), are other factors involved in liver disease. Indeed, deficiency of FXR in HSCs promotes hepatic inflammation and fibrosis (Kong et al. 2009). Furthermore, a dual PPARα-PPARδ agonist, GFT505, which may target both hepatocytes and HSCs, protects liver from steatosis, inflammation, and fibrosis in several mouse models (Staels et al. 2013). In line with this, inhibition of VDR signaling by sequestosome 1 (SQSTM1/p62) knockout attenuates liver inflammation and fibrosis in experimental steatohepatitis (Beilfuss et al. 2015; Duran et al. 2016).

The interaction between HSCs and neutrophils also seems to play an important role in the establishment of steatohepatitis through HSCs. Indeed, neutrophils stimulate the secretion of HSC-derived GM-CSF and IL-15, which in turn prolong neutrophil survival and may serve as a positive forward loop to promote liver damage and fibrosis (Zhou et al. 2018). Cholesterol, a component of high-fat diet, is another factor that can accumulate in HSCs and induce their activation (Tomita et al. 2014). Cholesterol-lowering drugs, such as ezetimibe or statins, attenuate steatohepatitis and fibrosis in a mouse model of NASH (Van Rooyen et al. 2013). Nevertheless, the effect of activated HSCs on recruiting pro-inflammatory cells is not fully understood and deserves further investigation.

9.4.3 HSCs and Hepatic Metastases

The combination of the unique microenvironment and its hemodynamics characteristics makes the liver one of the most targeted organs of cancer metastasis. Furthermore, liver metastases are dependent on the interactions between the tumor cells and the hepatic stromal cells, such as HSCs. HSCs have several important roles in liver tumors, such as promoting tumor growth, releasing growth factors and cytokines, regulating extracellular matrix (ECM) turnover and tumor-associated angiogenesis, and suppressing antitumor immune response (Kang et al. 2011). In 130 cases of hepatocellular carcinoma, peritumoral activated HSCs independently contributed to high recurrence or death rates (Ju et al. 2009). Similar to liver fibrosis, tumor-stimulated HSCs differentiate into myofibroblasts, upregulate αSMA expression, proliferate, migrate, and contribute significantly to collagen deposition (Vidal-Vanaclocha 2008). This behavior is stimulated by tumor cells through paracrine signaling, including secretion of TGFβ (Kang et al. 2011). In turn, activated HSCs induce tumor cell proliferation, migration, and invasion through secretion of several secreted growth factors and cytokines, such as TGFβ, HGF, PDGF, SDF-1, VEGF, and CXCL12 (Okabe et al. 2009; Amann et al. 2009; Zhao et al. 2011; Kang et al. 2011; Yaqoob et al. 2012; Dou et al. 2018), and by downregulating other molecules such as endosialin (Mogler et al. 2017). Indeed, conditioned media of activated HSCs promoted tumor cell invasion and proliferation in vitro (Okabe et al. 2009; Amann et al. 2009) and in vivo (Dou et al. 2018). Furthermore, HSCs cultured on a stiff environment induced a higher tumor growth than HSCs cultured on soft matrix (Dou et al. 2018), suggesting a role for stiffness in tumor progression.

Activated HSCs at the invasive front of liver metastasis can regulate the ECM by producing matrix metalloproteinase 2 (MMP2), TIMP2, and a disintegrin and metalloproteinase 9 (ADAM9) allowing cancer cells to increase their invasive behavior (Musso et al. 1997; Mazzocca et al. 2005). Furthermore, HSC-derived ECM components, such as collagen, fibronectin, laminin, and CCN family member 1 (CCN1), regulate adhesion, migration, and survival of tumor cells by activating the integrins on tumor cell surface (Kang et al. 2011; Azzariti et al. 2016; Li et al. 2018). For example, inhibiting α3 integrin on HCC cells abrogated the anti-apoptotic effect of laminin-332 (Azzariti et al. 2016). Moreover, the expression of neuropilin-1 by HSC-derived myofibroblasts promoted the secretion of fibronectin and therefore the increase of tumor microenvironment stiffness leading to a greater tumor growth (Yaqoob et al. 2012).

Activated HSCs can also promote tumor-associated angiogenesis since they release VEGF, angiopoietins, and IL-8 (Semela et al. 2008; Taura et al. 2008; Das et al. 2010; Zhu et al. 2015). Indeed, IL-8 derives mainly from activated HSCs, and IL-8 inhibition with a blocking antibody significantly reduced angiogenesis (Zhu et al. 2015). Moreover, activated HSCs protect the hepatic tumor by inhibiting antitumor cytotoxic T cells (Xia et al. 2017). Indeed, tumor-associated HSCs induce dendritic cell-derived immunoglobulin receptor 2 (DIgR2) in dendritic cells, which in turn inhibit the inflammatory response of T cells (Xia et al. 2017). It could be of interest to better understand the role of activated HSCs on the inflammatory cell behavior in the context of HCC.

9.5 Conclusion

In this chapter, we have identified HSCs as a crucial cell for liver development, homeostasis, and disease through paracrine and autocrine signaling, which includes growth factors, cytokines, and extracellular vesicles. While most of the studies describe HSC-dependent ECM regulation in response to microenvironment stimuli, less attention is brought to the understanding of other HSC-derived signals and their effect on the surrounding cell types. The HSC-derived signaling study remains incomplete, especially in the context of HSC heterogeneity. The isolation of different HSC subpopulations and specifically targeting the most fibrogenic ones are essential for developing novel therapies. Several encouraging anti-fibrotic strategies aiming at the prevention or reversal of HSC-induced liver fibrosis have been or are being developed (Schuppan et al. 2018). Recently, a dual C-C chemokine receptor type 2 and 5 (CCR2/CCR5) antagonist, cenicriviroc, is in phase 2 clinical trial and is being evaluated in liver inflammation and fibrosis in the context of NASH (Friedman et al. 2016). This is the first prospective study of an oral agent in patients with exclusively NASH and liver fibrosis, which aims to assess correlations between inflammation and fibrosis (Friedman et al. 2016). In parallel, other clinical trials are ongoing (clinicaltrials.gov) and throw a glimmer of hope in finding treatments for liver diseases.

References

(1996) Hepatic stellate cell nomenclature. Hepatology 23:193. No authors listed

Amann T, Bataille F, Spruss T, Mühlbauer M, Gäbele E, Schölmerich J, Kiefer P, Bosserhoff AK, Hellerbrand C (2009) Activated hepatic stellate cells promote tumorigenicity of hepatocellular carcinoma. Cancer Sci 100(4):646–653

Asahina K (2012) Hepatic stellate cell progenitor cells. J Gastroenterol Hepatol 27(Suppl 2):80–84

Asahina K, Sato H, Yamasaki C, Kataoka M, Shiokawa M, Katayama S, Tateno C, Yoshizato K (2002) Pleiotrophin/heparin-binding growth-associated molecule as a mitogen of rat hepatocytes and its role in regeneration and development of liver. Am J Pathol 160(6):2191–2205

Asahina K, Tsai SY, Li P, Ishii M, Maxson RE Jr, Sucov HM, Tsukamoto H (2009) Mesenchymal origin of hepatic stellate cells, submesothelial cells, and perivascular mesenchymal cells during mouse liver development. Hepatology 49(3):998–1011

Asahina K, Zhou B, Pu WT, Tsukamoto H (2011) Septum transversum-derived mesothelium gives rise to hepatic stellate cells and perivascular mesenchymal cells in developing mouse liver. Hepatology 53(3):983–995

Azzariti A, Mancarella S, Porcelli L, Quatrale AE, Caligiuri A, Lupo L, Dituri F, Giannelli G (2016) Hepatic stellate cells induce hepatocellular carcinoma cell resistance to sorafenib through the laminin-332/α3 integrin axis recovery of focal adhesion kinase ubiquitination. Hepatology 64(6):2103–2117

Beilfuss A, Sowa JP, Sydor S, Beste M, Bechmann LP, Schlattjan M, Syn WK, Wedemeyer I, Mathé Z, Jochum C, Gerken G, Gieseler RK, Canbay A (2015) Vitamin D counteracts fibrogenic TGF-β signalling in human hepatic stellate cells both receptor-dependently and independently. Gut 64(5):791–799

Bilzer M, Roggel F, Gerbes AL (2006) Role of Kupffer cells in host defense and liver disease. Liver Int 26(10):1175–1186

Blomhoff R, Green MH, Berg T, Norum KR (1990) Transport and storage of vitamin A. Science 250(4979):399–404

Boers W, Aarrass S, Linthorst C, Pinzani M, Elferink RO, Bosma P (2006) Transcriptional profiling reveals novel markers of liver fibrogenesis: gremlin and insulin-like growth factor-binding proteins. J Biol Chem 281(24):16289–16295

Brunt EM, Gouw AS, Hubscher SG, Tiniakos DG, Bedossa P, Burt AD, Callea F, Clouston AD, Dienes HP, Goodman ZD, Roberts EA, Roskams T, Terracciano L, Torbenson MS, Wanless IR (2014) Pathology of the liver sinusoids. Histopathology 64(7):907–920

Byun JS, Suh YG, Yi HS, Lee YS, Jeong WI (2013) Activation of toll-like receptor 3 attenuates alcoholic liver injury by stimulating Kupffer cells and stellate cells to produce interleukin-10 in mice. J Hepatol 58(2):342–349

Canbay A, Higuchi H, Bronk SF, Taniai M, Sebo TJ, Gores GJ (2002) Fas enhances fibrogenesis in the bile duct ligated mouse: a link between apoptosis and fibrosis. Gastroenterology 123(4):1323–1330
Canbay A, Taimr P, Torok N, Higuchi H, Friedman S, Gores GJ (2003) Apoptotic body engulfment by a human stellate cell line is profibrogenic. Lab Investig 83(5):655–663
Cassiman D, Barlow A, Vander Borght S, Libbrecht L, Pachnis V (2006) Hepatic stellate cells do not derive from the neural crest. J Hepatol 44(6):1098–1104
Chen M, Liu J, Yang W, Ling W (2017) Lipopolysaccharide mediates hepatic stellate cell activation by regulating autophagy and retinoic acid signaling. Autophagy 13(11):1813–1827
Chou MH, Huang YH, Lin TM, Du YY, Tsai PC, Hsieh CS, Chuang JH (2012) Selective activation of toll-like receptor 7 in activated hepatic stellate cells may modulate their profibrogenic phenotype. Biochem J 447(1):25–34
Dangi A, Huang C, Tandon A, Stolz D, Wu T, Gandhi CR (2016) Endotoxin-stimulated rat hepatic stellate cells induce autophagy in hepatocytes as a survival mechanism. J Cell Physiol 231(1):94–105
Das A, Shergill U, Thakur L, Sinha S, Urrutia R, Mukhopadhyay D, Shah VH (2010) Ephrin B2/EphB4 pathway in hepatic stellate cells stimulates Erk-dependent VEGF production and sinusoidal endothelial cell recruitment. Am J Physiol Gastrointest Liver Physiol 298(6):G908–G915
Ding BS, Cao Z, Lis R, Nolan DJ, Guo P, Simons M, Penfold ME, Shido K, Rabbany SY, Rafii S (2014) Divergent angiocrine signals from vascular niche balance liver regeneration and fibrosis. Nature 505(7481):97–102
Dou C, Liu Z, Tu K, Zhang H, Chen C, Yaqoob U, Wang Y, Wen J, van Deursen J, Sicard D, Tschumperlin D, Zou H, Huang WC, Urrutia R, Shah VH, Kang N (2018) P300 acetyltransferase mediates stiffness-induced activation of hepatic stellate cells into tumor-promoting myofibroblasts. Gastroenterology 154(8):2209–2221.e14
Drinane MC, Yaqoob U, Yu H, Luo F, Greuter T, Arab JP, Kostallari E, Verma VK, Maiers J, De Assuncao TM, Simons M, Mukhopadhyay D, Kisseleva T, Brenner DA, Urrutia R Lomberk G, Gao Y, Ligresti G, Tschumperlin DJ, Revzin A, Cao S, Shah VH. Synectin promotes fibrogenesis by regulating PDGFR isoforms through distinct mechanisms. JCI Insight. 2017;2(24). pii: 92821
Duran A, Hernandez ED, Reina-Campos M, Castilla EA, Subramaniam S, Raghunandan S, Roberts LR, Kisseleva T, Karin M, Diaz-Meco MT, Moscat J (2016) p62/SQSTM1 by binding to vitamin D receptor inhibits hepatic stellate cell activity, fibrosis, and liver Cancer. Cancer Cell 30(4):595–609
Ferré N, Martínez-Clemente M, López-Parra M, González-Périz A, Horrillo R, Planagumà A, Camps J, Joven J, Tres A, Guardiola F, Bataller R, Arroyo V, Clària J (2009) Increased susceptibility to exacerbated liver injury in hypercholesterolemic ApoE-deficient mice: potential involvement of oxysterols. Am J Physiol Gastrointest Liver Physiol 296(3):G553–G562
Friedman SL (2008) Hepatic stellate cells: protean, multifunctional, and enigmatic cells of the liver. Physiol Rev 88(1):125–172. https://doi.org/10.1152/physrev.00013.2007
Friedman S, Sanyal A, Goodman Z, Lefebvre E, Gottwald M, Fischer L, Ratziu V (2016) Efficacy and safety study of cenicriviroc for the treatment of non-alcoholic steatohepatitis in adult subjects with liver fibrosis: CENTAUR phase 2b study design. Contemp Clin Trials 47:356–365
Gäbele E, Mühlbauer M, Dorn C, Weiss TS, Froh M, Schnabl B, Wiest R, Schölmerich J, Obermeier F, Hellerbrand C (2008) Role of TLR9 in hepatic stellate cells and experimental liver fibrosis. Biochem Biophys Res Commun 376(2):271–276
Greuter T, Malhi H, Gores GJ, Shah VH. Therapeutic opportunities for alcoholic steatohepatitis and nonalcoholic steatohepatitis: exploiting similarities and differences in pathogenesis. JCI Insight. 2017;2(17). pii: 95354
Guo J, Loke J, Zheng F, Hong F, Yea S, Fukata M, Tarocchi M, Abar OT, Huang H, Sninsky JJ, Friedman SL (2009) Functional linkage of cirrhosis-predictive single nucleotide polymorphisms of toll-like receptor 4 to hepatic stellate cell responses. Hepatology 49(3):960–968
Guvendiren M, Perepelyuk M, Wells RG, Burdick JA (2014) Hydrogels with differential and patterned mechanics to study stiffness-mediated myofibroblastic differentiation of hepatic stellate cells. J Mech Behav Biomed Mater 38:198–208

Hendriks HF, Verhoofstad WA, Brouwer A, de Leeuw AM, Knook DL (1985) Perisinusoidal fat-storing cells are the main vitamin A storage sites in rat liver. Exp Cell Res 160(1):138–149
Huang H, Shiffman ML, Friedman S, Venkatesh R, Bzowej N, Abar OT, Rowland CM, Catanese JJ, Leong DU, Sninsky JJ, Layden TJ, Wright TL, White T, Cheung RC (2007) A 7 gene signature identifies the risk of developing cirrhosis in patients with chronic hepatitis C. Hepatology 46(2):297–306
Ibrahim SH, Hirsova P, Gores GJ (2018) Non-alcoholic steatohepatitis pathogenesis: sublethal hepatocyte injury as a driver of liver inflammation. Gut 67(5):963–972
Ito T, Nemoto M (1952) Kupfer's cells and fat storing cells in the capillary wall of human liver. Okajimas Folia Anat Jpn 24(4):243–258
Jiang JX, Venugopal S, Serizawa N, Chen X, Scott F, Li Y, Adamson R, Devaraj S, Shah V, Gershwin ME, Friedman SL, Török NJ (2010) Reduced nicotinamide adenine dinucleotide phosphate oxidase 2 plays a key role in stellate cell activation and liver fibrogenesis in vivo. Gastroenterology 139(4):1375–1384
Ju MJ, Qiu SJ, Fan J, Xiao YS, Gao Q, Zhou J, Li YW, Tang ZY (2009) Peritumoral activated hepatic stellate cells predict poor clinical outcome in hepatocellular carcinoma after curative resection. Am J Clin Pathol 131(4):498–510
Kang N, Gores GJ, Shah VH (2011) Hepatic stellate cells: partners in crime for liver metastases. Hepatology 54(2):707–713
Kiagiadaki F, Kampa M, Voumvouraki A, Castanas E, Kouroumalis E, Notas G (2018) Activin-a causes hepatic stellate cell activation via the induction of TNFα and TGFβ in Kupffer cells. Biochim Biophys Acta 1864(3):891–899
Kluwe J, Wongsiriroj N, Troeger JS, Gwak GY, Dapito DH, Pradere JP, Jiang H, Siddiqi M, Piantedosi R, O'Byrne SM, Blaner WS, Schwabe RF (2011) Absence of hepatic stellate cell retinoid lipid droplets does not enhance hepatic fibrosis but decreases hepatic carcinogenesis. Gut 60(9):1260–1268
Kong B, Luyendyk JP, Tawfik O, Guo GL (2009) Farnesoid X receptor deficiency induces nonalcoholic steatohepatitis in low-density lipoprotein receptor-knockout mice fed a high-fat diet. J Pharmacol Exp Ther 328(1):116–122
Kordes C, Häussinger D (2013) Hepatic stem cell niches. J Clin Invest 123(5):1874–1880
Kordes C, Sawitza I, Götze S, Herebian D, Häussinger D (2014) Hepatic stellate cells contribute to progenitor cells and liver regeneration. J Clin Invest 124(12):5503–5515
Kostallari E, Shah VH (2016) Angiocrine signaling in the hepatic sinusoids in health and disease. Am J Physiol Gastrointest Liver Physiol 311(2):G246–G251
Kostallari E, Hirsova P, Prasnicka A, Verma VK, Yaqoob U, Wongjarupong N, Roberts LR, Shah VH (2018) Hepatic stellate cell-derived PDGFRα-enriched extracellular vesicles promote liver fibrosis in mice through SHP2. Hepatology 68:333–348
Kupffer CV (1876) Ueber Sternzellen der Leber. Briefliche Mitteilung an Prof. Waldeyer. Arch Mikr Anat 12:353–358
Lan T, Li C, Yang G, Sun Y, Zhuang L, Ou Y, Li H, Wang G, Kisseleva T, Brenner D, Guo J (2018) Sphingosine kinase 1 promotes liver fibrosis by preventing miR-19b-3p-mediated inhibition of CCR2. Hepatology 68:1070–1086
Lee YS, Jeong WI (2012) Retinoic acids and hepatic stellate cells in liver disease. J Gastroenterol Hepatol 27(Suppl 2):75–79
Li ZQ, Wu WR, Zhao C, Zhao C, Zhang XL, Yang Z, Pan J, Si WK (2018) CCN1/Cyr61 enhances the function of hepatic stellate cells in promoting the progression of hepatocellular carcinoma. Int J Mol Med 41(3):1518–1528
Loo CK, Wu XJ (2008) Origin of stellate cells from submesothelial cells in a developing human liver. Liver Int 28(10):1437–1445
Marra F, Romanelli RG, Giannini C, Failli P, Pastacaldi S, Arrighi MC, Pinzani M, Laffi G, Montalto P, Gentilini P (1999) Monocyte chemotactic protein-1 as a chemoattractant for human hepatic stellate cells. Hepatology 29(1):140–148

Mazzocca A, Coppari R, De Franco R, Cho JY, Libermann TA, Pinzani M, Toker A (2005) A secreted form of ADAM9 promotes carcinoma invasion through tumor-stromal interactions. Cancer Res 65(11):4728–4738
Miura K, Yang L, van Rooijen N, Brenner DA, Ohnishi H, Seki E (2013) Toll-like receptor 2 and palmitic acid cooperatively contribute to the development of nonalcoholic steatohepatitis through inflammasome activation in mice. Hepatology 57(2):577–589
Mogler C, König C, Wieland M, Runge A, Besemfelder E, Komljenovic D, Longerich T, Schirmacher P, Augustin HG (2017) Hepatic stellate cells limit hepatocellular carcinoma progression through the orphan receptor endosialin. EMBO Mol Med 9(6):741–749
Musso O, Théret N, Campion JP, Turlin B, Milani S, Grappone C, Clément B (1997) In situ detection of matrix metalloproteinase-2 (MMP2) and the metalloproteinase inhibitor TIMP2 transcripts in human primary hepatocellular carcinoma and in liver metastasis. J Hepatol 26(3):593–605
Nejak-Bowen KN, Orr AV, Bowen WC Jr, Michalopoulos GK (2013) Gliotoxin-induced changes in rat liver regeneration after partial hepatectomy. Liver Int 33(7):1044–1055
Ohata M, Lin M, Satre M, Tsukamoto H (1997) Diminished retinoic acid signaling in hepatic stellate cells in cholestatic liver fibrosis. Am J Phys 272(3 Pt 1):G589–G596
Okabe H, Beppu T, Hayashi H, Horino K, Masuda T, Komori H, Ishikawa S, Watanabe M, Takamori H, Iyama K, Baba H (2009) Hepatic stellate cells may relate to progression of intrahepatic cholangiocarcinoma. Ann Surg Oncol 16(9):2555–2564
Patella S, Phillips DJ, Tchongue J, de Kretser DM, Sievert W (2006) Follistatin attenuates early liver fibrosis: effects on hepatic stellate cell activation and hepatocyte apoptosis. Am J Physiol Gastrointest Liver Physiol 290(1):G137–G144
Pérez-Pomares JM, Carmona R, González-Iriarte M, Macías D, Guadix JA, Muñoz-Chápuli R. (2004) Contribution of mesothelium-derived cells to liver sinusoids in avian embryos. Dev Dyn. Mar;229(3):465–74
Pinzani M, Milani S, De Franco R, Grappone C, Caligiuri A, Gentilini A, Tosti-Guerra C, Maggi M, Failli P, Ruocco C, Gentilini P (1996) Endothelin 1 is overexpressed in human cirrhotic liver and exerts multiple effects on activated hepatic stellate cells. Gastroenterology 110(2):534–548
Popper H (1944) Distribution of vitamin A in tissue as visualized by flourescence microscopy. Physiol Rev 24:205–224
Povero D, Panera N, Eguchi A, Johnson CD, Papouchado BG, de Araujo Horcel L, Pinatel EM, Alisi A, Nobili V, Feldstein AE (2015) Lipid-induced hepatocyte-derived extracellular vesicles regulate hepatic stellate cell via microRNAs targeting PPAR-γ. Cell Mol Gastroenterol Hepatol 1(6):646–663.e4
Preziosi ME, Monga SP (2017) Update on the mechanisms of liver regeneration. Semin Liver Dis 37(2):141–151
Ramadori G, Rieder H, Theiss F (1989) Meyer zum Büschenfelde KH. Fat-storing (Ito) cells of rat liver synthesize and secrete apolipoproteins: comparison with hepatocytes. Gastroenterology 97(1):163–172
Rangwala F, Guy CD, Lu J, Suzuki A, Burchette JL, Abdelmalek MF, Chen W, Diehl AM (2011) Increased production of sonic hedgehog by ballooned hepatocytes. J Pathol 224(3):401–410
Schuppan D, Ashfaq-Khan M, Yang AT, Kim YO (2018) Liver fibrosis: direct antifibrotic agents and targeted therapies. Matrix Biol 68:435–451. pii: S0945-053X(18)30160-4
Seki E, De Minicis S, Osterreicher CH, Kluwe J, Osawa Y, Brenner DA, Schwabe RF (2007) TLR4 enhances TGF-beta signaling and hepatic fibrosis. Nat Med 13(11):1324–1332
Seki E, de Minicis S, Inokuchi S, Taura K, Miyai K, van Rooijen N, Schwabe RF, Brenner DA (2009) CCR2 promotes hepatic fibrosis in mice. Hepatology 50(1):185–197
Semela D, Das A, Langer D, Kang N, Leof E, Shah V. (2008) Platelet-derived growth factor signaling through ephrin-b2 regulates hepatic vascular structure and function. Gastroenterology. Aug;135(2):671–9
Senoo H (2004) Structure and function of hepatic stellate cells. Med Electron Microsc 37(1):3–15

Shearer AM, Rana R, Austin K, Baleja JD, Nguyen N, Bohm A, Covic L, Kuliopulos A (2016) Targeting liver fibrosis with a cell-penetrating protease-activated Receptor-2 (PAR2) Pepducin. J Biol Chem 291(44):23188–23198
Shimizu H, Tsubota T, Kanki K, Shiota G (2018) All-trans retinoic acid ameliorates hepatic stellate cell activation via suppression of thioredoxin interacting protein expression. J Cell Physiol 233(1):607–616
Staels B, Rubenstrunk A, Noel B, Rigou G, Delataille P, Millatt LJ, Baron M, Lucas A, Tailleux A, Hum DW, Ratziu V, Cariou B, Hanf R (2013) Hepatoprotective effects of the dual peroxisome proliferator-activated receptor alpha/delta agonist, GFT505, in rodent models of nonalcoholic fatty liver disease/nonalcoholic steatohepatitis. Hepatology 58(6):1941–1952
Suskind DL, Muench MO (2004) Searching for common stem cells of the hepatic and hematopoietic systems in the human fetal liver: CD34+ cytokeratin 7/8+ cells express markers for stellate cells. J Hepatol 40(2):261–268
Taub R (2004) Liver regeneration: from myth to mechanism. Nat Rev Mol Cell Biol 5(10):836–847. Review
Taura K, De Minicis S, Seki E, Hatano E, Iwaisako K, Osterreicher CH, Kodama Y, Miura K, Ikai I, Uemoto S, Brenner DA (2008) Hepatic stellate cells secrete angiopoietin 1 that induces angiogenesis in liver fibrosis. Gastroenterology 135(5):1729–1738
Thompson KC, Trowern A, Fowell A, Marathe M, Haycock C, Arthur MJ, Sheron N (1998a) Primary rat and mouse hepatic stellate cells express the macrophage inhibitor cytokine interleukin-10 during the course of activation in vitro. Hepatology 28(6):1518–1524
Thompson K, Maltby J, Fallowfield J, McAulay M, Millward-Sadler H, Sheron N (1998b) Interleukin-10 expression and function in experimental murine liver inflammation and fibrosis. Hepatology 28(6):1597–1606
Toi M, Hayashi Y, Murakami I (2018) Hepatic stellate cells derived from the nestin-positive cells in septum transversum during rat liver development. Med Mol Morphol 51:199–207
Tomita K, Teratani T, Suzuki T, Shimizu M, Sato H, Narimatsu K, Usui S, Furuhashi H, Kimura A, Nishiyama K, Maejima T, Okada Y, Kurihara C, Shimamura K, Ebinuma H, Saito H, Yokoyama H, Watanabe C, Komoto S, Nagao S, Sugiyama K, Aosasa S, Hatsuse K, Yamamoto J, Hibi T, Miura S, Hokari R, Kanai T (2014) Acyl-CoA:cholesterol acyltransferase 1 mediates liver fibrosis by regulating free cholesterol accumulation in hepatic stellate cells. J Hepatol 61(1):98–106
Tsuchida T, Friedman SL (2017) Mechanisms of hepatic stellate cell activation. Nat Rev Gastroenterol Hepatol 14(7):397–411
Van Rooyen DM, Gan LT, Yeh MM, Haigh WG, Larter CZ, Ioannou G, Teoh NC, Farrell GC (2013) Pharmacological cholesterol lowering reverses fibrotic NASH in obese, diabetic mice with metabolic syndrome. J Hepatol 59(1):144–152
Vidal-Vanaclocha F (2008) The prometastatic microenvironment of the liver. Cancer Microenviron 1(1):113–129
Wake K (1971) "Sternzellen" in the liver: perisinusoidal cells with special reference to storage of vitamin A. Am J Anat 132(4):429–462
Wake K (1974) Development of vitamin A-rich lipid droplets in multivesicular bodies of rat liver stellate cells. J Cell Biol 63(2 Pt 1):683–691
Wang R, Ding Q, Yaqoob U, de Assuncao TM, Verma VK, Hirsova P, Cao S, Mukhopadhyay D, Huebert RC, Shah VH (2015) Exosome adherence and internalization by hepatic stellate cells triggers sphingosine 1-phosphate-dependent migration. J Biol Chem 290(52):30684–30696
Xia YH, Lu Z, Zhao M, Dai WT, Ding L, Hu LX, Jiang GL (2017) Tumor-specific hepatic stellate cells (tHSCs) induces DIgR2 expression in dendritic cells to inhibit T cells. Oncotarget 8(33):55084–55093
Yang L, Wang Y, Mao H, Fleig S, Omenetti A, Brown KD, Sicklick JK, Li YX, Diehl AM (2008a) Sonic hedgehog is an autocrine viability factor for myofibroblastic hepatic stellate cells. J Hepatol 48(1):98–106

Yang L, Jung Y, Omenetti A, Witek RP, Choi S, Vandongen HM, Huang J, Alpini GD, Diehl AM (2008b) Fate-mapping evidence that hepatic stellate cells are epithelial progenitors in adult mouse livers. Stem Cells 26(8):2104–2113

Yaqoob U, Cao S, Shergill U, Jagavelu K, Geng Z, Yin M, de Assuncao TM, Cao Y, Szabolcs A, Thorgeirsson S, Schwartz M, Yang JD, Ehman R, Roberts L, Mukhopadhyay D, Shah VH (2012) Neuropilin-1 stimulates tumor growth by increasing fibronectin fibril assembly in the tumor microenvironment. Cancer Res 72(16):4047–4059

Yin C, Evason KJ, Maher JJ, Stainier DY (2012) The basic helix-loop-helix transcription factor, heart and neural crest derivatives expressed transcript 2, marks hepatic stellate cells in zebrafish: analysis of stellate cell entry into the developing liver. Hepatology 56(5):1958–1970

Yoneda A, Sakai-Sawada K, Niitsu Y, Tamura Y (2016) Vitamin A and insulin are required for the maintenance of hepatic stellate cell quiescence. Exp Cell Res 341(1):8–17

Zaret KS (2002) Regulatory phases of early liver development: paradigms of organogenesis. Nat Rev Genet 3(7):499–512

Zhao W, Zhang L, Yin Z, Su W, Ren G, Zhou C, You J, Fan J, Wang X (2011) Activated hepatic stellate cells promote hepatocellular carcinoma development in immunocompetent mice. Int J Cancer 129(11):2651–2661

Zhou Z, Xu MJ, Cai Y, Wang W, Jiang JX, Varga ZV, Feng D, Pacher P, Kunos G, Torok NJ, Gao B (2018) Neutrophil-hepatic stellate cell interactions promote fibrosis in experimental steatohepatitis. Cell Mol Gastroenterol Hepatol 5(3):399–413

Zhu B, Lin N, Zhang M, Zhu Y, Cheng H, Chen S, Ling Y, Pan W, Xu R (2015) Activated hepatic stellate cells promote angiogenesis via interleukin-8 in hepatocellular carcinoma. J Transl Med 13:365

Chapter 10
Pericytes in the Periodontal Ligament

Motohiro Komaki

Abstract Teeth are exposed to hundreds of oral bacteria and also challenged by the mastication forces; because teeth are situated in oral cavity, the entrance of the digestive tract, and penetrates through the oral epithelium. The periodontal ligament is a noncalcified tissue that possesses abundant blood vessels, which exist between tooth root and alveolar bone. The ligament is thought to play an important role in absorbing the impact of mastication, in the maintenance of periodontal homeostasis, and in periodontal wound healing. We succeeded in isolating mesenchymal stem cells (MSCs), so-called periodontal stem cells (PDLSCs), with self-renewability and multipotency from the periodontal ligament. We also demonstrated that PDLSCs share some cell surface markers with pericytes and that PDLSCs distribute themselves to stay with the endothelial cell networks and that PDLSCs maintain the endothelial cell networks when added to endothelial cell network formation systems. Pericytes are located in the proximity of microvascular endothelial cells and thought to stabilize and supply nutrients to blood vessels. Recently, it was also reported that pericytes possess multipotency and can be the source of tissue stem cells and/or progenitor cells. This review explores the distinctive features of the periodontal ligament tissue and PDLSCs as well as the puzzling similarities between PDLSCs and pericytes.

Keywords Periodontal ligament · Mesenchymal stem cells · Pericytes · Periodontal ligament stem cells · Vascular network · Blood vessels · Regenerative therapy · Twist · Periostin · SDF-1 · Hypoxia

---

M. Komaki (✉)
Department of Highly Advanced Stomatology, Graduate School of Dentistry, Kanagawa Dental University, Yokohama City, Kanagawa, Japan
e-mail: m.komaki@kdu.ac.jp

A. Birbrair (ed.), *Pericyte Biology in Different Organs*, Advances in Experimental Medicine and Biology 1122,
https://doi.org/10.1007/978-3-030-11093-2_10

10.1 Anatomy, Function, and Distinctiveness of Periodontal Tissue

Teeth not only pulverize food brought to the mouth through mastication but also enable a person to perceive the texture of food through the periodontal ligament. It is also believed that mastication plays important role in the maintenance of periodontal ligament homeostasis. Teeth penetrate through the oral epithelium into the oral cavity. The root does not directly bind to the alveolar bone, rather it is drawn into the alveolar bone like a hammock by predominately collagen fibers of periodontal ligament. Periodontal ligament is situated in a space of approximately 0.25 mm between two calcified tissues, tooth root and alveolar bone. Thickening of the cementum due to aging gradually decreases this size; however, this space sustains the uncalcified zones throughout one's lifetime. Thus far, the inhibitory mechanism of periodontal ligament's calcification through homeobox protein Msx2 and extracellular matrix protein PLAP-1/asporin has been reported (Yamada et al. 2007; Yoshizawa et al. 2004). We also have reported the expression of Twist, a basic helix-loop-helix (bHLH) transcription factor and also known as a causal factor of Saethre-Chotzen syndrome, which causes the premature closure of cranial sutures, within the periodontal ligament (Komaki et al. 2007). Twist is also known as a regulatory factor of periostin that is involved in the maintenance of integrity of periodontal tissue in response to occlusal load. We demonstrated that Twist suppressed osteoblastic differentiation of periodontal ligament cells as a possible inhibitory mechanism in periodontal ligament (Komaki et al. 2007). Another feature of the periodontal ligament is abundant blood vessels that cover the root's surface akin to a bird's nest (Fig. 10.1). Periodontal ligament is a noncalcified tissue predominately composed of collagen fiber types I and III as well as small amounts of oxytalan fibers. A different type of gene expression in the periodontal ligament as compared to that of dermal connective tissue has been reported (Lallier et al. 2005). Compared to general connective tissue, which has a vascular volume of 5%, periodontal ligament has an abundant supply of blood vessels of ~20% (Lew 1987). The blood vessels of the periodontal ligament consist of routes that branch from the inferior alveolar artery inside the alveolar bone and reach the periodontal ligament through Volkmann's canal in alveolar bone, routes where anastomosis occurs with blood vessels of the gingiva near the tooth cervix, and routes where blood vessels, which diverge near the apex into the tooth pulp and periodontal ligament, diverge into the periodontal ligament (Berglundh et al. 1994; Matsuo and Takahashi 2002). The network of blood vessels formed a double-layered mesh on the root surface is made of terminal arterioles, capillaries, and postcapillary venules. Few valves that staunch regurgitation of blood are seen among the veins in the alveolar bone. Blood vessel system in periodontal ligament dissipates intermittently occurring occlusal loads because blood flows freely in and out between the periodontal ligament and alveolar bone. Fenestrated capillaries observed around the root apex enable efficient gas and substance exchange between blood vessels and surrounding tissues in response to high metabolic turnover of periodontal ligament (Lew 1987). It has been reported that vascular network in periodontal ligament can

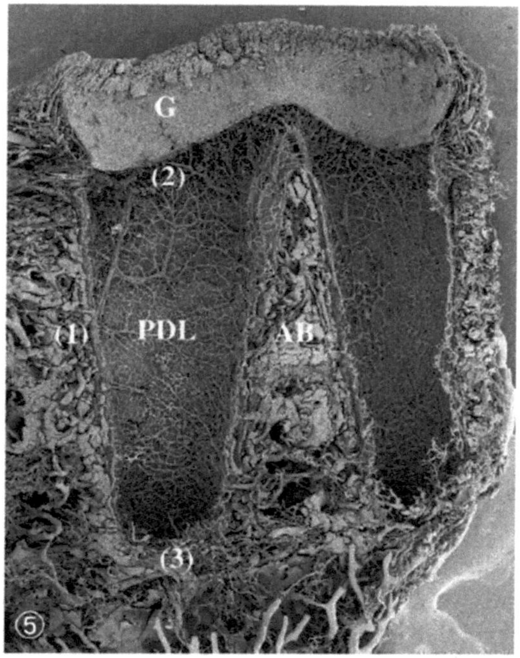

Fig. 10.1 Vascular network of the molar. The vascular network of periodontal ligament is supplied from three directions: (1) bifurcated in the alveolar bone (AB), (2) anastomosis with the gingival blood vessels, and (3) bifurcated into the apical area. (G) vascular network of gingiva [figures from Matsuo and Takahashi (2002)]

be influenced by external force including orthodontic tooth movement or local inflammation such as periodontitis (Rygh et al. 1986; Murrell et al. 1996; Zoellner et al. 2002). The compression of blood vessels due to orthodontic tooth movement causes local ischemia, which leads to angiogenesis via hypoxia inducible factor-1α (HIF1α) induction and alveolar bone resorption. The compression of the periodontal ligament due to orthodontic tooth movement and, in contrast, hypofunction of tooth due to the lack of adequate occlusal load may indirectly affect blood vessels via the temporarily altered expression of periostin (Wilde et al. 2003; Afanador et al. 2005; Lindner et al. 2005; Rios et al. 2008; Li et al. 2010). Besides, infections caused by periodontal pathogen such as *Porphyromonas gingivalis* (P.g.) cause disruption of blood vessels leading to local hypoxia via the production of protease, and chronic periodontitis further increases in the proportion of anaerobic bacteria. However, this mechanism is not fully understood.

10.2 The Importance of Periodontal Ligament in Periodontal Tissue Regeneration

Previous in vivo studies have indicated that the periodontal ligament has an important function in the restoration of periodontal tissue and the maintenance of homeostasis. In 1976, Melcher proposed that it is important to restore all periodontal tissue, not only the bone, when treating periodontal tissue destroyed by periodontitis, and that the periodontal ligament is important for that very purpose (Melcher 1976). In 1980, Nyman and Karring et al. conducted fervent in vivo research based on this hypothesis by Melcher and reported that the periodontal ligament plays an important role in the regeneration of periodontal tissue (Karring et al. 1980; Nyman et al. 1980). In other words, connective tissue attachment occurs continuing from the remaining periodontal ligament if there is any remaining healthy periodontal ligament on the surface of the tooth root, similar to that observed in normal tissues. In addition, it has been demonstrated that when periodontal ligament-derived cells are cultured in vitro and transplanted with decalcified tooth root, they restore the periodontal ligament (Boyko et al. 1981). In addition, even in tooth replantation studies on monkeys, it has been shown that root resorption does not occur when there is healthy periodontal ligament remaining on replanted teeth (Andreasen 1981). By contrast, when bone or gingival connective tissue contacts a root surface where the periodontal ligament is completely gone, tooth root resorption or ankylosis can take place due to the difference in adjacent tissues. In this regard, it is believed that the periodontal ligament can form collagen fibers that connect cementum to the alveolar bone called Sharpey's fibers, which are a supporting tissue of the teeth; in short, it may have the ability to regenerate periodontal tissue. In the 1990s, much research was published on the characteristics of fibroblastic cells derived from the periodontal ligament. According to Nojima and Basdra et al., fibroblasts derived from the periodontal ligament show osteoblast-related marker expression in vitro, and periodontal ligament fibroblasts possess mineralization capabilities similar to those of osteoblasts and can differentiate into cementoblasts or osteoblasts, the cells that form two hard tissues that flank the periodontal ligament, cementum, and alveolar bone (Nojima et al. 1990; Basdra and Komposch 1997). However, at this point, osteogenic activity of cells is derived from the periodontal ligament, but they are not described as being stem cells.

10.3 Periodontal Ligament Stem Cells

MSCs were first identified from bone marrow cells as plastic culture dish adhesive, colony-forming unit-fibroblastic (CFU-f). In the 1970s, Friedenstein et al. reported that CFU-f exist in bone marrow from low-density cultures of a bone marrow extract in vitro and demonstrated osteogenicity of the cells and that when the cells were transplanted into sub-renal capsule, fibrous tissue, bone, and bone marrow were formed where the recipient hematopoietic stem cells (HSCs) were homing, suggesting multipotency of CFU-f (Friedenstein et al. 1970). In the 1980s and 1990s, Pittenger et al. indicated that this cell type has the multipotency to differentiate into adipocytes, chondroblasts, and osteoblasts, suggesting that stem cells widely known as MSCs exist in postnatal adult tissue and bone marrow (Pittenger et al. 1999).

In the 1990s, it was shown that fibroblasts derived from the periodontal ligament possess osteogenic capacity (Nojima et al. 1990; Basdra and Komposch 1997).

Subsequently, it was confirmed from the results of gene expression analyses in fibroblasts by Lallier et al. that the gene expression of fibroblasts derived from the periodontal ligament is closer to that of osteoblasts than dermal fibroblasts (Lallier et al. 2005); however, the function or origin was not fully comprehended long afterward. McCulloch, using experiments involving mice with pulse labeled with H^3-thymidine, published results indicate that stem cells that have a slow rate of proliferation exist near the blood vessels in the periodontal ligament of mouse molars (McCulloch 1985). Chen et al., using antibodies against STRO-1, CD146, and CD44, investigated the localization of possible MSCs in healthy human teeth and periodontitis-affected teeth and reported the localization of putative MSCs near the blood vessels (Chen et al. 2006). In 2004, the first report came out regarding the existence of stem cells being implicated in the importance of periodontal ligament in the repair and maintenance of periodontal tissue homeostasis (Seo et al. 2004). This event happened 30 years after Friedenstein et al.'s quest for the discovery of bone marrow stem cells in the 1970s. MSCs were first identified as CFU-f from bone marrow. Currently, the minimum criteria of MSCs are (1) colony formation on plastic culture plates; (2) positive for mesenchymal cell markers such as CD73, CD90, and CD105 and negative for hematopoietic cell markers, such as CD11b, CD14, and CD45; and (3) differentiation capability into osteoblasts, chondrocytes, and adipocytes in vitro (Dominici et al. 2006). Seo et al. reported in vitro differentiation (into osteoblasts and adipocytes) of cells harvested from human periodontal ligaments using anti-STRO-1 antibodies that bind to an epitope on the cellular membrane of bone marrow MSCs; they reported the periodontal restorative ability in transplantations of these cells mixed with hydroxyapatite/tricalcium phosphate ceramic particles into periodontal tissue defects of immunocompromised rats (Seo et al. 2004). The anti-STRO-1 antibodies bind to CFU-f on plastic culture plates, one of minimum criteria for MSCs, but not to hematopoietic progenitor cells from human bone marrow. Therefore, compared to using whole bone marrow cells without fractionation, it is possible to increase the emergence rate of CFU-f 100-fold. Besides, it has been demonstrated that 50–70% of these cells bind to antibodies specific to smooth muscle actin (SMA) (Simmons and Torok-Storb 1991). Recently it has been reported that anti-STRO-1 antibodies cross-reacts with both heat shock cognate (HSC) 70 and heat shock protein (HSP) 70 and that STRO-1^{bright}/ $HSP70^{-}$ cell population demonstrates clonogenicity (Fitter et al. 2017). However, the function of HSC70 in the mesenchymal stem cell is still not fully understood. Nagatomo et al. demonstrated the presence of MSCS in the human periodontal ligament that are positive for markers of bone marrow stem cells, including STRO-1, and have colony-forming ability and differentiation capability into osteoblasts and adipocytes (Nagatomo et al. 2006). However, with this approach, the harvested MSCs represent a group of heterogeneous cells; and cells with clonogenicity and multipotency constitute a very small percentage of the cell group's population. Whether or not they are identical to stem cells that exist in tissues, their distribution in the adult organism and their origin are still not fully understood. With great changes in the notion that MSCs originate from fibroblasts due to their resemblance to clonogenic fibroblasts, along with reports that pericytes (which are localized near blood vessels and are instrumental for blood vessel nutrition and stabilization) possess mineralization activity (Doherty et al. 1998), multipotency (Pierantozzi et al. 2015), and tissue-reparative activity (Morgan and Muntoni 2007), it is now thought that MSCs also originate from pericytes (da Silva Meirelles et al. 2008; Lv et al. 2014; Sorrentino et al. 2008). It is believed that, because of their clonogenic properties and multipotency, MSCs are engaged in maintaining tissue homeostasis and restoration, but their true role in tissues remains unclear. It is unclear whether MSCs are involved in the repair process of periodontal tissue destruction associated with periodontitis. It is difficult to imagine, however, that MSCs dominantly contribute to the restoration of severe tissue loss because of the small number of cells isolated from the periodontal ligament. It is assumed that there is possibly another more biologically effective mechanism available to repair local tissue destruction in vivo. Sahara et al. demonstrated, using murine mechanically injured blood vessel models, that bone marrow-derived cells repair blood vessels via the blood circulation (Sahara et al. 2005). In murine spleen-injured model, it was reported that bone marrow cells enter peripheral blood and are mobilized to the wound site (Chen et al. 2010). We conducted an observational study using chimeric mice with green florescence protein (GFP)-labeled bone marrow cells (Fig. 10.2). The results showed that bone marrow-derived cells are present around the blood vessels of intact periodontal ligament. Furthermore, we demonstrated that the number of bone marrow-derived cells increases when the periodontal tissue is mechanically lesioned and that bone

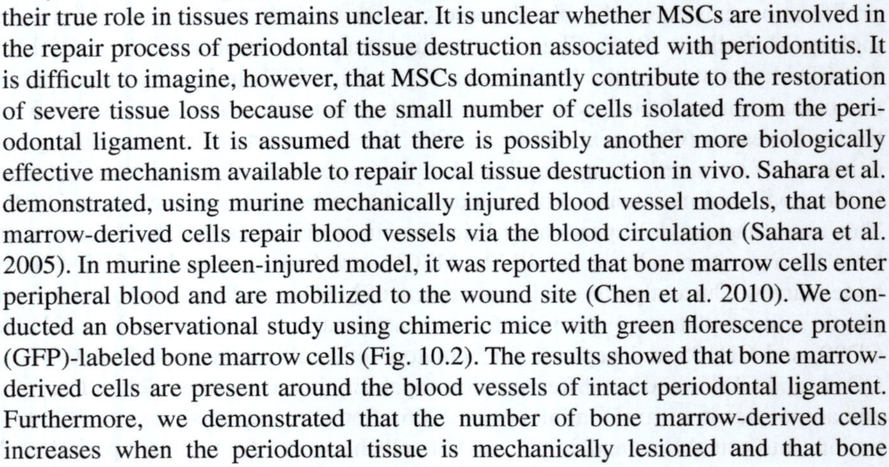

Fig. 10.2 Localization of GFP-positive cells in intact periodontal tissue. Visualization of bone marrow cells in vivo on periodontal tissue by using chimeric mice with GFP-labeled bone marrow cells. Arrows indicate osteoclast-like multinucleated giant cells (**a**) and macrophage-like cells in gingival epithelium (**b**) were GFP-positive, implying that GFP-positive images were confirmed to match bone marrow cells. GFP-positive cells were also observed at the periodontal ligament (**c**), blood vessels (**d**), and dental pulp (**e**). (**a**) Multinucleated giant cells in resorption lacunae of alveolar bone; (**b**) gingival epithelium; (**c**) periodontal ligament; (**d**) blood vessels in periodontal ligament; (**e**) dental pulp. (**g**), gingiva; *AB* alveolar bone, *P* periodontal ligament, *DP* dental pulp, *D* dentin

marrow-derived cells mobilize upon damage to the periodontal ligament tissue. In addition, we extract cells from bone marrow by enzymatic digestion and determine stem cell quantity using PDGFRα^+Sca-1$^+$CD45$^-$TER119$^-$ cell numbers according to Houlihan et al. method and realized that the number of MSCs in bone marrow decreases when the periodontal tissue is lesioned (Kimura et al. 2014). Although these results do not prove that they are pericytes, this evidence demonstrates that bone marrow cells are localized near the periodontal ligament's blood vessels and bone marrow cells, reacting to tissue destruction, are mobilized to the periodontal ligament as well (Fig. 10.3). Using FACS to compare the antigens on the surface of periodontal ligament cells with those of bone marrow stem cells and dermal fibroblasts, in addition to the mesenchymal marker expression, it was shown that PDLSCs have an expression pattern of marker CD34 similar to that of dermal fibroblasts and an expression pattern of marker CD166 similar to that of bone marrow stem cells (Iwasaki et al. 2013). This finding may suggest that stem cells harvested from the periodontal ligament may partly originate from bone marrow stem cells. Zhou et al. intravenously injected mice with GFP-labeled bone marrow cells and investigated the involvement of bone marrow cells in periodontal tissue regeneration. They reported that bone marrow stem cells are observed in periodontal reparative tissue and that stromal-derived factor 1 (SDF-1, also known as CXCL12) is involved in the

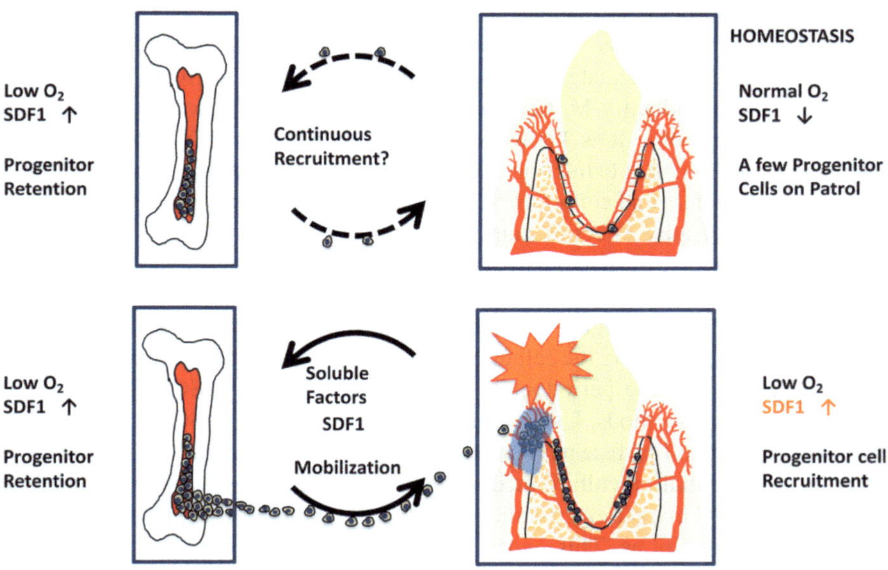

Fig. 10.3 Mobilization of bone marrow cells to periodontal ligament. Using chimeric mice with GFP-labeled bone marrow cells, we found that bone marrow cells are located around blood vessels in intact periodontal ligament. When periodontal lesion was mechanically made, hypoxia and/or inflammation was observed in periodontium, and the number of bone marrow cells increased in periodontal ligament. We found that bone marrow cells express SDF-1 receptor, CXCR4, and that bone marrow cells mobilize toward SDF in chemotaxis chamber. Taken together, the results imply that bone marrow cells were recruited to periodontal ligament via SDF-1 signal

mobilization of these cells (Zhou et al. 2011). This information is the essential proof needed for demonstrating that bone marrow cells are involved in the restoration of periodontal tissue. We also confirmed that bone marrow stem cells express SDF-1 receptor, CXCR4, and in experiments with a chemotaxis chamber to test cell migration, bone marrow stem cells migrate toward SDF-1 (Kimura et al. 2014).

10.4 Application of Mesenchymal Cell in Regenerative Periodontal Therapy and Related Findings

Based on in vitro functional assessments, MSCs have been harvested from various tissues including somatic tissues [bone marrow, fat, synovium (Hatakeyama et al. 2017), and cartilage (Xue et al. 2016)] and tooth and its surrounding tissues (dental pulp, periodontal ligament, gingiva), as well as fetal appendage [umbilical cord (Pirjali et al. 2013), cord blood (Zhang et al. 2011), and placenta (Komaki et al. 2017; Tooi et al. 2016)]. Stem cells are expected to find applications in tissue regenerative therapy because of their self-proliferation and differentiation potential. Various experimental evaluations of their ability to restore periodontal tissue have been conducted, and some clinical tests are being carried out (Monsarrat et al. 2014). However, there is some disparity seen in the restorative effect of stem cells, and it is still not safe to say at present whether they yield an adequate quantity of restored tissue or an adequate tissue regeneration efficiency. Currently, there is not a single marker to identify MSCs, and the cells are extracted according to in vitro functional evaluations. It is known that the phenotype of MSCs change easily depending on the tissue extraction method and culture condition such as cellular density, serum batch, and culture time. This situation may cause considerable variation in the regenerative effect of stem cells. Tanaka et al. used the enzymatic digestion and cell outgrowth methods to extract cells from the periodontal ligament to conduct a comparison. The periodontal ligament cells extracted via the enzymatic digestion method have higher colony-forming and cell differentiation potentials (Tanaka et al. 2011). Using the cell outgrowth and enzymatic digestion methods, we extracted cells from the periodontal ligament and obtained MSCs in accordance with conventional methods. Using commercial bone marrow MSCs as the control, we checked their cell surface markers again by FACS analysis after incubating these cells in a differentiation culture medium. We confirmed that there is no change in mesenchymal cell marker expression, even after differentiation of the stem cells due to incubation conditions. This experiment at least demonstrates that cell surface markers CD73, CD90, and CD105, used during in vitro identification of MSCs, are not affected by the cell differentiation (unpublished data). The heretofore seen periodontal tissue restoration via mesenchymal cells extracted from various sources [bone marrow (Kawaguchi et al. 2004; Li et al. 2009; Simsek et al. 2012), fat (Tobita and Mizuno 2013), or periodontal ligament (Dogan et al. 2002, 2003; Akizuki et al. 2005; Iwasaki et al. 2014)] has been demonstrated. We would like to cite an outstanding systematic review that presents a detailed comparison of periodontal ligament stem cells and bone marrow stem cells for periodontal tissue regeneration (Monsarrat et al. 2014; Yan et al. 2015). Tsumanuma et al. have conducted histological evaluations of MSCs extracted from bone marrow, synovium, and the periodontal ligament and transplanted them into experimentally created canine periodontal tissue defects. Cell sheets were manufactured before transplantation by taking cells incubated on temperature-responsive culture plates in a mineralization-inducing medium supplemented with ascorbic acid, β-glycerophosphate, and dexamethasone. Although no significant results were obtained with stem cells when used for alveolar bone regeneration, Tsumanuma et al. have reported superior results from the periodontal ligament stem cell transplantation, when using them to restore cementum, which makes up supporting dental tissue, and the periodontal ligament compared to stem cells derived from another tissue (Tsumanuma et al. 2011). This result may suggest tissue specificity of MSCs when they are used for regenerative therapy and/or the necessity to consider the histological idiosyncrasies of the lesion that is to be repaired.

MSCs taken from adults tissues are thought to include cells at varying stages of division and stemness depending on the local environmental factors such as inflammation or tissue metabolism. The establishment of a harvesting method that allows for the isolation of mesenchymal stem cell clones that retain their stemness, such as a single specific marker is expected.

It has been reported that stem cells in the periodontal ligament and bone marrow exist near blood vessels (McCulloch and Melcher 1983; Shi and Gronthos 2003). Furthermore, bone marrow cells have been observed near the blood vessels of the periodontal ligament (Kimura et al. 2014). Dohdrty et al. have found that bovine pericytes express osteoblastic markers and STRO-1 and can induce mineralization (Doherty et al. 1998). Although mesenchymal cells derived from avascular tissue such as cartilage have also been discovered (Xue et al. 2016), based on the previously discussed localization of periodontal ligament stem cells near blood vessels and stem cells being reported in the various vascular tissues (da Silva Meirelles et al. 2006), it is surmised that PDLSCs originate from pericytes and/or bone marrow stem cells located near capillaries. There are extremely small amounts of information regarding pericytes in the periodontal ligament tissues, and it remains unclear whether periodontal ligament stem cells and such cells near the blood vessels (e.g., pericytes) are identical.

10.5 Similarities Between Pericytes and Periodontal Ligament Stem Cells

It is thought that pericytes, along with endothelial cells, anatomically constitute the blood vessels of microvasculature including the terminal arterioles, capillaries, and postcapillary venules and are providing nutrient and stabilizing blood vessels. As with MSCs, there is no single marker that identifies pericytes; several markers such as alpha smooth muscle actin (αSMA), neuron-glial 2 (NG2), CD140a and CD140b also known as platelet-derived growth factor receptor (PDGFR) α and β, and CD146 are used in conjunction (Dore-Duffy and Cleary 2011). There are recent reports that pericytes possess multipotency, drawing focus to creating a new role for them (Yamazaki and Mukouyama 2018; Markovic et al. 2018). Shi et al. reported that STRO-1$^+$ bone marrow MSCs and dental pulp stem cells (DPSCs) express the markers widely used as pericyte markers: α smooth muscle action (αSMA) and CD146 (Shi and Gronthos 2003). As mentioned, thus far, MSCs and pericytes are localized near blood vessels and possess differentiation potentials. There is still debate about these cells' identities, their origins, their biological distribution, and their frequency. Meirelles et al. have conducted long-term incubation of MSC-like cells from various organs and tissues including the brain, pancreas, liver, bone marrow, and muscles to investigate their distribution in the adult organism. MSCs were observed in large and small blood vessels; thus, they concluded that MSCs exist in vascular niche (da Silva Meirelles et al. 2006). Xu et al. have conducted a review of studies on whether stem cells can differentiate into functional pericytes. Their criteria were (1) the study has an appropriate control; (2) it involves pericyte markers such as PDGFRβ, αSMA, NG2, and desmin; (3) the study functionally evaluates pericytes; and (4) it involves a differentiation efficiency protocol. They concluded that stem cells are the origins of pericytes (Xu et al. 2017). In contrast, based on the concept that the blood vessels in internal organs are reservoir for stem cells, Corselli et al. have isolated endothelial and perivascular cells by FACS and reported that microvascular pericytes and adventitial cells of large vessels express MSC markers (Corselli et al. 2010).

The periodontal ligament, supportive tissue for the teeth, is a connective tissue rich with blood vessels, but there is little information in relation to pericytes. Gould et al. conducted a cell kinetic analysis and cytology of murine injured molar model and reported that dividing cells can be observed around the blood vessels and on the outside of the blood vessel's basal lamina after wounding. Besides, they allude to the fact that accompanying wounding, the origins and functions of the dividing cells in the periodontal ligament are possibly different from those of classical pericytes (Gould et al. 1977, 1980). In 1985, McCulloch demonstrated, in autoradiographic investigation of mitotic cells with H^3-thymidine, that cells showing characteristics of stem cells exist perivascularly in the periodontal ligament in uninjured mice (McCulloch 1985). In 1986, Lew conducted a detailed study on the blood vessels of periodontal tissues in rats using scanning electron microscopy (SEM). There are three types of blood vessels in the periodontal ligament: terminal arterioles, capillaries, and the postcapillary venules. Using morphometry, Lew showed that the total vascular volume of the periodontal ligament root apex was approximately 20%: abundance of blood vessels compared to the 5% of general connective tissue. We found that MSCs in the periodontal ligament that form colonies are STRO-1-, CD105-, and CD166-positive and differentiate into osteoblasts, adipocytes, and chondrocytes (Nagatomo et al. 2006). We conducted additional experiments to test our hypothesis that bone marrow-derived cells were situated in periodontal ligament by using GFP transgenic mice to create bone marrow-substituted chimeric mice, during examination of the periodontal ligament tissue in both intact and injured sites, and found that GFP-positive cells were localized perivascularly in the intact periodontal ligament tissue, while damage leading to increase in the quantity of GFP-positive cells in the periodontal ligament. On the contrary, it was discovered that the number of MSCs (PDGFR-α^+Sca-1^+ and CD45^-TER119^-) extracted from bone marrow decreased (Kimura et al. 2014). These results possibly suggest that bone marrow stem cells are one source of the MSCs that exist in the periodontal ligament at least when the tissue was injured. Zhou et al. have shown that bone marrow MSCs are involved in periodontal ligament repair tissue via the chemotaxis of bone marrow stem cells toward SDF-1. Recently, it was reported that pericytes express CXCR4 much like bone marrow cells (Armulik et al. 2011); therefore, precaution is necessary when investigating the involvement of pericytes and bone marrow MSCs. We tested whether PDLSCs express stem cell and pericyte markers NG2, CD146, and CD140b (PDGFRβ), whether NG2- and CD146-positive cells can be observed near the vascular endothelium in rat periodontal ligament tissue, and whether periodontal ligament stem cells maintain and/or stabilize blood vessels formed by endothelial cells as do pericytes in the endothelial cell network formation assay or on cell pattern substrates we have reported previously (Akahori et al. 2010; Mukai et al. 2008). The results showed that PDLSCs express both stem cell and pericyte markers, including NG2, CD146, and CD140b and participate in the maintenance of blood vessel-like structures (Fig. 10.4). Furthermore, NG2- and CD146-positive cells are situated near the vascular endothelium in rat periodontal ligament (Iwasaki et al. 2013). There is a study stating that the blood vessels of the dental root apex are predominately postcapillary venules and that there is a full coverage of pericytes (Lew 1987). In contrast, there is a paper stating that pericytes exist in the periodontal ligament's blood vessels (Rhodin 1968). The content is not presented here because the paper is unobtainable. Whether or not PDLSCs and pericytes are the same must be investigated in more detail. Koike et al. have three-dimensionally cultured vascular endothelial cell with or without murine fetus-derived fibroblasts, 10 T1/2 cells in vitro, and then transplanted them into mice to investigate whether 10 T1/2 cells are involved in the maintenance of vascular network. They demonstrated that 10 T1/2 cells contribute to a long-term maintenance (56 days) of blood vessel density (Koike et al. 2004). The 10 T1/2 cells are a heterogeneous cell population whose phenotype can only be maintained under strict culture conditions. Some reports show their multipotency, and Koike et al. used 10 T1/2 cells as mesenchymal progenitor cells. The ontogenetic relation between fibroblasts and mural cells, including pericytes, is known (Armulik et al. 2011). The relation of bone marrow stem cells and pericytes needs to be clarified.

Cells that constitute the periodontal tissue are embryologically derived from the neural crest. We have shown that cells derived from bone marrow exist around the blood vessels of the periodontal ligament in 8- to 12-week-old mice. Using the lineage tracing method, Takashima et al. have revealed that MSCs in fetal mouse trunk are partially generated from the neuroepithelium through the intermediate stages of the neural crest but not from the mesoderm (Takashima et al. 2007). Subsequently, it was demonstrated that some of the MSCs within adult murine bone marrow come

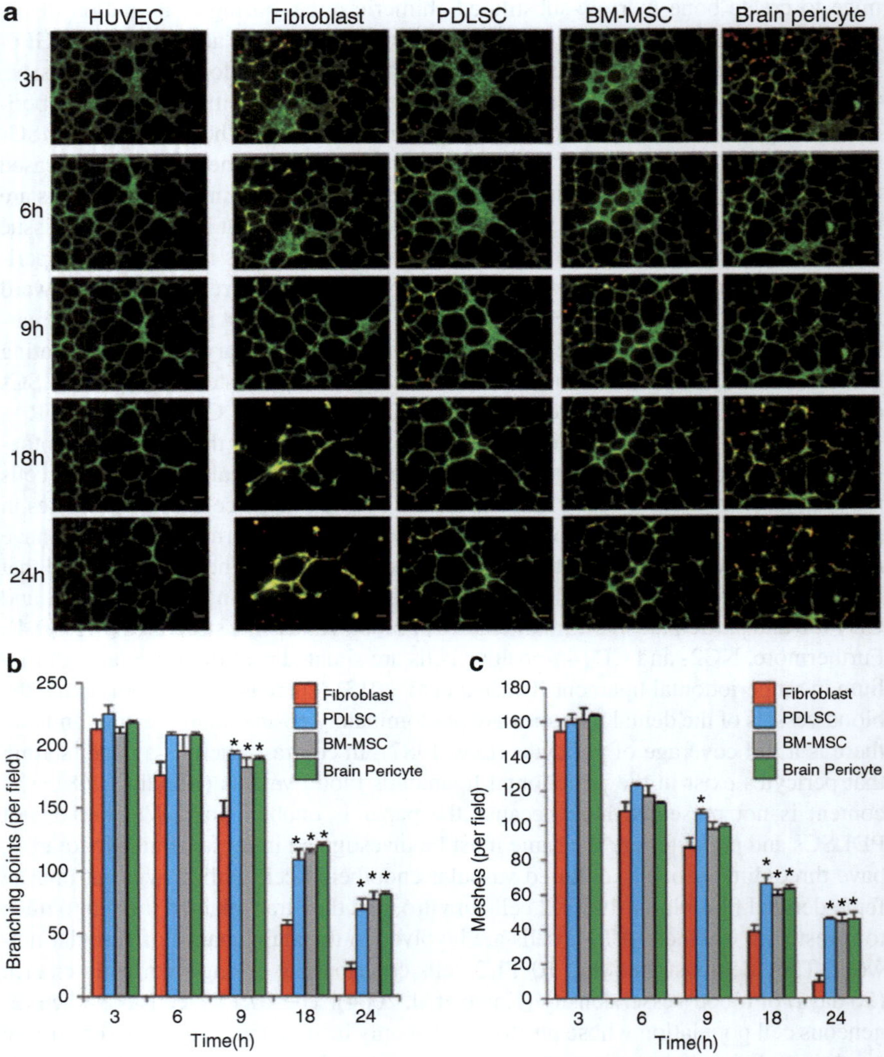

Fig. 10.4 Chronologic changes in capillary formation in co-culture of Human Umbilical Vein Endothelial Cells (HUVECs) with fibroblasts, PDLSCs, BM-MSCs, and brain pericytes. (**a**) PKH26-labeled fibroblasts, PDLSCs, BM-MSCs, and brain pericytes were co-cultured with GFP-HUVECs (cell-HUVEC ratio, 20:1) on a gel matrix. Photographs of red and green merged images from each experimental group are shown. Pictures in the leftmost column are from HUVECs alone. The capillary-like structures of HUVECs began to lose their shape at 9 hours after co-culture, and most of the structures had disappeared at 24 hours. However, the capillary structures were stably observed at 24 hours in the PDLSC, BM-MSC, and brain pericyte groups. Representative pictures from more than three independent experiments are shown. Scale bar = 300 μm (original magnification × 2) The time course changes in branching point (**b**) and mesh (**c**) counts are also indicated. (Asterisk) Statistically significant differences from the fibroblast group ($P < 0.05$)

from the neural crest, in which they were transiently observed during the developmental stage (Morikawa et al. 2009). From this information, it follows that the origin of MSCs in the adult organism is varied, much like that of pericytes. Unfortunately, there is no description regarding pericytes in any of the relevant reports. We must wait for more research to reveal the whole picture of the origins of MSCs and pericytes as well as their function in the adult organism.

10.6 Possibility of Treatments Targeting Pericytes

The blood vessels constitute a closed system that circulates throughout the body, and its length is two and a half that of the outer circumference of the Earth. It is known that blood vessels play both a physiologically and pathologically important role in tissues. Blood vessels are lined with a single layer of vascular endothelial cells along their lumen and are specialized according to the needs of the organs to which they deliver nutrients and oxygen (organ-specific endothelial cell differentiation). It is known that smooth muscle cells and pericytes join around the lumen formed by endothelial cells and work together with endothelial cells to stabilize and contract the blood vessels through receptors, including ephrin-B2, platelet-drived growth factor receptors, and epidermal growth factor receptor 2, and their ligands. Their diversity in abnormalities such as inflammation and their organ specificity are acknowledged (Armulik et al. 2011; Tomlinson and Topper 2003; Yoshida and Owens 2005; Sims 2000; Potente and Makinen 2017). In the adult, rapid metabolic turnover is known in tissues that face constant challenges, for example, the periodontal tissues, which are exposed to oral bacteria and mastication. The blood vessels in which pericytes and MSCs are believed to be localized are diversified as well. For this reason, it is thought that there are PDLSCs with various degrees of differentiation in the periodontal ligament. Currently, trials have been conducted using all sorts of cell surface antigens, including STRO-1, SSEA-4, CD271, and CD146, as indicators for enrichment of PDLSC subpopulations with high self-proliferation and multipotency (Lv et al. 2014). If pericytes and MSCs are tissue-specific cells that maintain and repair tissues, and if these cells ceaselessly self-proliferate and differentiate in response to a constantly changing environment, current MSC extraction protocols make highly efficient and stable extraction for clinical application difficult. Additionally, the quantity and quality of MSCs taken from the adult organism decrease with inflammation (Yang et al. 2013) and age (Zhang et al. 2012; Zheng et al. 2009). It is extremely difficult to maintain the phenotype of the obtained cells, in addition to securing the cell numbers needed for treatment, all the while avoiding senescent deterioration. Although still vague, expert use of lineage tracing is elucidating the true origin of pericytes and MSCs. On the basis of the plethora of information about pericytes and MSCs, if one could create pericytes or MSCs from induced pluripotent stem (iPS) cells (Fukuta et al. 2014; Chijimatsu et al. 2017), this advance would bring us closer to using these cells, which boosts such effective treatment results in various diseases and in periodontal regeneration.

10.7 Conclusion

MSCs and pericytes are cells thought to be harvestable from the adult organism and have multipotency. They have been used in research on disease models and have seen partial clinical application. On the other hand, the origin, the distribution in the adult organism, and the function of these perplexing cells are still not fully understood. There are still mixed results regarding the therapeutic effects, but the feasibility of the application of MSCs and pericytes to cell therapy has been demonstrated. Safe, stable, high-efficiency, and cheap cells will possibly be developed because of the abundance of accumulated information about them, leading to their application to therapy.

References

Afanador E et al (2005) Messenger RNA expression of periostin and twist transiently decrease by occlusal hypofunction in mouse periodontal ligament. Arch Oral Biol 50:1023–1031. https://doi.org/10.1016/j.archoralbio.2005.04.002

Akahori T et al (2010) Implantation of capillary structure engineered by optical lithography improves hind limb ischemia in mice. Tissue Eng Part A 16:953–959. https://doi.org/10.1089/ten.tea.2009.0097

Akizuki T et al (2005) Application of periodontal ligament cell sheet for periodontal regeneration: a pilot study in beagle dogs. J Periodontal Res 40:245–251. https://doi.org/10.1111/j.1600-0765.2005.00799.x

Andreasen JO (1981) Periodontal healing after replantation and autotransplantation of incisors in monkeys. Int J Oral Surg 10:54–61

Armulik A, Genove G, Betsholtz C (2011) Pericytes: developmental, physiological, and pathological perspectives, problems, and promises. Dev Cell 21:193–215. https://doi.org/10.1016/j.devcel.2011.07.001

Basdra EK, Komposch G (1997) Osteoblast-like properties of human periodontal ligament cells: an in vitro analysis. Eur J Orthod 19:615–621

Berglundh T, Lindhe J, Jonsson K, Ericsson I (1994) The topography of the vascular systems in the periodontal and peri-implant tissues in the dog. J Clin Periodontol 21:189–193

Boyko GA, Melcher AH, Brunette DM (1981) Formation of new periodontal ligament by periodontal ligament cells implanted in vivo after culture in vitro. A preliminary study of transplanted roots in the dog. J Periodontal Res 16:73–88

Chen SC, Marino V, Gronthos S, Bartold PM (2006) Location of putative stem cells in human periodontal ligament. J Periodontal Res 41:547–553. https://doi.org/10.1111/j.1600-0765.2006.00904.x

Chen Y et al (2010) Recruitment of endogenous bone marrow mesenchymal stem cells towards injured liver. J Cell Mol Med 14:1494–1508. https://doi.org/10.1111/j.1582-4934.2009.00912.x

Chijimatsu R et al (2017) Characterization of mesenchymal stem cell-like cells derived from human iPSCs via neural crest development and their application for osteochondral repair. Stem Cells Int 2017:1960965. https://doi.org/10.1155/2017/1960965

Corselli M, Chen CW, Crisan M, Lazzari L, Peault B (2010) Perivascular ancestors of adult multipotent stem cells. Arterioscler Thromb Vasc Biol 30:1104–1109. https://doi.org/10.1161/ATVBAHA.109.191643

da Silva Meirelles L, Chagastelles PC, Nardi NB (2006) Mesenchymal stem cells reside in virtually all post-natal organs and tissues. J Cell Sci 119:2204–2213. https://doi.org/10.1242/jcs.02932

da Silva Meirelles L, Caplan AI, Nardi NB (2008) In search of the in vivo identity of mesenchymal stem cells. Stem Cells 26:2287–2299. https://doi.org/10.1634/stemcells.2007-1122
Dogan A, Ozdemir A, Kubar A, Oygur T (2002) Assessment of periodontal healing by seeding of fibroblast-like cells derived from regenerated periodontal ligament in artificial furcation defects in a dog: a pilot study. Tissue Eng 8:273–282. https://doi.org/10.1089/107632702753725030
Dogan A, Ozdemir A, Kubar A, Oygur T (2003) Healing of artificial fenestration defects by seeding of fibroblast-like cells derived from regenerated periodontal ligament in a dog: a preliminary study. Tissue Eng 9:1189–1196. https://doi.org/10.1089/10763270360728099
Doherty MJ et al (1998) Vascular pericytes express osteogenic potential in vitro and in vivo. J Bone Miner Res 13:828–838. https://doi.org/10.1359/jbmr.1998.13.5.828
Dominici M et al (2006) Minimal criteria for defining multipotent mesenchymal stromal cells. The International Society for Cellular Therapy position statement. Cytotherapy 8:315–317. https://doi.org/10.1080/14653240600855905
Dore-Duffy P, Cleary K (2011) Morphology and properties of pericytes. Methods Mol Biol 686:49–68. https://doi.org/10.1007/978-1-60761-938-3_2
Fitter S, Gronthos S, Ooi SS, Zannettino AC (2017) The mesenchymal precursor cell marker antibody STRO-1 binds to cell surface heat shock cognate 70. Stem Cells 35:940–951. https://doi.org/10.1002/stem.2560
Friedenstein AJ, Chailakhjan RK, Lalykina KS (1970) The development of fibroblast colonies in monolayer cultures of Guinea-pig bone marrow and spleen cells. Cell Tissue Kinet 3:393–403
Fukuta M et al (2014) Derivation of mesenchymal stromal cells from pluripotent stem cells through a neural crest lineage using small molecule compounds with defined media. PLoS One 9:e112291. https://doi.org/10.1371/journal.pone.0112291
Gould TR, Melcher AH, Brunette DM (1977) Location of progenitor cells in periodontal ligament of mouse molar stimulated by wounding. Anat Rec 188:133–141. https://doi.org/10.1002/ar.1091880202
Gould TR, Melcher AH, Brunette DM (1980) Migration and division of progenitor cell populations in periodontal ligament after wounding. J Periodontal Res 15:20–42
Hatakeyama A et al (2017) Isolation and characterization of synovial mesenchymal stem cell derived from hip joints: a comparative analysis with a matched control knee group. Stem Cells Int 2017:9312329. https://doi.org/10.1155/2017/9312329
Iwasaki K et al (2013) Periodontal ligament stem cells possess the characteristics of pericytes. J Periodontol 84:1425–1433. https://doi.org/10.1902/jop.2012.120547
Iwasaki K et al (2014) Periodontal regeneration using periodontal ligament stem cell-transferred amnion. Tissue Eng Part A 20:693–704. https://doi.org/10.1089/ten.TEA.2013.0017
Karring T, Nyman S, Lindhe J (1980) Healing following implantation of periodontitis affected roots into bone tissue. J Clin Periodontol 7:96–105
Kawaguchi H et al (2004) Enhancement of periodontal tissue regeneration by transplantation of bone marrow mesenchymal stem cells. J Periodontol 75:1281–1287. https://doi.org/10.1902/jop.2004.75.9.1281
Kimura Y et al (2014) Recruitment of bone marrow-derived cells to periodontal tissue defects. Front Cell Dev Biol 2:19. https://doi.org/10.3389/fcell.2014.00019
Koike N et al (2004) Tissue engineering: creation of long-lasting blood vessels. Nature 428:138–139. https://doi.org/10.1038/428138a
Komaki M et al (2007) Twist negatively regulates osteoblastic differentiation in human periodontal ligament cells. J Cell Biochem 100:303–314. https://doi.org/10.1002/jcb.21038
Komaki M et al (2017) Exosomes of human placenta-derived mesenchymal stem cells stimulate angiogenesis. Stem Cell Res Ther 8:219. https://doi.org/10.1186/s13287-017-0660-9
Lallier TE, Spencer A, Fowler MM (2005) Transcript profiling of periodontal fibroblasts and osteoblasts. J Periodontol 76:1044–1055. https://doi.org/10.1902/jop.2005.76.7.1044
Lew KK (1987) The periodontal microvasculature—a morphological and morphometric study. J Nihon Univ Sch Dent 29:262–269

Li H, Yan F, Lei L, Li Y, Xiao Y (2009) Application of autologous cryopreserved bone marrow mesenchymal stem cells for periodontal regeneration in dogs. Cells Tissues Organs 190:94–101. https://doi.org/10.1159/000166547
Li G et al (2010) Periostin mediates vascular smooth muscle cell migration through the integrins alphavbeta3 and alphavbeta5 and focal adhesion kinase (FAK) pathway. Atherosclerosis 208:358–365. https://doi.org/10.1016/j.atherosclerosis.2009.07.046
Lindner V, Wang Q, Conley BA, Friesel RE, Vary C (2005) Vascular injury induces expression of periostin: implications for vascular cell differentiation and migration. Arterioscler Thromb Vasc Biol 25:77–83. https://doi.org/10.1161/01.ATV.0000149141.81230.c6
Lv FJ, Tuan RS, Cheung KM, Leung VY (2014) Concise review: the surface markers and identity of human mesenchymal stem cells. Stem Cells 32:1408–1419. https://doi.org/10.1002/stem.1681
Markovic BS et al (2018) Molecular and cellular mechanisms involved in mesenchymal stem cell-based therapy of inflammatory bowel diseases. Stem Cell Rev 14:153–165. https://doi.org/10.1007/s12015-017-9789-2
Matsuo M, Takahashi K (2002) Scanning electron microscopic observation of microvasculature in periodontium. Microsc Res Tech 56:3–14. https://doi.org/10.1002/jemt.10008
McCulloch CA (1985) Progenitor cell populations in the periodontal ligament of mice. Anat Rec 211:258–262. https://doi.org/10.1002/ar.1092110305
McCulloch CA, Melcher AH (1983) Continuous labelling of the periodontal ligament of mice. J Periodontal Res 18:231–241
Melcher AH (1976) On the repair potential of periodontal tissues. J Periodontol 47:256–260. https://doi.org/10.1902/jop.1976.47.5.256
Monsarrat P et al (2014) Concise review: mesenchymal stromal cells used for periodontal regeneration: a systematic review. Stem Cells Transl Med 3:768–774. https://doi.org/10.5966/sctm.2013-0183
Morgan J, Muntoni F (2007) Mural cells paint a new picture of muscle stem cells. Nat Cell Biol 9:249–251. https://doi.org/10.1038/ncb0307-249
Morikawa S et al (2009) Development of mesenchymal stem cells partially originate from the neural crest. Biochem Biophys Res Commun 379:1114–1119. https://doi.org/10.1016/j.bbrc.2009.01.031
Mukai N et al (2008) A comparison of the tube forming potentials of early and late endothelial progenitor cells. Exp Cell Res 314:430–440. https://doi.org/10.1016/j.yexcr.2007.11.016
Murrell EF, Yen EH, Johnson RB (1996) Vascular changes in the periodontal ligament after removal of orthodontic forces. Am J Orthod Dentofac Orthop 110:280–286
Nagatomo K et al (2006) Stem cell properties of human periodontal ligament cells. J Periodontal Res 41:303–310. https://doi.org/10.1111/j.1600-0765.2006.00870.x
Nojima N et al (1990) Fibroblastic cells derived from bovine periodontal ligaments have the phenotypes of osteoblasts. J Periodontal Res 25:179–185
Nyman S, Karring T, Lindhe J, Planten S (1980) Healing following implantation of periodontitis-affected roots into gingival connective tissue. J Clin Periodontol 7:394–401
Pierantozzi E et al (2015) Human pericytes isolated from adipose tissue have better differentiation abilities than their mesenchymal stem cell counterparts. Cell Tissue Res 361:769–778. https://doi.org/10.1007/s00441-015-2166-z
Pirjali T et al (2013) Isolation and characterization of human mesenchymal stem cells derived from human umbilical cord Wharton's jelly and amniotic membrane. Int J Organ Transplant Med 4:111–116
Pittenger MF et al (1999) Multilineage potential of adult human mesenchymal stem cells. Science 284:143–147
Potente M, Makinen T (2017) Vascular heterogeneity and specialization in development and disease. Nat Rev Mol Cell Biol 18:477–494. https://doi.org/10.1038/nrm.2017.36
Rhodin JA (1968) Ultrastructure of mammalian venous capillaries, venules, and small collecting veins. J Ultrastruct Res 25:452–500

Rios HF et al (2008) Periostin is essential for the integrity and function of the periodontal ligament during occlusal loading in mice. J Periodontol 79:1480–1490. https://doi.org/10.1902/jop.2008.070624
Rygh P, Bowling K, Hovlandsdal L, Williams S (1986) Activation of the vascular system: a main mediator of periodontal fiber remodeling in orthodontic tooth movement. Am J Orthod 89:453–468
Sahara M et al (2005) Comparison of various bone marrow fractions in the ability to participate in vascular remodeling after mechanical injury. Stem Cells 23:874–878. https://doi.org/10.1634/stemcells.2005-0012
Seo BM et al (2004) Investigation of multipotent postnatal stem cells from human periodontal ligament. Lancet 364:149–155. https://doi.org/10.1016/S0140-6736(04)16627-0
Shi S, Gronthos S (2003) Perivascular niche of postnatal mesenchymal stem cells in human bone marrow and dental pulp. J Bone Miner Res 18:696–704. https://doi.org/10.1359/jbmr.2003.18.4.696
Simmons PJ, Torok-Storb B (1991) Identification of stromal cell precursors in human bone marrow by a novel monoclonal antibody, STRO-1. Blood 78:55–62
Sims DE (2000) Diversity within pericytes. Clin Exp Pharmacol Physiol 27:842–846
Simsek SB, Keles GC, Baris S, Cetinkaya BO (2012) Comparison of mesenchymal stem cells and autogenous cortical bone graft in the treatment of class II furcation defects in dogs. Clin Oral Investig 16:251–258. https://doi.org/10.1007/s00784-010-0486-7
Sorrentino A et al (2008) Isolation and characterization of CD146+ multipotent mesenchymal stromal cells. Exp Hematol 36:1035–1046. https://doi.org/10.1016/j.exphem.2008.03.004
Takashima Y et al (2007) Neuroepithelial cells supply an initial transient wave of MSC differentiation. Cell 129:1377–1388. https://doi.org/10.1016/j.cell.2007.04.028
Tanaka K et al (2011) Comparison of characteristics of periodontal ligament cells obtained from outgrowth and enzyme-digested culture methods. Arch Oral Biol 56:380–388. https://doi.org/10.1016/j.archoralbio.2010.10.013
Tobita M, Mizuno H (2013) Adipose-derived stem cells and periodontal tissue engineering. Int J Oral Maxillofac Implants 28:e487–e493. https://doi.org/10.11607/jomi.te29
Tomlinson JE, Topper JN (2003) New insights into endothelial diversity. Curr Atheroscler Rep 5:223–229
Tooi M et al (2016) Placenta mesenchymal stem cell derived exosomes confer plasticity on fibroblasts. J Cell Biochem 117:1658–1670. https://doi.org/10.1002/jcb.25459
Tsumanuma Y et al (2011) Comparison of different tissue-derived stem cell sheets for periodontal regeneration in a canine 1-wall defect model. Biomaterials 32:5819–5825. https://doi.org/10.1016/j.biomaterials.2011.04.071
Wilde J, Yokozeki M, Terai K, Kudo A, Moriyama K (2003) The divergent expression of periostin mRNA in the periodontal ligament during experimental tooth movement. Cell Tissue Res 312:345–351. https://doi.org/10.1007/s00441-002-0664-2
Xu J, Gong T, Heng BC, Zhang CF (2017) A systematic review: differentiation of stem cells into functional pericytes. FASEB J 31:1775–1786. https://doi.org/10.1096/fj.201600951RRR
Xue K, Zhang X, Qi L, Zhou J, Liu K (2016) Isolation, identification, and comparison of cartilage stem progenitor/cells from auricular cartilage and perichondrium. Am J Transl Res 8:732–741
Yamada S et al (2007) PLAP-1/asporin, a novel negative regulator of periodontal ligament mineralization. J Biol Chem 282:23070–23080. https://doi.org/10.1074/jbc.M611181200
Yamazaki T, Mukouyama YS (2018) Tissue specific origin, development, and pathological perspectives of Pericytes. Front Cardiovasc Med 5:78. https://doi.org/10.3389/fcvm.2018.00078
Yan XZ, Yang F, Jansen JA, de Vries RB, van den Beucken JJ (2015) Cell-based approaches in periodontal regeneration: a systematic review and meta-analysis of periodontal defect models in animal experimental work. Tissue Eng Part B Rev 21:411–426. https://doi.org/10.1089/ten.TEB.2015.0049
Yang N et al (2013) Tumor necrosis factor alpha suppresses the mesenchymal stem cell osteogenesis promoter miR-21 in estrogen deficiency-induced osteoporosis. J Bone Miner Res 28:559–573. https://doi.org/10.1002/jbmr.1798

Yoshida T, Owens GK (2005) Molecular determinants of vascular smooth muscle cell diversity. Circ Res 96:280–291. https://doi.org/10.1161/01.RES.0000155951.62152.2e
Yoshizawa T et al (2004) Homeobox protein MSX2 acts as a molecular defense mechanism for preventing ossification in ligament fibroblasts. Mol Cell Biol 24:3460–3472
Zhang X et al (2011) Isolation and characterization of mesenchymal stem cells from human umbilical cord blood: reevaluation of critical factors for successful isolation and high ability to proliferate and differentiate to chondrocytes as compared to mesenchymal stem cells from bone marrow and adipose tissue. J Cell Biochem 112:1206–1218. https://doi.org/10.1002/jcb.23042
Zhang J et al (2012) The effect of aging on the pluripotential capacity and regenerative potential of human periodontal ligament stem cells. Biomaterials 33:6974–6986. https://doi.org/10.1016/j.biomaterials.2012.06.032
Zheng W et al (2009) Loss of proliferation and differentiation capacity of aged human periodontal ligament stem cells and rejuvenation by exposure to the young extrinsic environment. Tissue Eng Part A 15:2363–2371. https://doi.org/10.1089/ten.tea.2008.0562
Zhou J et al (2011) Role of bone marrow-derived progenitor cells in the maintenance and regeneration of dental mesenchymal tissues. J Cell Physiol 226:2081–2090. https://doi.org/10.1002/jcp.22538
Zoellner H, Chapple CC, Hunter N (2002) Microvasculature in gingivitis and chronic periodontitis: disruption of vascular networks with protracted inflammation. Microsc Res Tech 56:15–31. https://doi.org/10.1002/jemt.10009

Chapter 11
Pericytes in the Heart

Linda L. Lee and Vishnu Chintalgattu

Abstract Mural cells known as pericytes envelop the endothelial layer of microvessels throughout the body and have been described to have tissue-specific functions. Cardiac pericytes are abundantly found in the heart, but they are relatively understudied. Currently, their importance is emerging in cardiovascular homeostasis and dysfunction due to their pleiotropism. They are known to play key roles in vascular tone and vascular integrity as well as angiogenesis. However, their dysfunctional presence and/or absence is critical in the mechanisms that lead to cardiac pathologies such as myocardial infarction, fibrosis, and thrombosis. Moreover, they are targeted as a therapeutic potential due to their mesenchymal properties that could allow them to repair and regenerate a damaged heart. They are also sought after as a cell-based therapy based on their healing potential in preclinical studies of animal models of myocardial infarction. Therefore, recognizing the importance of cardiac pericytes and understanding their biology will lead to new therapeutic concepts.

Keywords Cardiac pericyte · Heart · Cardiovascular physiology · Cardiovascular pathophysiology · Myocardial infarction · Myocardial ischemic disease · Vascular stem cell · Perivascular cell · Mural cell · Vascular tone · Vascular integrity · Vascular biology

Abbreviations

3g5	Ganglioside 3g5
αSMA	Alpha-smooth muscle actin
Alk-p	Alkaline phosphatase
Angpt-1	Angiopoietin-1

L. L. Lee · V. Chintalgattu (✉)
Department of CardioMetabolic Disorders, Amgen Research and Discovery, Amgen Inc., South San Francisco, CA, USA
e-mail: vishnuc@amgen.com

A. Birbrair (ed.), *Pericyte Biology in Different Organs*, Advances in Experimental Medicine and Biology 1122,
https://doi.org/10.1007/978-3-030-11093-2_11

Angpt-2	Angiopoietin-2
BDNF	Endothelial brain-derived neurotrophic factor
CNP	C-type natriuretic peptide
cPC	Cardiac pericyte
CTGF/CCN2	Connective tissue growth factor
Cx43	Connexin-43
CXCR3	Chemokine receptor 3
ECM	Extracellular matrix
Gli1	Glioma-associated oncogene 1
HB-EGF	Heparin-binding epidermal growth factor
HGF	Hepatocyte growth factor
LRP	Low-density lipoprotein receptor-related protein 6
MI	Myocardial infarction
MMPs	Matrix metalloproteinases
MSC	Mesenchymal stem cells
NANOG	Homeobox transcription factor Nanog
NG2	Neural glial 2
OCT4	Octamer-binding transcription factor 4
p75 NTR	Neurotrophin receptor
PDGFbb	Platelet-derived growth factor bb
PDGFRβ	Platelet-derived growth factor receptor beta
pro-NGF	Pro-nerve growth factor
RGS5	Regulator of G protein signaling 5
S1P	Sphingosine-1-phosphate
SIRT3	Sirtuin 3
SorCS2	Sortilin-related VPS10 domain-containing receptor 2
SOX2	Sex-determining region box
SVPs	Saphenous vein-derived pericyte progenitor cells
Tbx18	T-box protein 18
TF	Tissue factor
TGFβ	Transforming growth factor beta
TrkB	Tropomyosin receptor kinase B
VCAF	Vascular calcification-associated factor
VEGF-A	Vascular endothelial growth factor A
VEGFR2	Vascular endothelial growth factor receptor 2
vWF	von Willebrand factor

11.1 Introduction

Pericytes are perivascular cells that are present throughout various vascular beds of the body including the heart. Pericytes are found on small blood vessels including arterioles, venules, and capillaries where they are known to have tissue-specific functions. Pericyte number also differs in different tissue beds, and their number appears to determine the tissue-specific endothelial barrier function. For example, the brain and retina need stringent barrier integrity as compared to other tissues like skeletal muscle. The ratio of pericytes to endothelial cells in the brain is found as high as 1:1 compared to places like the skeletal muscle which is found to be as low as 1:100 (Sims 1986; Armulik et al. 2011). Moreover, pericyte dysfunction and/or absence has been implicated in the development of fibrosis, atherosclerosis, osteogenesis, tumor angiogenesis, dementia, and Alzheimer's disease (Armulik et al. 2011, 2010; Hellstrom et al. 2001; Sengillo et al. 2013; Sagare et al. 2013; Collett and Canfield 2005; Fang and Salven 2011; Henderson et al. 2013; Matthews et al. 2016). Despite these critical roles in various tissue physiology and disease, relatively little is known about cardiac pericytes. Roughly 1 in every 5000 cardiac-related publications mentions pericytes. Due to this gap in the literature and emerging importance of the vascular system in pathophysiology of cardiovascular diseases, cardiac pericyte biology has recently gained momentum and attention in the cardiac field.

This chapter will first focus on the current known properties of cardiac pericytes (cPC) in terms of origin, location, structure, and characterization. Next, we will discuss their physiological function and their role in cardiac homeostasis. Through their intimate relationship with endothelial cells, cPC participate in angiogenesis, are important in vascular integrity and perfusion, and promote vessel stability and vessel maturation. Cardiac pericytes have also been shown to play a part in hemodynamics by altering vascular tone (O'Farrell and Attwell 2014; O'Farrell et al. 2017). We will also dive into the cardiac pericyte's role in cardiovascular pathophysiology. Various cardiovascular diseases such as myocardial ischemia, fibrosis, calcification, atherosclerosis, and thrombosis all involve cPC at different stages of disease progression. From there, we will discuss their plasticity as they have been shown to be part of the regeneration and reparative process of a diseased heart as well as their use in cell therapy. Understanding cPC biology and their role under pathophysiological conditions could pave a path for development of novel therapies for various forms of cardiovascular diseases.

11.2 Properties of Cardiac Pericytes

11.2.1 Origin

Due to the vascularization of the heart, it has been said that pericytes are the second most abundant cell type after endothelial cells in the heart where the ratio of pericyte to endothelial cells is 1:2 or 1:3 (Nees et al. 2012a). Others have said that they occupy 5% of the total non-myocyte population (Pinto et al. 2016). Cardiac pericytes are of mesothelial origin and are derived from the mesenchymal angioblast during development (Kovacic et al. 2012; Kumar et al. 2017). Due to their origin, they are seen as vascular stem cells (Nees et al. 2012a). Through linage studies, Cai et al. elegantly showed that cPC originate from the epicardium (Cai et al. 2008a). However, the mechanism to how they are recruited to the heart is currently unknown.

11.2.2 Anatomical Location

Anatomical location is one of the main properties that distinguish pericytes from the closely related smooth muscle cells where both are mural cells derived from the mesenchymal angioblast (Kumar et al. 2017; Feng et al. 2011). As the heart pumps blood to feed the tissues of the body, the heart also feeds itself through its coronary arteries, and it is here where the bulk of the cPCs lie. There are adventitial pericytes that are found in the intima of arterioles (where the microvessel widens to the arteries) and venules (where the microvessels widens to the veins), and there are pericytes that are found on the capillaries (Campagnolo et al. 2010; Nees et al. 2012b; Klein et al. 2011; Corselli et al. 2012; Avolio et al. 2015a). In the heart, endocardial endothelial cells have been shown to be progenitors for cPCs and smooth muscle cells (Chen et al. 2016). Also specific to the heart, smooth muscle cells derive from pericyte progenitors (Volz et al. 2015). Smooth muscle cells sit on the higher arteries and veins where they regulate blood flow through vasoconstriction and vasodilation. Pericytes sit on the abluminal side of a vessel and envelop a single layer of endothelial cells. The unique location of the pericytes allows them to be the liaison between the endothelium and myocytes and the liaison between the intima and media in the higher vascular branches (Nees et al. 2012b).

11.2.3 Characterization

Morphologically, pericytes have a single round nucleus with extended processes that reach out and wrap around endothelial cells. Visually, one can spot a pericyte as a small "bump-on-a-log" on a microvessel. In culture, they are quite flat and less spindle-like compared to their smooth muscle cell counterpart. Their cellular architecture has been defined, but there is currently no one marker that can unequivocally identify pericytes; rather, multiple pericytic markers must be used. Phenotypically, the most popular and accepted markers are neural glial-2 (NG2), platelet-derived growth factor receptor beta (PDGFRβ), and alpha-smooth muscle actin (αSMA) (Armulik et al. 2010). These markers are expressed ten times higher in pericytes compared to smooth muscle cells which also express the same markers (Nees et al. 2013). Microvascular pericytes are usually NG2^{+} αSMA^{-}; venular pericytes are NG2^{-} αSMA^{+}; arteriolar pericytes are NG2^{+} αSMA^{-} (Wanjare et al. 2013). However, a recent study showed that microvascular pericytes are also αSMA^{+}; they were previously undetected due to the quick depolymerization of F-actin (Alarcon-Martinez et al. 2018). Other markers include tissue factor (TF), ganglioside 3g5 (3g5), vimentin, desmin, calponin, caldesmon, and alkaline phosphatase (Alk-p) (Nees et al. 2012a; Avolio et al. 2015a; Shepro and Morel 1993; Hughes and Chan-Ling 2004; Crisan et al. 2008; Juchem et al. 2010; Chen et al. 2015). Most recently, glioma-associated oncogene 1 (Gli1) and T-box protein 18 (Tbx18) have also been identified as molecular markers for pericytes (Kramann et al. 2015a, 2017; Guimaraes-Camboa et al. 2017). They have been reported to express stemness markers such as CD44, CD90, and CD105 (Chen et al. 2015; Covas et al. 2008) and genetically express homeobox transcription factor Nanog (NANOG), sex-determining region box (SOX2), and octamer-binding transcription factor 4 (OCT4) (Avolio et al. 2015a). They are also negative for endothelial markers such as CD31 and von Willebrand factor (vWF) as well as for hematopoietic markers such as CD45 (Nees et al. 2012a; Avolio et al. 2015a; Juchem et al. 2010; Chen et al. 2015). Regulator of G protein signaling 5 (RGS5) has been described as a specific marker for brain pericytes, but the progress of research in the heart is behind (Cho et al. 2003; Mitchell et al. 2008; Nisancioglu et al. 2008).

The homogeneity of pericytes in the heart (as well as in other tissues) is under debate. Pericytes have been isolated and characterized from different species and tissues, and their marker expression varies (Dias Moura Prazeres et al. 2017). To further complicate their characterization, their marker expression is also dynamic within the same tissue as an indication of their activation state and/or maturity (van Dijk et al. 2015). However, their functional characteristics are more homogeneous within the tissues. Functional characteristics of pericytes include migration toward endothelial platelet-derived growth factor bb (PDGFbb) signaling and the ability to mediate tubule formation in co-culture with endothelial cells (Chen et al. 2015; Zhou et al. 2016). Cardiac pericytes are pro-angiogenic and procoagulatory (Juchem et al. 2010). Cardiac pericytes will also connect quickly with other cells (endothelial or other pericytes) via gap junctions like connexin-43 (Cx43) as shown by cell dye transfer studies (Nees et al. 2012a). Enzymatically, they have been shown to have alkaline phosphatase activity (Juchem et al. 2010). However, few groups have been able to isolate and characterize cPCs, and each group found a different subtype. They all varied in origin, location, and molecular markers. Nees et al. used NG2, PDGFRβ, αSMA, calponin, Cx43, and 3g5 as molecular markers as well as functional characterization to identify their isolated pericyte population from eight different species (Nees et al. 2012a). Couple of years later, Chen et al. followed up with their primary isolation from human fetal and adult hearts. They isolated a population that was $NG2^+$ $CD146^+$ $CD34^-$ $CD45^-$ $CD56^-$ $CD117^-$ (Chen et al. 2015). They claim that this population can be cultivated in homogeny by FACS. Around the same time, Avolio et al. were also able to isolate and characterize isolated primary pericytes from human neonatal hearts. Their unique population consisted of pericytes expressing $CD34^+$ $NG2^+$ $CD146^-$ $cKit^-$. This population was found on both the arterioles and capillaries (Avolio et al. 2015a). Table 11.1 provides all of the markers that can be used to identify cPCs. Work remains to identify a cPC-specific marker that is homogenously expressed in all cPC subtypes. For now, identification of pericytes requires immunogenic markers, gene and protein expression profiles, functional characteristics, and localization in situ.

Table 11.1 Markers for cardiac pericytes (table adopted from Murray et al. 2017)

Gene symbol	Gene name	References
3G5	3g5-defined ganglioside	Nees et al. (2012a)
αSMA	Alpha-smooth muscle actin	Chen et al. (2015) and Crisan et al. (2008)
ALPL	Alkaline phosphatase	Juchem et al. (2010)
CD10	Neural endopeptidase	Crisan et al. (2008)
CD13	Alanine aminopeptidase	Crisan et al. (2008)
CD29	Integrin beta	Dar et al. (2012)
CD34	CD34 molecule	Chen et al. (2013)
CD44	Receptor for hyaluronic acid	Chen et al. (2015) and Crisan et al. (2008)
CD73	5′nucleotidase, ecto	Chen et al. (2015) and Crisan et al. (2008)
CD90	Thy-1	Chen et al. (2015) and Crisan et al. (2008)
CD105	Endoglin	Chen et al. (2015) and Crisan et al. (2008)
CD108	Sema L	Chen et al. (2015) and Crisan et al. (2008)
CD109	Platelet activation factor	Crisan et al. (2008)
CD140b	Platelet-derived growth factor beta (PDGFRβ)	Chen et al. (2015) and Crisan et al. (2008)
CD140a	Platelet-derived growth factor alpha (PDGFRα)	Psaltis et al. (2010)
CD142	Tissue factor	Juchem et al. (2010)
CD146	Melanoma cell adhesion molecule	Chen et al. (2015) and Crisan et al. (2008)
CD164	Sialomucin core protein 24	Crisan et al. (2008)
CD166	ALCAM	Crisan et al. (2008)
CD318	CUB domain-containing protein 1	Crisan et al. (2008)
CD340	Human epidermal growth factor receptor 2	Crisan et al. (2008)
CD349	Frizzled-9	Crisan et al. (2008)
CNN1	Calponin	Nees et al. (2012a)
Cx43	Connexin-43	Nees et al. (2012a)
Gli1	Glioma-associated oncogene 1	Kramann et al. (2015b, 2017)
HLA-CLI	Human leukocyte antigen class 1	Crisan et al. (2008)
NG2	Neurol/glial antigen 2	Chen et al. (2015) and Crisan et al. (2008)
SM-MHC	Smooth muscle myosin heavy chain	Chen et al. (2015)
SSEA-4	Stage-specific embryonic antigen-4	Crisan et al. (2008)
STRO-1	N/A	Psaltis et al. (2010)
Tbx18	T-BOX protein 18	Guimaraes-Camboa et al. (2017)
VIM	Vimentin	Avolio et al. (2015a)

11.3 Role in Cardiac Homeostasis

11.3.1 Vascular Integrity and Perfusion

To help maintain cardiac homeostasis, cPCs contribute to the integrity of the endothelial layer by regulating vascular permeability (Armulik et al. 2005). Pericytes and endothelial cells have an intimate relationship and together they form a functional intercommunicating unit. They share a basement membrane where multiple intracellular connections are made. The two cell types form gap junctions, tight junctions, and peg-socket contacts to communicate and strengthen the barrier (Fig. 11.1). These gap junctions are formed by proteins such as connexins, cadherins, occludins, and claudins that are integrated into the cytoskeletal structure (Winkler et al. 2011). In cell culture models, co-culture models of pericytes and endothelial cells show a decrease in barrier permeability compared to endothelial cells alone (Al Ahmad et al. 2009; Nakagawa et al. 2009). PDGFRβ is not only used as a pericyte molecular marker, it is also part of a signaling pathway essential to pericyte survival. As shown by Kaminski et al., PDGFRβ$^{-/-}$ mice are embryonically lethal. The embryos have vascular malformations and hemorrhaging (Hellstrom et al. 2001; Kaminski et al. 2001). In a conditional adult PDGFRβ$^{-/-}$ mouse, the mice exhibit high vascular permeability and leakage, and their perfusion is compromised (Bell et al. 2010; Winkler et al. 2010). Bjarnegard et al. showed that deletion of endothelial-derived PDGFbb causes pericyte loss and cardiac deficits (Bjarnegard et al. 2004). While studying the cardiotoxicity of cancer drugs, sunitinib malate decreased the number of pericyte coverage of vessels on the heart, caused microvascular dysfunction, decreased cardiac function (coronary flow), and induced hypertrophy (Chintalgattu et al. 2013). Interestingly, using an anticancer agent thalidomide reversed the physiological effects that they found. All together, these observations demonstrate that cPCs play a critical role which is vascular function and homeostasis in the heart.

11.3.2 Angiogenesis

Another physiological aspect is that pericytes participate in is angiogenesis (Gerhardt and Betsholtz 2003). They do so by presenting growth factors that stimulate the formation of new vessels (Stapor et al. 2014; Matsuki et al. 2015; Caporali et al. 2017). Under angiogenic conditions, the cross talk between endothelial cells and pericytes allows for both cell types to migrate and proliferate (Darland et al. 2003; Bergers and Song 2005; Ribatti et al. 2011; Fuxe et al. 2011). There have been several key processes that have been identified. First, angiopoietin-1/2 (angpt-1/2) signaling from endothelial cell binding to Tie-2 receptors on pericytes causes dissociation and aids in survival of pericytes from the vessels (Benest et al. 2006;

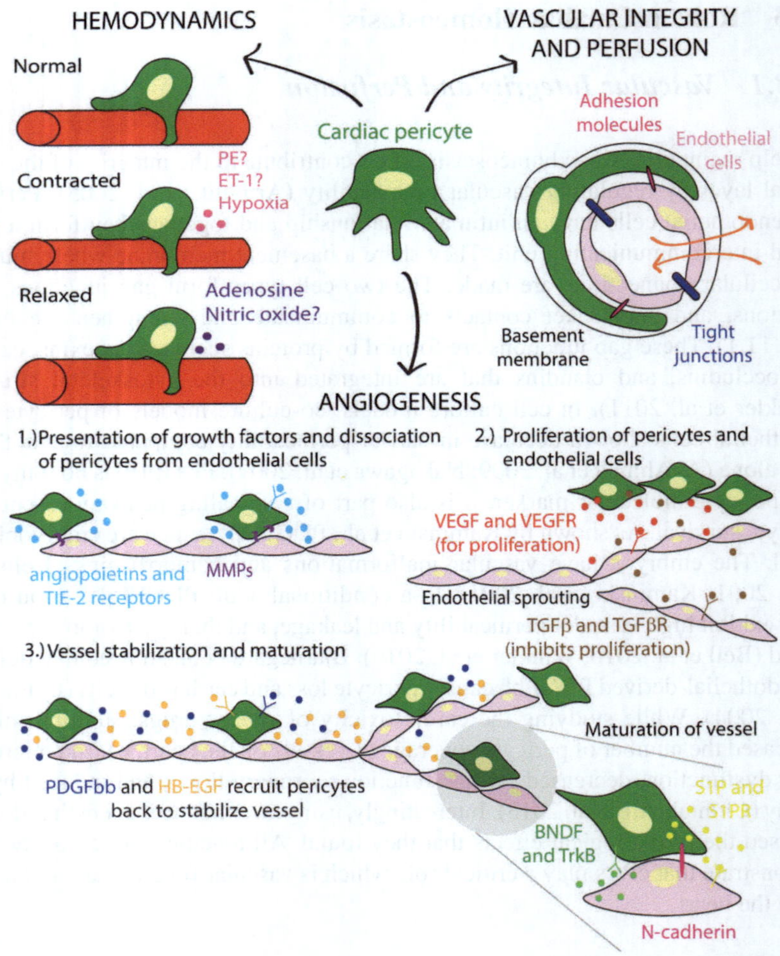

Fig. 11.1 Summary of the roles of cardiac pericytes in cardiac homeostasis. They are known to play a part in hemodynamics, vascular integrity, and perfusion and angiogenesis

Wakui et al. 2006; Cai et al. 2008b; Maisonpierre et al. 1997; Dewi et al. 2018). Matrix metalloproteinases (MMPs) and proteases alter the extracellular matrix (ECM) to allow for pericyte migration (Stratman et al. 2010; Stratman and Davis 2012; He and Spiro 1995). Pericyte-derived vascular endothelial growth factor A (VEGF-A) binding to endothelial VEGF receptor 2 (VEGFR2) allows for endothelial proliferation and survival and leads to sprouting and formation of new vessels (Darland et al. 2003; Benest et al. 2006; Wakui et al. 2006; Franco et al. 2011). Next, pericyte-derived transforming growth factor beta (TGFβ) is secreted to inhibit continuous proliferation of endothelial cells (Winkler et al. 2011; Bergers and Song 2005). Lastly, endothelial-derived PDGFbb and heparin-binding epidermal growth factor (HB-EGF) recruit pericytes back to stabilize the new vessels (Stratman et al. 2010; Magnusson et al. 2007; Nadal et al. 2002). Jagged1 expression and Notch-1 signaling are required for vessel maturation for both the newly proliferated endothelium and pericytes (Boscolo et al. 2011; Tattersall et al. 2016; Tao et al. 2017).

N-cadherin is an adhesion molecule that is required for maturation and stabilization of the newly sprouted vessel due to its interaction with pericytes (Tillet et al. 2005). To induce the expression of adhesion molecules such as N-cadherin where its absence would impede pericyte recruitment, pericyte sphingosine-1-phosphate (S1P) signaling is required to bind to the S1P receptor on endothelial cells (Paik et al. 2004; McGuire et al. 2011). Blocking N-cadherin expression decreased endothelial barrier properties. Found at the edges of sprouting vessels, pericyte-derived angpt-1 binds to the endothelial Tie-2 receptor to signal vessel stabilization, and pericytes spatially regulate sprouting due to their expression of VEGFR1 (Zeng et al. 2016; Eilken et al. 2017; Teichert et al. 2017). Once the vessels are stabilized, endothelial brain-derived neurotrophic factor (BDNF) binds to tropomyosin receptor kinase B (TrkB) on the pericytes for vessel maturation; however, the downstream signaling cascade is currently unknown. In the heart, only pericytes and smooth muscle cells express TrkB. Anastasia et al. showed that in trk$^{-/-}$ mice are embryonically lethal (Anastasia et al. 2014). The mice also had cardiac vascular abnormalities, increased vascular permeability, and compromised vascular integrity. Lastly, pericytes signal through chemokine receptor 3 (CXCR3) to regulate angiogenesis by inhibiting vessel formation and mediate vessel dissociation (Bodnar et al. 2013). As evidenced by the pathways discussed above, cross talk between endothelial and pericytes is required for angiogenesis in the heart (Fig. 11.1).

11.3.3 Vascular Tone

Scientists have hypothesized that pericytes can control capillary function and act as pre- and postcapillary sphincters due to their anatomical location and expression of contractile proteins similar to their close relatives—the smooth muscle cells. Multiple studies have investigated the contractile function of pericytes. Using optical and atomic force microscopy, retinal pericytes have been shown to generate contractile forces that can wrinkle elastomeric substrata in vitro studies (Lee et al. 2010). Renal pericytes have been shown intravitally that they regulate blood flow and pressure through endothelial-derived C-type natriuretic peptide (CNP) (Spiranec et al. 2018). Moreover, renal smooth muscle cells and pericytes have been shown to have calcium signals that are integrated across the vascular bed and control vascular tone (Borysova et al. 2013). Under in vitro conditions, brain pericytes in culture have been shown to contract by measuring their impedance changes due to dose-dependent responses to contraction and relaxation stimulants such as endothelin-1 and adenosine (Neuhaus et al. 2017). Calpain and talin are proteins that have been shown to play a part in pericyte contraction and stiffness (Kotecki et al. 2010). Multiple groups have contributed to showing that brain capillary pericytes do contract post-ischemia (Hamilton et al. 2010; Mazzoni et al. 2015; Hall et al. 2014; Mishra et al. 2014; Yemisci et al. 2009; Peppiatt et al. 2006; Fernandez-Klett et al. 2010). Using state-of-the-art intravital and ex vivo imaging techniques on the brain, they showed that pericytes contract during ischemia and even after reperfusion, the pericytes do not relax and they prevent flow back into the ischemic area (Hamilton et al. 2010; Mazzoni et al. 2015; Hall et al. 2014; Mishra et al. 2014; Yemisci et al. 2009; Peppiatt et al. 2006; Fernandez-Klett et al. 2010). This phenomenon is known as no-reflow (O'Farrell and Attwell 2014; Watanabe et al. 2004). Due to their tissue-specific properties, it was unknown if cPCs behave the same way during a myocardial infraction (MI) or under ischemic conditions. Most recently, O'Farrell et al. showed that pericytes can alternate vascular tone and cardiac hemodynamics by constricting capillaries post-MI and reperfusion (O'Farrell et al. 2017). Adenosine is known as a pericyte relaxant (Matsugi et al. 1997). Treating the mice with adenosine showed a 30% improvement in reperfusion in the infracted area. However, the mechanism to this no-reflow phenomenon in the heart remains to be elucidated. Due to the evidence provided by the study from O'Farrell et al., we are left wondering if cPCs have the ability to alter vascular tone under normal physiological conditions and not just under ischemic conditions (Fig. 11.1).

11.4 Role in Cardiovascular Pathophysiology

We have discussed how cPC help maintain cardiac homeostasis under normal physiological conditions, but what about their role in cardiac pathophysiology? Because cPC play multiple roles in cardiac homeostasis, their dysfunction could also contribute to cardiovascular pathologies. Cells respond to their environment such as mechanical and chemical stimuli, and any perturbations and/or stress can alter their molecular and biological functions that govern their normal physiological function. Reformation of their cell biology can render the cell to transform into subsidies contributing to cardiac pathogenesis. Dysfunctional cPC have been implicated in many cardiovascular diseases and pathologies such as myocardial infarction (MI), atherosclerosis, fibrosis, calcification, thrombosis, and inflammation (Fig. 11.2).

11.4.1 Myocardial Ischemic Disease

Myocardial ischemic disease develops when there is an occlusion to the coronary arteries most likely due to a blockage in the lumen of the vessel that is built up over time. The heart is deprived of oxygen, and the surrounding muscle becomes ischemic and eventually necrotic. Post-MI, the heart will remodel to try to compensate for the damaged tissue and this will eventually lead to a failing heart. As discussed earlier in their ability to control vascular tone under ischemic conditions, cPC play a role in the no-reflow phenomena during ischemia/reperfusion injury. Pericytes

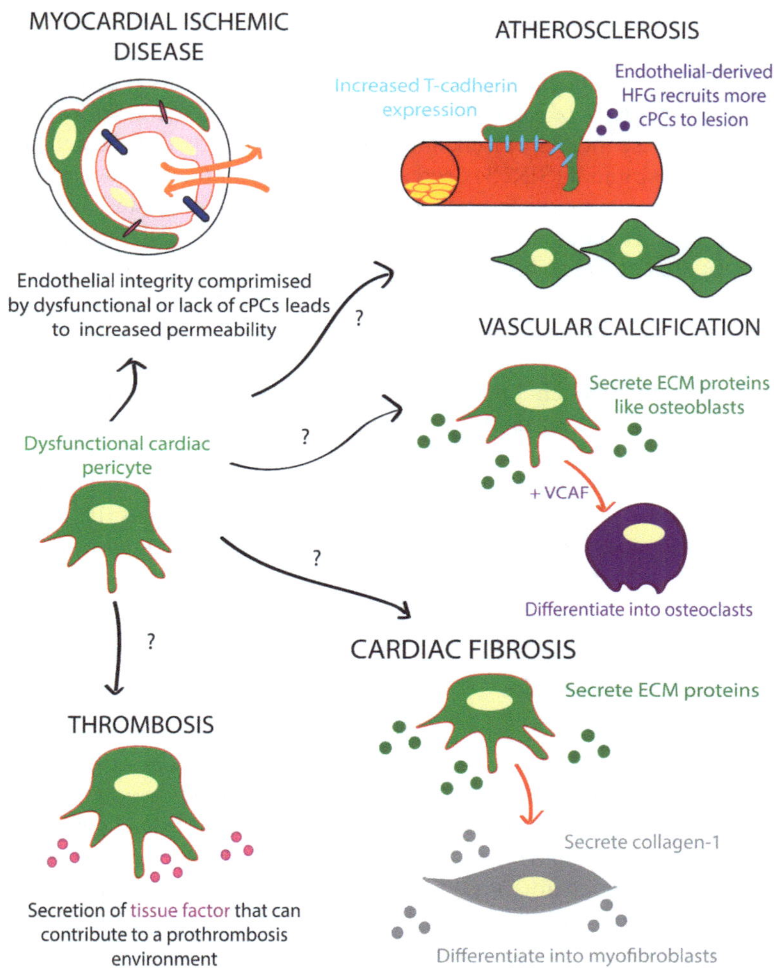

Fig. 11.2 Summary of the roles and implicated roles of cardiac pericytes in cardio pathophysiology. They have been shown to play a part in myocardial ischemic disease and have been implicated in atherosclerosis, thrombosis, vascular calcification, and fibrosis

contract and close off the microvessels that could potentially reperfuse the oxygen-deprived area due to the acute MI. Because of cPC constriction, inadequate blood flow is prevented from being reestablished (O'Farrell et al. 2017; Costa et al. 2018). Under hypoxic conditions, hypoxia-induced long noncoding RNAs were identified in pericytes, and HypERlnc was characterized to regulate pericyte cell proliferation, viability, and endothelial interaction. They also found that HypERlnc expression was significantly decreased in human cardiac tissue from heart failure patients (Bischoff et al. 2017). Moreover, in a human and mouse study, they found that pro-nerve growth factor (pro-NGF), an inflammatory cytokine, accumulates at the site of a MI secreted by cardiomyocytes and binds to neurotrophin receptor (p75 NTR) and sortilin-related VPS10 domain-containing receptor 2 (SorCS2) on cPCs which mediates a proapoptotic pathway (Siao et al. 2012). This eventually leads to the downstream effects of rescinded pericyte processes and increased vascular permeability. In mouse models of ischemia-reperfusion injury studies, genetic ablation of Notch-3 and sirtuin 3 (SIRT3) decreased the ability of the heart to recover post-MI because the proteins are important in microvascular homeostasis and function (Tao et al. 2017; He et al. 2016). Notch-3 knockout mice, which have been shown to be essential in vessel stability and maturation, showed a significant reduction in cPC vessel coverage leading to larger infarct sizes and higher mortality rate post-MI. The knockout mice also had a significant difference in cardiac function post-MI recovery compared to their wild-type counterparts due to their vascular dysfunction predisposed by their Notch-3 mutation (Tao et al. 2017). In another study utilizing a SIRT3 knockout mouse model, they also show that the genetic manipulation caused microvascular dysfunction where there was a decrease of cPC coverage of capillaries and exacerbated the MI injury. This study also showed that cardiac function post-MI in the KO was decreased compared to the WT animals (He et al. 2016). These studies in mouse model of ischemia-reperfusion injury demonstrate that cPCs contribute to microvascular dysfunction in myocardial ischemic disease.

11.4.2 Atherosclerosis

Atherosclerosis is defined as the narrowing and hardening of the vessel walls caused by fatty depositions in form of plaques which impede blood flow to tissues. In the heart, complete blockage in the coronary arteries leads to a MI, and cPCs have been implicated in the development and/or progression of atherosclerosis (Ivanova and Orekhov 2016). The atherosclerotic process causes changes in adhesion molecule expression during the different stages of lesion progression. T-cadherin is a surface adhesion molecule found in endothelial cells, smooth muscle cells, and pericytes of the aorta. The expression of T-cadherin in smooth muscle cells and pericytes is directly correlated with the severity of disease development (Ivanov et al. 2001). Secondly, human aortic endothelial cells and smooth muscle cells produce hepatocyte growth factor (HGF) (Nakamura et al. 1995). HGF has been implicated in the atherosclerotic process and shown to mediate pericyte migration in vitro. A study in human femoral atherosclerotic arteries shows that c-Met receptor expression on pericytes is activated by endothelial-derived HGF (Liu et al. 2007). Activation of c-Met by HGF triggers the PI3K/Akt pathway which mediates pericyte migration and recruitment to atherosclerotic lesions. Moreover, microvessels within the atherosclerotic lesions co-localized expression of c-Met with smooth muscle cells and pericytes (Liu et al. 2007). Finally, a mutation in the low-density lipoprotein receptor-related protein 6 (LRP) causes early atherosclerosis. This mutation increases the proliferation of smooth muscle cells by activating the PDGFRβ pathway, whereas in the normal functioning LRP6, PDGFRβ is targeted for degradation (Keramati et al. 2011). Because pericytes have high expression of PDGFRβ, they can be targeted by the same mechanism to cause early atherosclerosis. These studies taken together implicate cPCs in atherogenesis.

11.4.3 Calcification

Another cardiovascular pathology where dysfunctional pericytes contribute to its calcific vasculopathy, which is also known as vascular calcification, has been well reviewed (Collett and Canfield 2005; Canfield et al. 2000; Bardeesi et al. 2017; Leszczynska and Murphy 2018). There are two types: intimal vascular calcification which contributes to the complicated pathology of atherogenesis or medial vascular calcification which contributes to vascular stiffness (Wu et al. 2013). Under pathological conditions, pericytes can utilize its ability to secrete ECM proteins and its alkaline phosphatase activity to cause calcification by mirroring what osteoblasts do during bone formation (Murshed and McKee 2010). Under normal physiological conditions, pericytes synthesize and release pyrophosphatase that will inhibit calcification. Therefore, a dysfunctional pericyte can contribute to calcification by becoming chondrogenic (Collett and Canfield 2005; Crisan et al. 2008; Murshed and McKee 2010; Farrington-Rock et al. 2004). Furthermore, a mechanism has been described for pericyte chondrogenic differentiation involving the Wnt/β-catenin pathway in retinal pericytes (Kirton et al. 2007). Pericytes have been shown to express several Wnt receptors such as LDL-related receptor 5 and 6 (Keramati et al. 2011; Kirton et al. 2007). The TGFβ3 ligand activates the Wnt/β-catenin signaling pathway and induces chondrogenic characteristics such as accumulation of collagen and SOX-9 expression (Kirton et al. 2007). In the heart, TGFβ3 has been implicated in aortic atherosclerotic lesion development since several of the cell types (smooth muscle cells, foam cells, and macrophages) in the lesions have been found to secrete high levels of the TGFβ3 ligand (Bobik et al. 1999; Toma and McCaffrey 2012). Not only are pericytes chondrogenic, but they can also differentiate into osteoclasts (Collett and Canfield 2005; Crisan et al. 2008; Farrington-Rock et al. 2004; Kirton et al. 2006; Davaine et al. 2014). Glucocorticoid therapy such as treatment with dexamethasone enhances osteogenic differentiation and decreases the inhibition of mineralization in pericytes (Kirton et al. 2006). A novel gene, vascular calcification-associated factor (VCAF), is found to be upregulated during vascular calcification and during the osteogenic differentiation of pericytes in femoral arteries (Alexander et al. 2005). Calcified arterial lesions contain VCAF, while their non-calcified counterpart did not. On the other hand, knockdown of VCAF in cultured pericytes increased the amount of calcified lesions and the time it took for the mineralization to occur (Alexander et al. 2005). Together, these studies thus provide evidence where cPC can behave similarly in the development of calcific vasculopathy in the heart.

11.4.4 Cardiac Fibrosis

Cardiac fibrosis is defined as a reduction in cardiac muscle compliance due to an excessive deposition of ECM proteins which leads to myocardial stiffness and decreased cardiac function. Resident fibroblast maintains ECM homeostasis, but upon injury, they convert into myofibroblast and contribute to cardiac fibrosis. Cardiac fibrosis is necessary post-MI to help preserve cardiac structure, the scarring process, and healing (van Amerongen et al. 2008). However, if the process is not governed, it becomes pathogenic and consequential. How do pericytes contribute to cardiac fibrosis? The exact role and mechanism of pericytes in fibrosis still remain to be determined; however, recent studies in the heart and studies from other organs such as the kidney have shed some light (Travers et al. 2016). Most recently, a mouse model of spinal cord injury demonstrated that a subtype of pericytes was responsible for fibrosis and scar tissue formation. Inhibiting pericyte proliferation and their contribution to the ECM showed a reduction in fibrosis and facilitated healing (Dias et al. 2018). Under pathological conditions, dysfunctional pericytes can detach from the endothelium, migrate, differentiate into myofibroblasts, and compromise the integrity of the vasculature (Schrimpf and Duffield 2011; Greenhalgh et al. 2013). Cardiac pericytes have already been shown to secrete ECM proteins (Nees et al. 2012a). Under culture conditions, pericytes can become fibroblast-like cells by induction of type-1 collagen mRNA and protein expression as well as protein expression of connective tissue growth factor (CTGF/CCN2) and fibroblast marker ASO2 (Shiwen et al. 2009). Shown in vitro, fibroblast-like cells are derived from pericytes (Ivarsson et al. 1996). Moreover, studies have shown that pericytes can differentiate into fibrotic cells that secrete collagen-1 under inflammatory and wound healing conditions in vivo including the heart (Sundberg et al. 1996; Lin et al. 2008; Weiss et al. 2013). Galectin-3 is secreted by macrophages in inflamed myocardial tissue. This protein has been shown to stimulate pericyte proliferation and secrete procollagen I which can thus lead to fibrosis (McCullough et al. 2011). Conversely, type-1 pericytes accumulate at the site of injury in multiple organs, but they do not secrete collagen-I in the heart (Birbrair et al. 2014). Moreover, Gli1$^+$ cells were identified as a subpopulation pericytes in the mouse heart where they differentiate into myofibroblast and are a source of fibrotic tissue (Kramann et al. 2015b, 2017; Kim and Braun 2015). Interestingly, genetic ablation of Gli$^+$ pericytes decreased the observed fibrosis and recovered organ function. Future studies are required to resolve the conflicting results found in the heart.

11.4.5 Thrombosis

The development of a blood clot that can occlude and impede blood flow through the vascular system is known as a thrombosis, and a thrombosis usually happens when a vessel is injured. The blood clot is formed by thrombocytes to prevent further blood loss. The role of pericytes has yet to be elucidated in not just the heart but other organs as well. In the heart, a subset of pericytes discovered by Juchem et al. secrete high levels of TF which is a key initiator in coagulation (Juchem et al. 2010). Interestingly, pericytes are the only cells in the vascular wall secreting TF, whereas smooth muscle cells and monocytes that migrate to the area of plaques don't secrete TF unless activated (Osterud and Bjorklid 2006). Endothelial cells also do not secrete TF. This study suggests that pericytes can secrete procoagulatory factors that can create a thrombotic environment and can lead to plaque formation (Juchem et al. 2010). Saphenous veins are used in the transplantation of coronary artery bypass grafting. However, there is a high rate of acute occlusion post-surgery due to the treatment of the grafts that leaves them de-endothelialized and causes an increase in the release of TF (Weiss et al. 2009). Therefore, grafted vessels fail due to pro-thrombosis environment that is created partially due to intimal pericytes.

11.5 Therapeutic Potential

11.5.1 Repair and Regeneration

During a MI, the coronary artery is blocked. This will cause a deprivation of oxygen to the area; the myocytes and vessels downstream of the blockage will experience ischemia and die. The heart will therefore have to work harder to compensate for the damaged muscle. In order to heal this damage, first there needs to be repair—meaning myofibroblast needs to perform fibrosis to help retain structure by replacing the necrotic tissue. As discussed previously, pericytes can differentiate into myofibroblasts to aid in preserving structure (Crisan et al. 2008; Kramann et al. 2015b, 2017; Sundberg et al. 1996; Lin et al. 2008; Kim and Braun 2015). Next, there needs to be revascularization to perfuse the ischemic zones as the ischemic vessels are no longer able to deliver oxygen and nutrient. As discussed above, pericytes play a central role in angiogenesis and, in this case, even collateralization. Lastly, there needs to be regeneration because new muscle mass needs to be formed to keep the heart from becoming hypertrophic due to ventricular remodeling triggered by the acute MI. Pericytes can differentiate into myoblast and provide the foundation for new muscle development (Crisan et al. 2008; Dellavalle et al. 2007; Stallcup 2013; Cappellari et al. 2013). Therefore, pericytes are able to address all of the above requirements for cardiac repair and regeneration and have high therapeutic potential.

11.5.2 Mesenchymal Properties

Mesenchymal stem cells (MSCs) are at the forefront of regenerative therapy. Pericytes share several characteristics with MSCs, and their plasticity toward other cell phenotypes implies great healing potential. In addition to pericytes being of mesenchymal origin and the characterized subpopulations of isolated pericytes expressing “stemness” markers (CD44, CD73, CD90, CD105), pericytes are highly proliferative and have multilineage capabilities as discussed in previous sections (Volz et al. 2015; Crisan et al. 2008, 2012; Covas et al. 2008; Tintut et al. 2003). Interestingly, pericytes are also integral to stem cell niches and hematopoietic stem cells (Crisan et al. 2008; Sa da Bandeira et al. 2017). This brings us to the pericyte-MSC conundrum. Some believe that all MSCs are pericytes as it has been shown that human pluripotent stem cells gave rise to pericytes (Dar et al. 2012). Pericytes have also been referred to as a vascular stem cell or vascular progenitor cells. Whether they are pluripotent or progenitors or both have yet to be elucidated, but their potential has been rigorously reviewed and discussed (Geevarghese and Herman 2014; Wong et al. 2015; Murray et al. 2017). Another way pericytes have healing potential is their secretome which is also similar to stem cells. Many ways that pericytes contribute to vascular remodeling and stabilization are through their paracrine signaling rather than their direct physical contact (Ellison-Hughes and Madeddu 2017). Their secretome has been found to contain pro-angiogenic factors and anti-inflammatory cytokines as well as a plethora of other enzymes and factors that can contribute to healing and regeneration (Juchem et al. 2010; Chen et al. 2015). More specifically, pericytes are targeted for angiogenic therapies because they are bi-directional (Stallcup 2013; Kelly-Goss et al. 2014). In diseases such as cancer, inhibition of angiogenesis would be beneficial, while in cases like MI, promoting angiogenesis would help improve cardiac function. However, Guimaraes-Camboa et al. reported the opposite—pericytes in vivo do not behave like MSCs under normal physiological nor pathophysiological conditions (Guimaraes-Camboa et al. 2017).

11.5.3 Cell-Based Therapeutics

There have been several studies using pericyte-like cells and pericytes as cell-based therapeutics for regenerative medicine. Many of these studies were done by a group based in the United Kingdom led by Paolo Madeddu where they utilized leftover human saphenous vein grafts from coronary bypass surgeries to isolate pericyte-like progenitor cells. The first study used human saphenous vein-derived pericyte progenitor cells (SVPs) injected into the ischemic limb of immunodeficient mice, and they showed improvement in blood flow due to reparative angiogenesis (Campagnolo et al. 2010). Several years later, they transplanted human adventitial progenitor cells and showed improved reperfusion in the same ischemic limb model along with an epigenetic expression profile (Gubernator et al. 2015). The same group also transplanted SVPs into the infract zone of mouse MI model. The study showed that the transplantation reduced scarring, apoptosis, and fibrosis; it also increased angiogenesis through a mechanism involving microRNA-132 as well as improved overall cardiac function (Katare et al. 2011; Katare and Madeddu 2013). Due to the complexity of the heart, the group also did a study using a combined therapy of SVPs and cardiac stem cells to see if they could optimize their treatment. They found the combination therapy to be complimentary by stimulating repair of vessels and muscle. As previous studies, they saw improvement in cardiac function and perfusion, increased angiogenesis, and reduced infarct size (Avolio et al. 2015b). In a study where human pluripotent stem cells were transformed into pericytes and injected into an ischemic limb mouse model, they found that the pericytes significantly improved limb perfusion and vascular density (Dar et al. 2012). Next, Chen et al. investigated the therapeutic potential of purified human skeletal pericytes on mouse model of MI. They found that their transplantation improved cardiac contractility, increased angiogenesis, reduced fibrosis, decreased inflammation at the infract zones, and attenuated left ventricular dilation (Chen et al. 2013). Within that same year, Yannarelli et al. came out with a study showing the reparative ability of human umbilical cord perivascular cells improving cardiac function in an immunodeficient mouse model of MI (Yannarelli et al. 2013). Most recently, a study was published where they intramyocardially injected both human and swine allogeneic pericytes post-acute MI in a swine model. They found that transplantation of human pericytes only decreased fibrosis and no other endpoints they measured due to an immunological response. As an alternative, the transplantation of swine allogeneic pericytes resulted in an increase in angiogenesis and decrease in fibrosis, but did not improve overall cardiac function (Alvino et al. 2018). The first and only study using cPC as a therapy was done by Chen and colleagues. They isolated human cPCs and injected them into mouse models of MI, and they found that the cPCs demonstrated cardiomyogenic potential (Chen et al. 2015). This could be because the cells are derived from a cardiac source. As the popularity of using pericytes as cell-based therapies for cardiovascular regenerative medicine continues to grow, the findings from the studies above are encouraging and demonstrate its feasibility.

11.6 Future Directions

In the past 15–20 years, research on cPCs have accelerated. With the latest knowledge gained from recent studies, the compilation of cPC biology is beginning to unravel. We know that cPCs play a fundamental role in vascular integrity, stability, and remodeling, and their dysfunction proves consequential to global cardiac function. However, plenty of questions remain to be answered. As we saw in Table 11.1, there are numerous markers found in cPCs but with no specificity. The current publications that have isolated cPCs have different expression profiles. Under in vivo conditions, this difference in expression profile could be due to their localization along the vessel, whether they are microvascular, arteriole, or venular. This begs us to answer “What is a cardiac pericyte?” In order to truly understand cPC biology as a whole and differentiate among the subtypes, we need to rigorously define them.

There are plenty of functions pericytes have been described to participate in, but do they have a main role? Most of the signaling and functions have translated well across different tissues. Because of their heterogeneity, do the subtypes interact with endothelial cells and influence cardiac homeostasis differently? It’s possible that each subtype has a different role due to where they are found in the vascular bed. We have discussed how pericyte dysfunction and/or lack of pericytes can disrupt and decrease cardiac efficiency and function and contribute to cardiac pathophysiology. However, different subtypes of cPCs can contribute to each pathology differently. For example, type-1 pericytes in the heart do not secrete collagen-1 and contribute to cardiac fibrosis, but perhaps another subpopulation does. Moreover, the pericyte-MSC conundrum needs to be resolved. Although studies have started on targeting pericytes as a cell-based therapeutic agent and have seen improvements in the ischemic limb and MI mouse models, we wonder if the regenerative capabilities are different among the different subtypes, tissue they are derived from, or species they are derived from. For example, most studies have used pericytes derived from human saphenous vein grafts and transplanted into the hearts of mice. Would we see a bigger difference if the cells came from mice or cardiac derived? In the swine model, there was an immunoreaction to the human cells. In the study where they transplanted human cPCs into the heart, they showed that the pericytes had myogenic potential. It is possible that this subpopulation of pericytes was preprogrammed to gravitate toward a myogenic phenotype. We briefly mentioned that the regenerative potential of pericytes come from their paracrine signaling. Perhaps instead of transplanting pericytes themselves, therapies can just include the secretome of pericytes. It's possible that this could restore the signaling pathways that were disrupted due to pericyte dysfunction or death associated with a particular disease. There are many directions where the field of cPC biology can take, but we believe that we must first critically and rigorously define cardiac pericytes and their subpopulations before we are able to truly draw any conclusions on their capabilities as a therapeutic agent.

References

Al Ahmad A, Gassmann M, Ogunshola OO (2009) Maintaining blood-brain barrier integrity: pericytes perform better than astrocytes during prolonged oxygen deprivation. J Cell Physiol 218(3):612–622

Alarcon-Martinez L et al (2018) Capillary pericytes express alpha-smooth muscle actin, which requires prevention of filamentous-actin depolymerization for detection. Elife 7

Alexander MY et al (2005) Identification and characterization of vascular calcification-associated factor, a novel gene upregulated during vascular calcification in vitro and in vivo. Arterioscler Thromb Vasc Biol 25(9):1851–1857

Alvino VV et al (2018) Transplantation of allogeneic pericytes improves myocardial vascularization and reduces interstitial fibrosis in a swine model of reperfused acute myocardial infarction. J Am Heart Assoc 7(2)

Anastasia A et al (2014) Trkb signaling in pericytes is required for cardiac microvessel stabilization. PLoS One 9(1):e87406

Armulik A, Abramsson A, Betsholtz C (2005) Endothelial/pericyte interactions. Circ Res 97(6):512–523

Armulik A et al (2010) Pericytes regulate the blood-brain barrier. Nature 468(7323):557–561

Armulik A, Genove G, Betsholtz C (2011) Pericytes: developmental, physiological, and pathological perspectives, problems, and promises. Dev Cell 21(2):193–215

Avolio E et al (2015a) Expansion and characterization of neonatal cardiac pericytes provides a novel cellular option for tissue engineering in congenital heart disease. J Am Heart Assoc 4(6):e002043
Avolio E et al (2015b) Combined intramyocardial delivery of human pericytes and cardiac stem cells additively improves the healing of mouse infarcted hearts through stimulation of vascular and muscular repair. Circ Res 116(10):e81–e94
Bardeesi ASA et al (2017) A novel role of cellular interactions in vascular calcification. J Transl Med 15(1):95
Bell RD et al (2010) Pericytes control key neurovascular functions and neuronal phenotype in the adult brain and during brain aging. Neuron 68(3):409–427
Benest AV et al (2006) VEGF and angiopoietin-1 stimulate different angiogenic phenotypes that combine to enhance functional neovascularization in adult tissue. Microcirculation 13(6):423–437
Bergers G, Song S (2005) The role of pericytes in blood-vessel formation and maintenance. Neuro Oncol 7(4):452–464
Birbrair A et al (2014) Type-1 pericytes accumulate after tissue injury and produce collagen in an organ-dependent manner. Stem Cell Res Ther 5(6):122
Bischoff FC et al (2017) Identification and functional characterization of hypoxia-induced endoplasmic reticulum stress regulating lncRNA (HypERlnc) in Pericytes. Circ Res 121(4):368–375
Bjarnegard M et al (2004) Endothelium-specific ablation of PDGFB leads to pericyte loss and glomerular, cardiac and placental abnormalities. Development 131(8):1847–1857
Bobik A et al (1999) Distinct patterns of transforming growth factor-beta isoform and receptor expression in human atherosclerotic lesions. Colocalization implicates TGF-beta in fibrofatty lesion development. Circulation 99(22):2883–2891
Bodnar RJ et al (2013) Pericyte regulation of vascular remodeling through the CXC receptor 3. Arterioscler Thromb Vasc Biol 33(12):2818–2829
Borysova L et al (2013) How calcium signals in myocytes and pericytes are integrated across in situ microvascular networks and control microvascular tone. Cell Calcium 54(3):163–174
Boscolo E et al (2011) JAGGED1 signaling regulates hemangioma stem cell-to-pericyte/vascular smooth muscle cell differentiation. Arterioscler Thromb Vasc Biol 31(10):2181–2192
Cai CL et al (2008a) A myocardial lineage derives from Tbx18 epicardial cells. Nature 454(7200):104–108
Cai J et al (2008b) The angiopoietin/Tie-2 system regulates pericyte survival and recruitment in diabetic retinopathy. Invest Ophthalmol Vis Sci 49(5):2163–2171
Campagnolo P et al (2010) Human adult vena saphena contains perivascular progenitor cells endowed with clonogenic and proangiogenic potential. Circulation 121(15):1735–1745
Canfield AE et al (2000) Role of pericytes in vascular calcification: a review. Z Kardiol 89(Suppl 2):20–27
Caporali A et al (2017) Contribution of pericyte paracrine regulation of the endothelium to angiogenesis. Pharmacol Ther 171:56–64
Cappellari O et al (2013) Dll4 and PDGF-BB convert committed skeletal myoblasts to pericytes without erasing their myogenic memory. Dev Cell 24(6):586–599
Chen CW et al (2013) Human pericytes for ischemic heart repair. Stem Cells 31(2):305–316
Chen WC et al (2015) Human myocardial pericytes: multipotent mesodermal precursors exhibiting cardiac specificity. Stem Cells 33(2):557–573
Chen Q et al (2016) Endothelial cells are progenitors of cardiac pericytes and vascular smooth muscle cells. Nat Commun 7:12422
Chintalgattu V et al (2013) Coronary microvascular pericytes are the cellular target of sunitinib malate-induced cardiotoxicity. Sci Transl Med 5(187):187ra69
Cho H et al (2003) Pericyte-specific expression of Rgs5: implications for PDGF and EDG receptor signaling during vascular maturation. FASEB J 17(3):440–442
Collett GD, Canfield AE (2005) Angiogenesis and pericytes in the initiation of ectopic calcification. Circ Res 96(9):930–938

Corselli M et al (2012) The tunica adventitia of human arteries and veins as a source of mesenchymal stem cells. Stem Cells Dev 21(8):1299–1308
Costa MA et al (2018) Pericytes constrict blood vessels after myocardial ischemia. J Mol Cell Cardiol 116:1–4
Covas DT et al (2008) Multipotent mesenchymal stromal cells obtained from diverse human tissues share functional properties and gene-expression profile with CD146+ perivascular cells and fibroblasts. Exp Hematol 36(5):642–654
Crisan M et al (2008) Purification and long-term culture of multipotent progenitor cells affiliated with the walls of human blood vessels: myoendothelial cells and pericytes. Methods Cell Biol 86:295–309
Crisan M et al (2012) Perivascular cells for regenerative medicine. J Cell Mol Med 16(12):2851–2860
Dar A et al (2012) Multipotent vasculogenic pericytes from human pluripotent stem cells promote recovery of murine ischemic limb. Circulation 125(1):87–99
Darland DC et al (2003) Pericyte production of cell-associated VEGF is differentiation-dependent and is associated with endothelial survival. Dev Biol 264(1):275–288
Davaine JM et al (2014) Osteoprotegerin, pericytes and bone-like vascular calcification are associated with carotid plaque stability. PLoS One 9(9):e107642
Dellavalle A et al (2007) Pericytes of human skeletal muscle are myogenic precursors distinct from satellite cells. Nat Cell Biol 9(3):255–267
Dewi NA et al (2018) Mechanism of retinal pericyte migration through angiopoietin/Tie-2 signaling pathway on diabetic rats. Int J Ophthalmol 11(3):375–381
Dias Moura Prazeres PH et al (2017) Pericytes are heterogeneous in their origin within the same tissue. Dev Biol 427(1):6–11
Dias DO et al (2018) Reducing pericyte-derived scarring promotes recovery after spinal cord injury. Cell 173(1):153–165. e22
Eilken HM et al (2017) Pericytes regulate VEGF-induced endothelial sprouting through VEGFR1. Nat Commun 8(1):1574
Ellison-Hughes GM, Madeddu P (2017) Exploring pericyte and cardiac stem cell secretome unveils new tactics for drug discovery. Pharmacol Ther 171:1–12
Fang S, Salven P (2011) Stem cells in tumor angiogenesis. J Mol Cell Cardiol 50(2):290–295
Farrington-Rock C et al (2004) Chondrogenic and adipogenic potential of microvascular pericytes. Circulation 110(15):2226–2232
Feng J et al (2011) Dual origin of mesenchymal stem cells contributing to organ growth and repair. Proc Natl Acad Sci U S A 108(16):6503–6508
Fernandez-Klett F et al (2010) Pericytes in capillaries are contractile in vivo, but arterioles mediate functional hyperemia in the mouse brain. Proc Natl Acad Sci U S A 107(51):22290–22295
Franco M et al (2011) Pericytes promote endothelial cell survival through induction of autocrine VEGF-A signaling and Bcl-w expression. Blood 118(10):2906–2917
Fuxe J et al (2011) Pericyte requirement for anti-leak action of angiopoietin-1 and vascular remodeling in sustained inflammation. Am J Pathol 178(6):2897–2909
Geevarghese A, Herman IM (2014) Pericyte-endothelial crosstalk: implications and opportunities for advanced cellular therapies. Transl Res 163(4):296–306
Gerhardt H, Betsholtz C (2003) Endothelial-pericyte interactions in angiogenesis. Cell Tissue Res 314(1):15–23
Greenhalgh SN, Iredale JP, Henderson NC (2013) Origins of fibrosis: pericytes take Centre stage. F1000Prime Rep 5:37
Gubernator M et al (2015) Epigenetic profile of human adventitial progenitor cells correlates with therapeutic outcomes in a mouse model of limb ischemia. Arterioscler Thromb Vasc Biol 35(3):675–688
Guimaraes-Camboa N et al (2017) Pericytes of multiple organs do not behave as mesenchymal stem cells in vivo. Cell Stem Cell 20(3):345–359. e5
Hall CN et al (2014) Capillary pericytes regulate cerebral blood flow in health and disease. Nature 508(7494):55–60

Hamilton NB, Attwell D, Hall CN (2010) Pericyte-mediated regulation of capillary diameter: a component of neurovascular coupling in health and disease. Front Neuroenergetics 2

He Q, Spiro MJ (1995) Isolation of rat heart endothelial cells and pericytes: evaluation of their role in the formation of extracellular matrix components. J Mol Cell Cardiol 27(5):1173–1183

He X, Zeng H, Chen JX (2016) Ablation of SIRT3 causes coronary microvascular dysfunction and impairs cardiac recovery post myocardial ischemia. Int J Cardiol 215:349–357

Hellstrom M et al (2001) Lack of pericytes leads to endothelial hyperplasia and abnormal vascular morphogenesis. J Cell Biol 153(3):543–553

Henderson NC et al (2013) Targeting of alphav integrin identifies a core molecular pathway that regulates fibrosis in several organs. Nat Med 19(12):1617–1624

Hughes S, Chan-Ling T (2004) Characterization of smooth muscle cell and pericyte differentiation in the rat retina in vivo. Invest Ophthalmol Vis Sci 45(8):2795–2806

Ivanov D et al (2001) Expression of cell adhesion molecule T-cadherin in the human vasculature. Histochem Cell Biol 115(3):231–242

Ivanova EA, Orekhov AN (2016) Cellular model of Atherogenesis based on pluripotent vascular wall pericytes. Stem Cells Int 2016:7321404

Ivarsson M et al (1996) Recruitment of type I collagen producing cells from the microvasculature in vitro. Exp Cell Res 229(2):336–349

Juchem G et al (2010) Pericytes in the macrovascular intima: possible physiological and pathogenetic impact. Am J Physiol Heart Circ Physiol 298(3):H754–H770

Kaminski WE et al (2001) Basis of hematopoietic defects in platelet-derived growth factor (PDGF)-B and PDGF beta-receptor null mice. Blood 97(7):1990–1998

Katare RG, Madeddu P (2013) Pericytes from human veins for treatment of myocardial ischemia. Trends Cardiovasc Med 23(3):66–70

Katare R et al (2011) Transplantation of human pericyte progenitor cells improves the repair of infarcted heart through activation of an angiogenic program involving micro-RNA-132. Circ Res 109(8):894–906

Kelly-Goss MR et al (2014) Targeting pericytes for angiogenic therapies. Microcirculation 21(4):345–357

Keramati AR et al (2011) Wild-type LRP6 inhibits, whereas atherosclerosis-linked LRP6R611C increases PDGF-dependent vascular smooth muscle cell proliferation. Proc Natl Acad Sci U S A 108(5):1914–1918

Kim J, Braun T (2015) Targeting the cellular origin of organ fibrosis. Cell Stem Cell 16(1):3–4

Kirton JP et al (2006) Dexamethasone downregulates calcification-inhibitor molecules and accelerates osteogenic differentiation of vascular pericytes: implications for vascular calcification. Circ Res 98(10):1264–1272

Kirton JP et al (2007) Wnt/beta-catenin signaling stimulates chondrogenic and inhibits adipogenic differentiation of pericytes: potential relevance to vascular disease? Circ Res 101(6):581–589

Klein D et al (2011) Vascular wall-resident CD44+ multipotent stem cells give rise to pericytes and smooth muscle cells and contribute to new vessel maturation. PLoS One 6(5):e20540

Kotecki M et al (2010) Calpain- and Talin-dependent control of microvascular pericyte contractility and cellular stiffness. Microvasc Res 80(3):339–348

Kovacic JC et al (2012) Epithelial-to-mesenchymal and endothelial-to-mesenchymal transition: from cardiovascular development to disease. Circulation 125(14):1795–1808

Kramann R et al (2015a) Pharmacological GLI2 inhibition prevents myofibroblast cell-cycle progression and reduces kidney fibrosis. J Clin Invest 125(8):2935–2951

Kramann R et al (2015b) Perivascular Gli1+ progenitors are key contributors to injury-induced organ fibrosis. Cell Stem Cell 16(1):51–66

Kramann R et al (2017) Gli1(+) Pericyte loss induces capillary rarefaction and proximal tubular injury. J Am Soc Nephrol 28(3):776–784

Kumar A et al (2017) Specification and diversification of Pericytes and smooth muscle cells from Mesenchymoangioblasts. Cell Rep 19(9):1902–1916

Lee S et al (2010) Pericyte actomyosin-mediated contraction at the cell-material interface can modulate the microvascular niche. J Phys Condens Matter 22(19):194115

Leszczynska A, Murphy JM (2018) Vascular calcification: is it rather a stem/progenitor cells driven phenomenon? Front Bioeng Biotechnol 6:10

Lin SL et al (2008) Pericytes and perivascular fibroblasts are the primary source of collagen-producing cells in obstructive fibrosis of the kidney. Am J Pathol 173(6):1617–1627

Liu Y et al (2007) Hepatocyte growth factor and c-met expression in pericytes: implications for atherosclerotic plaque development. J Pathol 212(1):12–19

Magnusson PU et al (2007) Platelet-derived growth factor receptor-beta constitutive activity promotes angiogenesis in vivo and in vitro. Arterioscler Thromb Vasc Biol 27(10):2142–2149

Maisonpierre PC et al (1997) Angiopoietin-2, a natural antagonist for Tie2 that disrupts in vivo angiogenesis. Science 277(5322):55–60

Matsugi T, Chen Q, Anderson DR (1997) Adenosine-induced relaxation of cultured bovine retinal pericytes. Invest Ophthalmol Vis Sci 38(13):2695–2701

Matsuki M et al (2015) Ninjurin1 is a novel factor to regulate angiogenesis through the function of pericytes. Circ J 79(6):1363–1371

Matthews BG et al (2016) Osteogenic potential of alpha smooth muscle actin expressing muscle resident progenitor cells. Bone 84:69–77

Mazzoni J, Cutforth T, Agalliu D (2015) Dissecting the role of smooth muscle cells versus Pericytes in regulating cerebral blood flow using in vivo optical imaging. Neuron 87(1):4–6

McCullough PA, Olobatoke A, Vanhecke TE (2011) Galectin-3: a novel blood test for the evaluation and management of patients with heart failure. Rev Cardiovasc Med 12(4):200–210

McGuire PG et al (2011) Pericyte-derived sphingosine 1-phosphate induces the expression of adhesion proteins and modulates the retinal endothelial cell barrier. Arterioscler Thromb Vasc Biol 31(12):e107–e115

Mishra A et al (2014) Imaging pericytes and capillary diameter in brain slices and isolated retinae. Nat Protoc 9(2):323–336

Mitchell TS et al (2008) RGS5 expression is a quantitative measure of pericyte coverage of blood vessels. Angiogenesis 11(2):141–151

Murray IR et al (2017) Skeletal and cardiac muscle pericytes: functions and therapeutic potential. Pharmacol Ther 171:65–74

Murshed M, McKee MD (2010) Molecular determinants of extracellular matrix mineralization in bone and blood vessels. Curr Opin Nephrol Hypertens 19(4):359–365

Nadal JA et al (2002) Angiotensin II stimulates migration of retinal microvascular pericytes: involvement of TGF-beta and PDGF-BB. Am J Physiol Heart Circ Physiol 282(2):H739–H748

Nakagawa S et al (2009) A new blood-brain barrier model using primary rat brain endothelial cells, pericytes and astrocytes. Neurochem Int 54(3–4):253–263

Nakamura Y et al (1995) Expression of local hepatocyte growth factor system in vascular tissues. Biochem Biophys Res Commun 215(2):483–488

Nees S et al (2012a) Isolation, bulk cultivation, and characterization of coronary microvascular pericytes: the second most frequent myocardial cell type in vitro. Am J Physiol Heart Circ Physiol 302(1):H69–H84

Nees S et al (2012b) Wall structures of myocardial precapillary arterioles and postcapillary venules reexamined and reconstructed in vitro for studies on barrier functions. Am J Physiol Heart Circ Physiol 302(1):H51–H68

Nees S et al (2013) Abundant Pericytes in the venous intima and the vasa Venarum: evidence for their key role in venous thrombosis. J Vasc Surg Venous Lymphat Disord 1(1):113

Neuhaus AA et al (2017) Novel method to study pericyte contractility and responses to ischaemia in vitro using electrical impedance. J Cereb Blood Flow Metab 37(6):2013–2024

Nisancioglu MH et al (2008) Generation and characterization of rgs5 mutant mice. Mol Cell Biol 28(7):2324–2331

O'Farrell FM, Attwell D (2014) A role for pericytes in coronary no-reflow. Nat Rev Cardiol 11(7):427–432

O'Farrell FM et al (2017) Capillary pericytes mediate coronary no-reflow after myocardial ischaemia. Elife 6

Osterud B, Bjorklid E (2006) Sources of tissue factor. Semin Thromb Hemost 32(1):11–23
Paik JH et al (2004) Sphingosine 1-phosphate receptor regulation of N-cadherin mediates vascular stabilization. Genes Dev 18(19):2392–2403
Peppiatt CM et al (2006) Bidirectional control of CNS capillary diameter by pericytes. Nature 443(7112):700–704
Pinto AR et al (2016) Revisiting cardiac cellular composition. Circ Res 118(3):400–409
Psaltis PJ et al (2010) Enrichment for STRO-1 expression enhances the cardiovascular paracrine activity of human bone marrow-derived mesenchymal cell populations. J Cell Physiol 223(2):530–540
Ribatti D, Nico B, Crivellato E (2011) The role of pericytes in angiogenesis. Int J Dev Biol 55(3):261–268
Sa da Bandeira D, Casamitjana J, Crisan M (2017) Pericytes, integral components of adult hematopoietic stem cell niches. Pharmacol Ther 171:104–113
Sagare AP et al (2013) Pericyte loss influences Alzheimer-like neurodegeneration in mice. Nat Commun 4:2932
Schrimpf C, Duffield JS (2011) Mechanisms of fibrosis: the role of the pericyte. Curr Opin Nephrol Hypertens 20(3):297–305
Sengillo JD et al (2013) Deficiency in mural vascular cells coincides with blood-brain barrier disruption in Alzheimer's disease. Brain Pathol 23(3):303–310
Shepro D, Morel NM (1993) Pericyte physiology. FASEB J 7(11):1031–1038
Shiwen X et al (2009) Pericytes display increased CCN2 expression upon culturing. J Cell Commun Signal 3(1):61–64
Siao CJ et al (2012) ProNGF, a cytokine induced after myocardial infarction in humans, targets pericytes to promote microvascular damage and activation. J Exp Med 209(12):2291–2305
Sims DE (1986) The pericyte—a review. Tissue Cell 18(2):153–174
Spiranec K et al (2018) Endothelial C-type natriuretic peptide acts on Pericytes to regulate microcirculatory flow and blood pressure. Circulation 138(5):494–508
Stallcup WB (2013) Bidirectional myoblast-pericyte plasticity. Dev Cell 24(6):563–564
Stapor PC et al (2014) Pericyte dynamics during angiogenesis: new insights from new identities. J Vasc Res 51(3):163–174
Stratman AN, Davis GE (2012) Endothelial cell-pericyte interactions stimulate basement membrane matrix assembly: influence on vascular tube remodeling, maturation, and stabilization. Microsc Microanal 18(1):68–80
Stratman AN et al (2010) Endothelial-derived PDGF-BB and HB-EGF coordinately regulate pericyte recruitment during vasculogenic tube assembly and stabilization. Blood 116(22):4720–4730
Sundberg C et al (1996) Pericytes as collagen-producing cells in excessive dermal scarring. Lab Investig 74(2):452–466
Tao YK et al (2017) Notch3 deficiency impairs coronary microvascular maturation and reduces cardiac recovery after myocardial ischemia. Int J Cardiol 236:413–422
Tattersall IW et al (2016) In vitro modeling of endothelial interaction with macrophages and pericytes demonstrates notch signaling function in the vascular microenvironment. Angiogenesis 19(2):201–215
Teichert M et al (2017) Pericyte-expressed Tie2 controls angiogenesis and vessel maturation. Nat Commun 8:16106
Tillet E et al (2005) N-cadherin deficiency impairs pericyte recruitment, and not endothelial differentiation or sprouting, in embryonic stem cell-derived angiogenesis. Exp Cell Res 310(2):392–400
Tintut Y et al (2003) Multilineage potential of cells from the artery wall. Circulation 108(20):2505–2510
Toma I, McCaffrey TA (2012) Transforming growth factor-beta and atherosclerosis: interwoven atherogenic and atheroprotective aspects. Cell Tissue Res 347(1):155–175
Travers JG et al (2016) Cardiac fibrosis: the fibroblast awakens. Circ Res 118(6):1021–1040

van Amerongen MJ et al (2008) Bone marrow-derived myofibroblasts contribute functionally to scar formation after myocardial infarction. J Pathol 214(3):377–386

van Dijk CG et al (2015) The complex mural cell: pericyte function in health and disease. Int J Cardiol 190:75–89

Volz KS et al (2015) Pericytes are progenitors for coronary artery smooth muscle. Elife 4

Wakui S et al (2006) Localization of Ang-1, −2, Tie-2, and VEGF expression at endothelial-pericyte interdigitation in rat angiogenesis. Lab Investig 86(11):1172–1184

Wanjare M, Kusuma S, Gerecht S (2013) Perivascular cells in blood vessel regeneration. Biotechnol J 8(4):434–447

Watanabe N et al (2004) Three-dimensional microstructural abnormality of the coronary capillary network after myocardial reperfusion—comparison between 'reflow' and 'no-reflow'. Circ J 68(9):868–872

Weiss DR et al (2009) Extensive deendothelialization and thrombogenicity in routinely prepared vein grafts for coronary bypass operations: facts and remedy. Int J Clin Exp Med 2(2):95–113

Weiss RM, Miller JD, Heistad DD (2013) Fibrocalcific aortic valve disease: opportunity to understand disease mechanisms using mouse models. Circ Res 113(2):209–222

Winkler EA, Bell RD, Zlokovic BV (2010) Pericyte-specific expression of PDGF beta receptor in mouse models with normal and deficient PDGF beta receptor signaling. Mol Neurodegener 5:32

Winkler EA, Bell RD, Zlokovic BV (2011) Central nervous system pericytes in health and disease. Nat Neurosci 14(11):1398–1405

Wong SP et al (2015) Pericytes, mesenchymal stem cells and their contributions to tissue repair. Pharmacol Ther 151:107–120

Wu M, Rementer C, Giachelli CM (2013) Vascular calcification: an update on mechanisms and challenges in treatment. Calcif Tissue Int 93(4):365–373

Yannarelli G et al (2013) Human umbilical cord perivascular cells exhibit enhanced cardiomyocyte reprogramming and cardiac function after experimental acute myocardial infarction. Cell Transplant 22(9):1651–1666

Yemisci M et al (2009) Pericyte contraction induced by oxidative-nitrative stress impairs capillary reflow despite successful opening of an occluded cerebral artery. Nat Med 15(9):1031–1037

Zeng H et al (2016) LPS causes pericyte loss and microvascular dysfunction via disruption of Sirt3/angiopoietins/Tie-2 and HIF-2alpha/Notch3 pathways. Sci Rep 6:20931

Zhou Z et al (2016) Induction of initial steps of angiogenic differentiation and maturation of endothelial cells by pericytes in vitro and the role of collagen IV. Histochem Cell Biol 145(5):511–525

Chapter 12
Pericytes in the Umbilical Cord

Andrée Gauthier-Fisher, Peter Szaraz, and Clifford L. Librach

Abstract The structural components of the umbilical cord, including two arteries and one vein, the stromal region/Wharton's jelly, and amniotic epithelial membrane, are well described at various time points of gestation. Over the last two decades, evidence has emerged that multipotent cells sharing properties of mesenchymal stromal cell and pericytes/mural cells can be isolated from multiple regions of the umbilical cord, including the perivascular region of the umbilical cord arteries and vein, Wharton's jelly, and subamnion. These cells have increasingly gained interest for their potential use in regenerative and immunomodulatory medicine. Recent studies suggest that obstetrical complications including gestational diabetes mellitus and preeclampsia may alter the yield, properties, and potency of mesenchymal stromal cells isolated from the umbilical cord. The role that pericytes or pericyte-like cells play in the development of the human umbilical cord and associated pathologies, however, remains to be investigated.

Keywords Human umbilical cord · Perivascular cells · Fetal development · Gestation · Body stalk · Umbilical artery · Umbilical vein · Wharton's jelly · Placenta · Obstetrical complication · Vasculature · Preeclampsia · Mesenchymal stromal cells · Regenerative medicine

---

Part of "*Biology of Pericytes: Development, Homeostasis and Disease*", a forthcoming volume which will be in the Springer Nature series "Advances in Experimental Medicine and Biology (ISSN:0065-2598)".

A. Gauthier-Fisher (✉) · P. Szaraz
CReATe Fertility Centre, University of Toronto, Toronto, ON, Canada
e-mail: andree@createivf.com

C. L. Librach
CReATe Fertility Centre, University of Toronto, Toronto, ON, Canada

Department of Obstetrics and Gynecology, University of Toronto, Toronto, ON, Canada

Department of Physiology, University of Toronto, Toronto, ON, Canada

Institute of Medical Sciences, University of Toronto, Toronto, ON, Canada

Department of Obstetrics and Gynecology, Women's College Hospital, Toronto, ON, Canada

A. Birbrair (ed.), *Pericyte Biology in Different Organs*,
Advances in Experimental Medicine and Biology 1122,
https://doi.org/10.1007/978-3-030-11093-2_12

12.1 Introduction

In placental mammals, the umbilical cord is a vital structure that forms early during fetal development, allowing for oxygen- and nutrient-rich blood to pass from the placenta to the fetus and for the return of waste and oxygen-depleted blood back to the placenta. Normally, human umbilical cords at term are composed of one vein and two arteries that are embedded in Wharton's jelly, a connective tissue consisting of a network of myofibroblasts and collagen that protect the blood vessels from compression. Wharton's jelly is abundant in mucopolysaccharides that provide the jellylike substance and is covered by an amniotic epithelial cell membrane (Fig. 12.1). An understanding of pericytes and related perivascular cells in the human umbilical cord, the focus of this chapter, necessitates an understanding of how this seemingly simple organ, and especially the vasculature and perivascular tissue within it, develop and change during gestation. How mesenchymal cells or pericyte-like cells in the umbilical cord tissue and cord blood participate in the growth, maintenance, and pathological aspects of the umbilical cord structures and fetal development throughout gestation remains largely unclear. However, these cells with demonstrated plasticity have increasingly gained interest in the field of regenerative medicine and immunomodulatory medicine, as they represent one of the youngest and richest known sources of mesenchymal stromal cells. In this chapter, we will review:

- The development of the umbilical cord, focusing on vascular structures and perivascular connective tissue.
- In situ analyses of pericyte-like cells and perivascular cells in the umbilical cord.
- Analyses of pericyte-associated markers in cells isolated from various regions of the perivascular regions of the umbilical cord as sources of stem cells for regenerative and immunomodulatory medicine.
- The potential involvement of pericytes/perivascular cells in umbilical cord pathologies as well as the impact of obstetrical complications on perivascular cells isolated from the umbilical cord.

12.2 Development of the Umbilical Cord, Focusing on Vascular Structures and Connective Tissue

Early embryonic processes involved in the formation of the human umbilical cord are described and illustrated in detail in eminent textbooks (Cullen 1916; Benirschke and Kaufmann 1995) and have been summarized in recent scientific reviews (Spurway et al. 2012; Corrao et al. 2013; Davies et al. 2017). The following is an overview of early human umbilical cord development described in these references, with the inclusion of occasional references to studies performed in animal models, as these have improved our knowledge of umbilical cord vascularization (Inman and Downs 2007). The connective tissue of the umbilical cord is derived from the

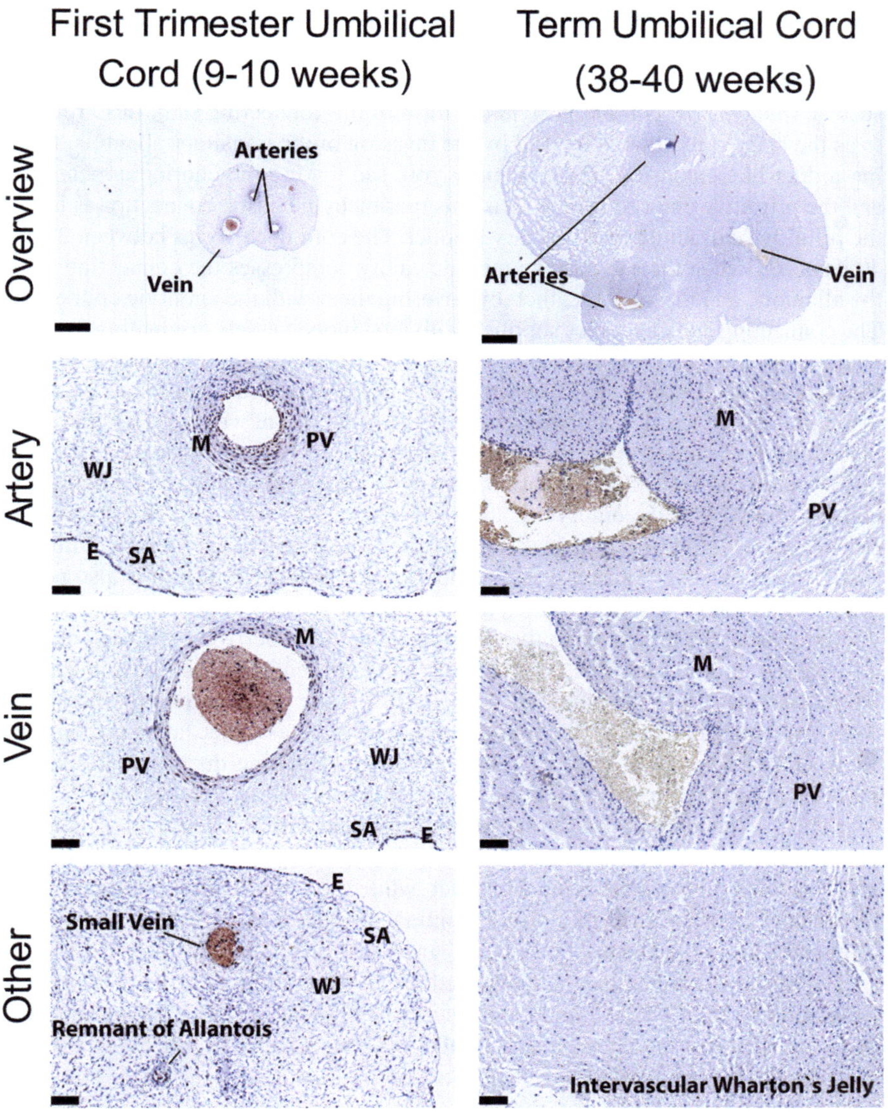

Fig. 12.1 Structural regions of the umbilical cord during the first trimester (FTM) (left panels) and at term (right panel). In both stages, the umbilical cord is mainly comprised of two arteries (shown for each stage in the second row) and one vein (shown for each stage in the third row). Additional structures, including the second vein which normally atrophies during the first trimester and remnants of the allantois, can also be observed, as shown here in FTM (fourth row). *E* epithelium, *SA* subamnion, *C* clefts, *WJ* Wharton's jelly, *PV* perivascular zone, *M* media, *I* tunica intima. Slides used to illustrate umbilical cord regions were stained with OCT4 and counterstained with hematoxylin (Librach lab, unpublished). One can appreciate the expansion of the media (M), perivascular zone (PV), and intervascular WJ between FTM and term cords. OCT4+ cells are found in FTM, mainly in the subendothelial regions of the arteries and vein, with occasional cells staining positive for OCT4 in WJ, but are not observed prevalently in term cords

extraembryonic mesoblast, the layer of mesoderm that surrounds the amniotic vesicle and primary yolk sac in the 2- to 3-week-old embryo, shortly after trophoblastic invasion of the maternal endometrium. The earliest sign of the umbilical cord can be seen at 18 days *post coitum* (*p.c*). in the form of the connecting stalk (also referred to as the body stalk) that is formed by the invasion of the transitory allantois, forming a duct-like extension of the primary yolk sac linking the chorionic mesoderm and the primitive embryo (Fig. 12.2). This mesenchyme tissue connection is in fact the primitive extraembryonic urinary bladder. The cord itself forms between 28 and 40 days *p.c.* when the expanding amniotic cavity compresses the connecting stalk, the allantois, and the yolk sac, thereby covering them with the amniotic epithelium. The connecting stalk becomes supplied with fetal blood vessels originating from the allantois around the third week *p.c*. The blood vessels are formed through de novo vasculogenesis (Downs 1998; Downs et al. 1998) and invade the placenta at the distal end of the cord, becoming connected to the villous vessels and conveying placental blood to and from the embryo. It remains poorly understood how this is orchestrated and how allantoic vasculature at the proximal end (at the caudal part of embryo) fuses with the embryonic and yolk sac vessels (Downs 1998; Inman and Downs 2007). How the primitive vasculature develops into the mature umbilical cord vessels observed during the remainder of gestation and at birth is also poorly described. During the first trimester, the omphalomesenteric duct (also known as the vitelline duct), the precursor to the gastrointestinal tract derived from the yolk sac, also passes through the cord (Fig. 12.2b). The cord lengthens as the embryo prolapses backward into the amniotic sac. The full fusion of the amniotic membrane with the chorionic membrane/mural trophoblasts takes place at approximately 12 weeks of fetal development. The allantois and vitelline duct typically recede within the umbilical cord during this period, thinning out and losing their patency (Fig. 12.3). The allantois and yolk sac are important structures in the formation of the cord as they form the initial link with the body through the connecting stalk and are the source of vascularization. However, while the allantois-derived ducts recede in the cord as they form the intra-abdominal bladder and gut, the blood vessels originating from the allantois persist and grow with the umbilical cord.

The umbilical cord tissue is composed of six structural and functional regions: from outer to inner UC, we find (1) the surface epithelial membrane (amniotic epithelium), (2) subamniotic stroma, (3) clefts, (4) intervascular stroma (WJ), (5) perivascular stroma, and (6) blood vessels, where regions (2–5) are often, and some would argue ambiguously, referred to collectively as the umbilical cord stromal region or WJ (Figs. 12.1 and 12.4) (Nanaev et al. 1997; Can and Karahuseyinoglu 2007; Corrao et al. 2013; Davies et al. 2017). There are several ways in which human umbilical cord vessels differ from major vessels in the body. First, the umbilical cord vessels' structure includes an arterial and venous tunica intima (layer of endothelial cells) and tunica media (cell-dense layers of vascular smooth muscle cells) but lacks a distinct tunica adventitia and external elastic lamina (Blanco et al. 2011), which provide contractile and vascular support functions in corporal blood

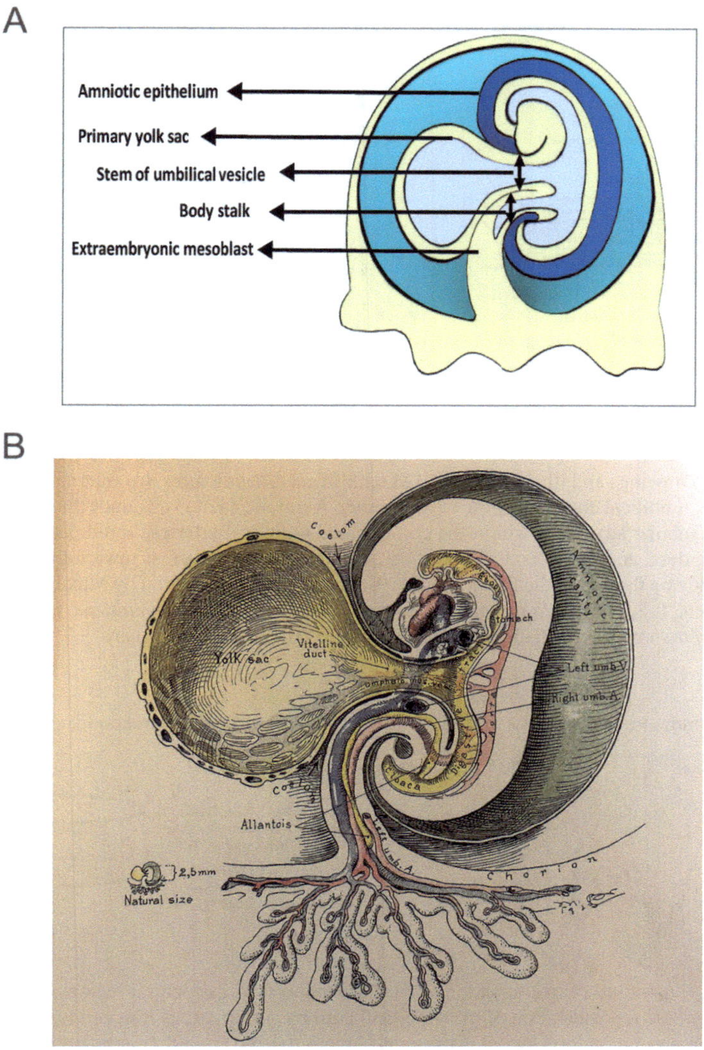

Fig. 12.2 Illustrations describing early development of the umbilical cord and structures. (**a**) Body stalk and stem of umbilical vesicle are identified as primitive components of the developing umbilical cord. Continuity and invagination of chorionic and amniotic cavities are illustrated. Amniotic cavity surrounds the developing cord structure on the fetal side, while the chorionic cavity surrounds it on the yolk sac side. (**b**) Hand-drawn illustration of early embryonic and umbilical cord development from Cullen (1916). Illustration by Max Brödel, reprinted from Cullen, T. S. (1916). *Embryology, anatomy, and diseases of the umbilicus, together with the diseases of the urachus.* Philadelphia, and London, W. B. Saunders company

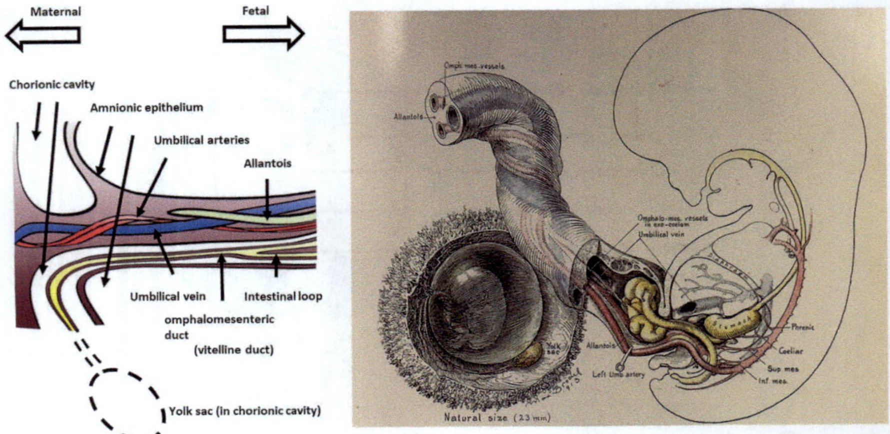

Fig. 12.3 Drawings and illustration of FTM umbilical cord structures. (**a**) Internal structure of the developing umbilical cord during the first trimester. Amniotic cavity surrounds the developing cord structure. Tubular structures within the cord include umbilical arteries, umbilical vein, omphalomesenteric duct, and allantois, the latter two of which normally regress toward the fetus and lose their patency by the end of the first trimester. (**b**) Hand-drawn illustration by Max Brödel, reprinted from Cullen, T. S. (1916). *Embryology, anatomy, and diseases of the umbilicus, together with the diseases of the urachus*. Philadelphia and London, W. B. Saunders company

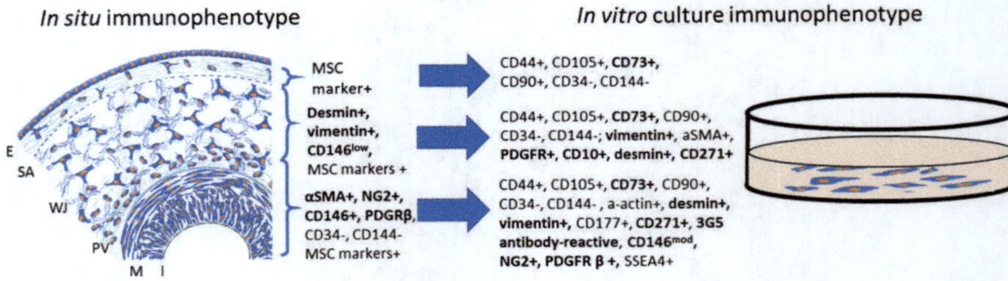

Fig. 12.4 Illustration of in situ and in vitro cell phenotypes in structural regions of the umbilical cord. Schematic representation of the umbilical cord tissue structures illustrating cell density and localization around a blood vessel. Concentric layers starting from the umbilical cord epithelium toward the umbilical blood vessel are as follows: *E* epithelium, *SA* subamnion, *WJ* Wharton's jelly, *PV* perivascular region, *M* tunica media, and *I* tunica intima. A non-exhaustive list of markers associated with mesenchymal and pericyte-like cells (*pericyte-associated markers are bolded*) from each structural region in cells in situ or in culture is presented

vessels. From placental to fetal ends, the umbilical cord is largely considered not to be innervated (Hoyes 1969). A few controversial studies, however, have reported the presence of neural elements supplying the umbilical arteries at the proximal end (Pearson and Sauter 1970; Ellison 1971). Wharton's jelly or umbilical stroma, the perivascular mucoid connective tissue described by Thomas Wharton in 1656, and its dispersed myofibroblasts and collagen fibers are thought to fulfill the roles of the adventitia. This source of perivascular cells will be discussed in the next section. Umbilical cord endothelial and perivascular cells also differ from those of typical blood vessels, with changing characteristics throughout gestation (Parry and Abramovich 1972; Sexton et al. 1996; Nanaev et al. 1997; Stehbens et al. 2005). Umbilical cord vessel endothelial cells are rich in organelles such as mitochondria, endoplasmic reticulum, and Golgi and, in the sheep at least, have been described as resembling those of growing or of regenerating tissues (Sheppard and Bishop 1973). Finally, unlike blood flow and exchange in other tissues, oxygen- and nutrient-rich blood passes from the placenta to fetal circulation through one vein (which goes to the liver), and depleted blood/waste is returned to the placenta through two large arteries that are branches of the internal iliac arteries, with anastomosis about 3 centimeters from the placental insertion site (Cullen 1916, Benirschke and Kaufmann 1995, Spurway et al. 2012). It is important to note that in some cords, there are deviations from this most often described structure and that other structures can be observed in first trimester or even second trimester umbilical cords (Figs. 12.1, 12.2 and 12.3). The human umbilical cord initially contains two veins, the right and the left, where the left normally disappears during the second month of pregnancy. However, two veins (four vessels) or one artery (two vessels) or, in rare cases, five vessels (Singh et al. 2012) can be observed in about one percent of human umbilical cords at term. In twenty percent of umbilical cords at term, short and typically non-patent remnants of allantois containing epithelial cells can be observed at the distal end of the cord and would be expected to be located between the two arteries during earlier stages of gestation, as it normally regresses by the eighth week of gestation (Cullen 1916, Spurway et al. 2012). Similarly, as other umbilical cord lengthens, the omphalomesenteric duct and the umbilical coelom regress by the tenth and 12th week of gestation, respectively (Khati et al. 1998).

The human umbilical cord is generally described as lacking microvessels (Hoyes 1969). Transmission electron microscopy studies of developing human umbilical arteries and veins at various time points of gestation suggest that human umbilical cord blood vessels are devoid of vasa vasorum (Sexton et al. 1996), networks of smaller blood vessels that supply the walls of large blood vessels such as arteries and veins in many other tissues and where classically defined pericytes would be found as discussed below. However, the intra-abdominal part of the umbilical cord artery and vein do contain vasa vasorum, and these appear to increase in number between the first trimester and term (Clarke 1965; Malas et al. 2003). It should be noted that in many mammals, including sheep, cow, horses, and the dromedary camel, the umbilical cord contains four blood vessels, including two umbilical veins, two umbilical arteries, and a large allantoic duct. In the sheep umbilical cord, vasa vasorum, including fenestrated endothelial cells and pericytes, were described in the outer layers of the media of the artery in electron microscopy studies (Sheppard and Bishop 1973). There have been at least two published reports of microvessels in the umbilical cord. Montemurro et al. reported the presence of venules and arterioles with diameters ranging from 10 to 100 μm in umbilical cords resected from 23–32-week preterm births but not in term umbilical cords (Montemurro et al. 2011). In addition, Blanco et al. reported the frequent finding of one or two scarcely developed but erythrocyte-filled vessels in Wharton's jelly, away from the three major vessels, in segments of cord isolated specifically approximately 2 centimeters from the placental insertion site (29–41 weeks gestational age) (Blanco et al. 2011). It is important to note that our understanding of the structural organization of the human umbilical cord has been developed from ultrastructural and histological assessments of short segments of the cord (representing less than ten to twenty-five percent of the total length) that are accessible after resection of the cord at birth (from both distended and collapsed specimens) or from abortions. Furthermore, the proximal, medial, and distal distance of these segments to the fetus are almost never specified in publications. Specifically, vascular structures at the proximal and intra-abdominal regions of the umbilical cord as well as the rostral region that links to the placenta throughout fetal development have not been extensively studied. This, however, may be relevant to our understanding of potential sources of pericyte-like cells throughout umbilical cord development.

In a manner that is crucial to proper fetal development and offspring health, the length and diameter of the human umbilical cord increase from a few millimeters in the first weeks of pregnancy to an average length of 40–65 cm and 1–2 cm girth at birth. The steepest growth curve of the umbilical cord, including vascular structures, occurs before the end of the second trimester (Cullen 1916; Malas et al., 2003). This implies the presence and regulation of a rich source of active progenitors that supply the growth of the amniotic epithelium, of endothelial cells, as well as of the subendothelial smooth muscle and connective tissue, which will be the focus of the next section. Many assumptions have been made, but not empirically tested, with regard to the types of cells that act as progenitors and feed the growth of the perivascular and vascular components of the cord and with regard to the source of mitogenic factors. However, these and the mechanisms regulating progenitors are of great relevance, both for the understanding of the development and pathologies of the umbilical cord, as well as for the application of umbilical cord-derived cells in many areas of regenerative and immunomodulatory medicine. For example, cords that are abnormally short or too long and defects in umbilical vasculature or connective tissue can each result in inadequate nutrient and oxygen delivery to the fetus, leading to fetal demise or neurological and other severe birth defects (Benirschke and Kaufmann 1995). As discussed at the end of this chapter, the potential association between pericyte-like cells in the umbilical cord and these umbilical cord pathologies remains largely unexplored.

12.3 In Situ Analysis of Pericytes and Related Perivascular Cells in the Umbilical Cord

Pericytes have been classically defined as contractile cells that wrap around endothelial cells in venules, capillaries, and arterioles throughout the body. Over the last two decades, the definition of pericytes has been extended to include a heterogeneous, specialized population of multipotent mural cells residing in the subendothelial layer, adventitia, and vasa vasorum of the vasculature, including in small, medium, and large arteries and veins (Andreeva et al. 1998). Pericytes are now thought to function not only as hemostasis regulators but also as a source of cells for repair and maintenance of the tissues in which they reside. For instance, pericytes can differentiate into osteoblasts, chondrocytes, adipocytes, fibroblasts, myofibroblasts, and smooth muscle cells (Diaz-Flores et al. 2009). Pericytes or pericyte-like cells have thus emerged as critical components of the vascular bed in blood vessels of all sizes, not only covering endothelial cell junctions to regulate cellular processes such as inflammation but also as a mesenchymal progenitor cells with metabolic, signaling, and mechanical roles to support endothelial cells and tissue hemostasis as well homeostasis (Sims 2000).

In addition, many perivascular cell types across the vascular wall, including smooth muscle cells found in the intima and myofibroblasts found in the stroma/adventitia, are thought to have close relationships with pericytes, and as such the reference to pericytes, mural cells, adventitial cells, and perivascular cells has become somewhat fluid in the literature. These cells may represent a continuum of cell phenotypes along growing and maturing vasculature (Holm et al. 2018). Immunophenotypically, pericytes are defined by their anatomical localization in subendothelial layers of vessels and the expression of at least one of the following markers, neither one of which is specific to, or representative of, all pericytes and may be associated with their quiescent or activated states: CD146 (MCAM), platelet-derived growth factor receptor beta (PDGRβ, CD140b), alkaline phosphatase (AP), regulator of G protein signaling (RGS5), alpha-smooth muscle actin (αSMA), desmin, and CD271 and the absence of CD31, CD34, and von Willebrand factor (VWF) and CD144 or CD45 which are endothelial and pan-hematopoietic markers, respectively (Andreeva et al. 1998, Diaz-Flores et al. 2009). Adventitial cells in turn are typified as CD34+, CD31−, CD146−, and CD45−. In vitro, adventitial cells can acquire a pericyte-like cell phenotype, suggesting that adventitial cells may be a source of pericyte progenitors (Corselli et al. 2012).

In a groundbreaking research study, Crisan et al. showed that the perivascular region of blood vessels in many tissues, including the umbilical cord, harbors a reservoir of CD146+, NG2+, and PDGRβ+ progenitor cells that may be the source of mesenchymal stromal cells (MSCs) in all vascularized bodily tissues (Crisan et al. 2008). This has led MSC researchers to conclude that at least a subset of MSCs are pericytes (Caplan 2008). MSCs are currently defined as plastic adherent cells that express CD90, CD73, and CD105; do not express CD45, CD34, CD14, CD11b, CD79a, CD19, and HLA-DR; and can differentiate into osteocytes, adipocytes, and chondrocytes in appropriate culture conditions (Dominici et al. 2006). The co-expression of pericyte-associated markers in a subset of cells expressing MSC-associated markers is well described in various sources of MSCs including the umbilical cord, both in vivo (as discussed in this section) and in vitro (as discussed in the next section).

With regard to the study of perivascular cells in situ in the developing and term human umbilical cord, ultrastructural and histological studies have been performed on tissue sections of term human umbilical cord (both distended and collapsed) following their resection in preterm or term births (Parry 1970; Takechi et al. 1993; Nanaev et al. 1997; Schugar et al. 2009; Coskun and Can 2015). Additional studies have been performed on umbilical cords spanning fetal development (Parry and Abramovich 1972; Sexton et al. 1996; Stehbens et al. 2005; Montemurro et al. 2011; Spurway et al. 2012; Hong et al. 2013).

It was noted a long time ago that cells in the subamnion, Wharton's Jelly, and perivascular regions of the arteries and veins differ from each other phenotypically and that the perivascular cells had increased proliferative activity (Parry and Abramovich 1970; Nanaev et al. 1997, Friedman et al. 2007). The umbilical cord epithelium is characterized by epithelial-like cells expressing various cytokeratins, while the stromal regions include mesenchymal cells expressing vimentin, desmin, and alpha-smooth muscle actin (αSMA). Here, we will specifically discuss the pericyte-like properties of umbilical cord stromal cells identified in the perivascular compartments and Wharton's jelly.

Pericyte-associated markers NG2, CD146, and PDGFRβ have been detected exclusively in perivascular cells of the umbilical cord arteries and vein in first trimester (FTM), second trimester (STM), and term umbilical cords (Sarugaser et al. 2005; Baksh et al. 2007; Crisan et al. 2008; Schugar et al. 2009; Montemurro et al. 2011; Hong et al. 2013). CD146 (MCAM, S-Endo 1) is expressed by some pericytes, but endothelial, smooth muscle cells, and stromal and perivascular stem cells are also known to express this antigen. A subpopulation of cells surrounding the large umbilical arteries and vein as well as small vessels in mid-gestation fetal cords co-express CD146 and αSMA, as well as MSC-associated markers (Montemurro et al. 2011). These cells, consisting mainly of vascular smooth muscle cells, form the intima media, a layer around the endothelial cells that becomes more prominent during the second trimester (Nanaev et al. 1997; Stehbens et al. 2005). In term cords, the two arteries present a similar expression pattern for these markers, but only rare CD146^{high} and αSMA-positive cells were detected around the vein, where the vSMCs are αSMA-positive (Schugar et al. 2009; Montemurro et al. 2011). Moderate CD146 expression (relative to the high expression in endothelial cells) was detected in perivascular zone cell (outside of media) of term but not first trimester umbilical cord (Davies et al. 2017) (Fig. 12.5a). The multipotency-associated marker, Oct4A, however was detected in the perivascular region of FTM and second trimester human umbilical cord perivascular cells (HUCPVCs), but not in term (Hong et al. 2013, 2013; Montemurro et al. 2011) (Fig. 12.1). Flow cytometric analysis of freshly isolated fetal versus term umbilical cord cells suggest that there is a higher frequency of perivascular cell populations with pericyte-like properties (described as either CD146+ CD34− CD45− CD56−; CD146+ PDGFRb+; or CD146+) in early gestation (Montemurro et al. 2011; Hong et al. 2013). While the exact proportion of such cells varies across studies based on tissue digestion and immunophenotyping methods used, this suggests that a perivascular cell population with increased multipotential is present in the fetal umbilical cord perivascular region. The functional significance of these cell phenotypes during fetal development is unclear but likely contributes to the growth of the blood vessels and expan-

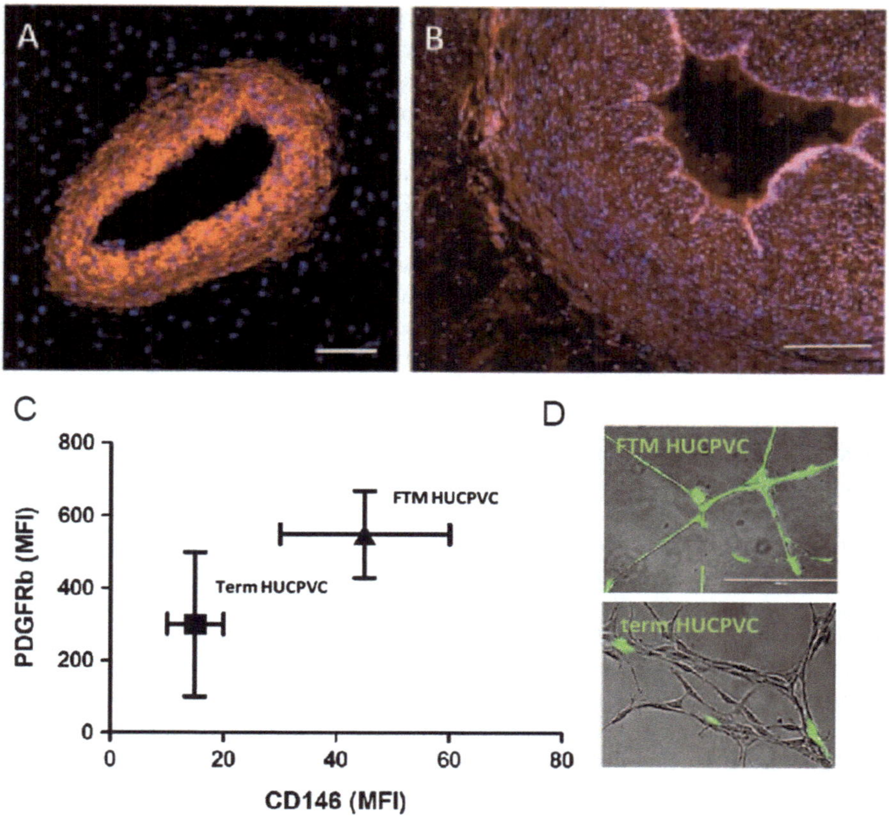

Fig. 12.5 Comparison of FTM and term human umbilical cord perivascular cells (HUCPVCs). Expression of pericyte-associated markers and physical interaction with endothelial cells in tube formation assay. (**a,b**) CD146 immunostaining in first trimester (FTM) and term perivascular regions, respectively. This image was obtained with permission from Davies, J. E., J. T. Walker and A. Keating (2017). "Concise Review: Wharton's Jelly: The Rich, but Enigmatic, Source of Mesenchymal Stromal Cells." *Stem Cells Transl Med* **6**(7): 1620–1630. (**c**) Flow cytometry-based cell surface expression levels of PDGFRβ and CD146 in perivascular cell populations isolated from FTM and term cords ($n = 4$, passage 4–5). Marker levels are expressed as mean fluorescence intensity. Both markers show higher expression in FTM-derived cell population compared to term. (**d**): Aortic endothelial cell co-cultures demonstrating a pericyte-like characteristic of FTM HUCPVC in the extent of adhesion and elongation of these cells in endothelial networks. FTM HUCPVC exhibits higher endothelial adhesion and network coverage than term HUCPVCs, in agreement with the pericyte-associated immunophenotype corresponding to each cell type (**c**) (obtained from unpublished data from C Librach laboratory)

sion of perivascular cells (Holm et al. 2018). The properties of these HUCPVCs when isolated and propagated in culture will be discussed in more details in the next section.

The intervascular or distal perivascular stromal region or Wharton's jelly (WJ), which can be distinguished from the proximal perivascular region by a reduction in cell density, with highest density near perivascular region and lower density near the subamniotic region, is another compartmentalized source of cells with MSC-like properties in the umbilical cord (Fig. 12.4). WJ is largely composed of myofibroblasts (Parry 1970; Takechi et al. 1993), and occasional mast cells dispersed in a mucoid substance composed of water (90%), sulfated glycosaminoglycans, (HA) proteoglycans (decorin, biglycan), and collagens in which the blood vessels are embedded. WJ is a crucial structure, with contractile function and elasticity which protects the vessels from compression, allowing for adequate blood flow from the maternal placenta to fetus. Electron microscopy studies of WJ cells and smooth vascular cells first described WJ cells as "unusual fibroblasts," displaying partial morphological similarities with fibroblasts and smooth muscle cells. A stellate cell morphology and extended inter-cell connections were described in collapsed cord segments, with cells on the outer periphery displaying longer processes (Takechi et al. 1993; Ryu et al. 2013). In general, these cells resemble fibroblasts, with organelle-rich cytoplasm, with well-developed ER, abundant mitochondria, and Golgi, suggestive of active biosynthesis and secretory activity. Myofibroblasts are thought to produce the extracellular matrix molecules of the WJ. Small elongated and indented nuclei, with prominent nucleoli, were observed. Gap junctions are observed between interconnecting cell processes. Unlike fibroblasts which are not covered by basal lamina, and smooth muscle cells which are fully covered by basal lamina, WJ cell surfaces are partially covered by basal lamina. Unlike fibroblasts, WJ cells express contractile proteins actin and non-muscle myosin, as well as cytoskeletal protein desmin and vimentin. Furthermore, unlike WJ cells, vSMCs express muscle-associated myosin. It has been suggested, but not demonstrated, that WJ myofibroblasts are derived from either vSMC or fibroblasts. Multiple immunohistological studies have shown that, in the first trimester, second trimester, and term umbilical cord, the bulk of WJ myofibroblasts are CD146-negative but do stain positive for MSC-associated markers and other pericyte-associated markers such as αSMA, desmin, and PDGFR (Takechi et al. 1993; Baksh et al. 2007; Schugar et al. 2009; Montemurro et al. 2011; Ryu et al. 2013; Coskun and Can 2015). In the term umbilical cord at least, CD34 and CD144 only mark the endothelial cell layer; the perivascular and WJ regions are negative for these markers (Schugar et al. 2009).

Altogether, these data suggest that at least two cell populations expressing MSC-associated markers and differential expression of pericyte-associated markers are present in the umbilical cord. The origin and function of these cells remain unclear, and lineage tracing studies using animal models with tagged pericyte markers have not been performed on the umbilical cord. The differential expression of pericyte marker CD146 between perivascular and WJ cells could be attributed to the proximity of perivascular cells to endothelial cells, as it was shown that direct MSC-endothelial cell contact induces the upregulation of pericyte genes (CD146, NG2, PDGFRβ) (Loibl et al. 2014).

12.4 Analysis of Pericyte-Associated Markers in Cells Isolated from the Perivascular Regions of the Umbilical Cord as Sources of Mesenchymal Stromal Cells for Regenerative Medicine

Early microscopy studies focusing on describing the ultrastructural properties of stromal cells in the umbilical cord have improved our understanding of the cellular mechanisms associated with umbilical cord anomalies. Scientific interest in the cellular composition of each structural region of the umbilical cord was re-ignited in the last 20 years when it was found that they contained stem cell-like cells that could be propagated in vitro. Cells exhibiting stem cell-like properties can be derived from many compartments in the cord including amniotic epithelial membrane (AM), subamnion (SA) or "cord lining", intervascular Wharton's jelly (WJ), and perivascular region (PV) surrounding the three blood vessels, as well as cord blood.

It seems apparent, based on the description of cell phenotypes observed in the defined structural regions of the umbilical cord tissue at different developmental time points, that the compartment of umbilical cord tissue included during cell isolation would yield cells with the molecular phenotypes and biological functions associated with those regions, as described above (Fig. 12.4). MSC-like cells have been isolated and propagated from whole umbilical cord or using protocols that include the subamnion, WJ, perivascular region, and vessels, suggesting that there is a close and plastic relationship, at least in culture, between myofibroblast, pericytes, and smooth muscle cells in the umbilical cord. According to the description of cells in the umbilical cord in situ, protocols which include the harvest of cells from the whole cord or perivascular regions (and these are not consistently described as discussed in detail in Davies et al. 2017), as opposed to removing blood vessels and isolating cells from Wharton's jelly, would yield corresponding proportions of CD146+ cells. Published isolation protocols vary based on the inclusion/exclusion of epithelial membrane, Wharton's jelly and vessels, enzymatic digestion, or explant culture methods. Schugar et al. demonstrated that the protocol used to isolate cells from whole umbilical cord can have a significant impact on cell yield, morphology, and immunophenotype. Collagenase digestion yielded an increased proportion of freshly isolated cells expressing MSC-associated markers (CD44, CD105, CD73, CD90) where about half of the cells expressed moderate levels of CD146 ($CD146^{mod}$) and decreased proportion of endothelial cell-associated markers (CD34, CD144, and $CD146^{high}$) when compared to dispase digestion, which yielded an increased proportion of cells with an endothelial phenotype and decreased proportion of cells with an MSC phenotype. These differences in immunophenotype persisted in culture for 1 to 3 weeks, and only the collagenase-isolated population showed the ability to differentiate in vitro toward the osteogenic and chondrogenic lineage (Schugar et al. 2009). This group used fluorescence-activated cell sorting (FACS) technology to enrich CD146+ cells in the collagenase digestion population and found that within 48 hours, the proportion of CD146+ cells returned to fifty percent of the total cells, suggesting that CD146+ cells may give rise to a CD146-negative population of cells expressing MSC markers. This also suggests that the distinct in situ cell phenotypes observed between Wharton's jelly myofibroblasts and perivascular tissue pericytes/vascular smooth muscle cells persist following isolation and propagation in culture (at least in the early passages). This will be discussed further below.

MSC-like cells were first isolated from the umbilical cord stroma, by focusing on protocols that removed the blood vessels, thereby presumably varying amounts of the perivascular tissue (McElreavey et al. 1991; Karahuseyinoglu et al. 2007; Subramanian et al. 2015). A few of these studies have demonstrated that when WJ cells are isolated and plated in culture, they generally maintain the ultrastructural and molecular properties of myofibroblasts, including stellate-shaped cells with filopodia, small nuclei in situ (relatively larger in culture), prominent nucleoli, numerous Golgi apparatus and filaments, and cytoplasmic filopodia. Differences in the arrangements of myofilaments (random in culture) and lipid inclusion (not observed in culture) were however observed. The cells maintain expression of vimentin, αSMA and PDGFR, and CD10 but were desmin-positive in early passages only (Ryu et al. 2013). Comparison of MSC derived from proximal, middle, and distal segments of the umbilical cord WJ using explant cultures yielded cells expressing a similar immunophenotype, morphology, and tri-lineage differentiation potential, in addition to the ability to transdifferentiate toward osteocytes and hepatocyte-like cells following induction in culture, suggesting that WJ cells with similar therapeutic potential can be isolated throughout the length of the cord (Ishige et al. 2009). An independent study, where an enzymatic digestion protocol was used, showed similar findings, with the exception of decreased differentiation potential in cells derived from the middle segment when compared to those derived from maternal and fetal ends of the cord (Lim et al. 2016). Neither of these studies investigated pericyte-like properties in WJ-derived MSC.

When plated in culture, cells isolated from the perivascular region using a method that excludes Wharton's Jelly, subamnion, and epithelium have a fibroblast-like appearance with a stellate shape and long cytoplasmic processes that extend 100–300 μm. In vitro, HUCPVCs express alpha-actin, desmin, and vimentin and can be stained with the 3G5 antibody (used to mark pericytes). In addition, they are positive for MSC markers but negative for hematopoietic markers, and a subpopulation of HUCPVC expresses low levels of CD177 (c-kit) and HLA-A, HLA-B, and HLA-C (Sarugaser et al. 2005). The bulk population of these cells shows tri-lineage differentiation potential in vitro and in vivo. When compared to cord blood or bone marrow, and whole cord tissue, the perivascular-targeted isolation of cells allowed for a higher frequency of MSC-like cells and CFU-F (1:200 million vs 1:100,000 vs 1:300, respectively) (Sarugaser et al. 2005), making the perivascular region of the umbilical cord one of the richest sources of MSC. Importantly, and in contrast to other studies, Davies' group used clonal experiments, starting from single cells, to demonstrate that HUCPVCs represent a heterogeneous, hierarchical population of self-renewing MSCs where 1:3 cells have self-renewing and multi-differentiation capacity toward five mesenchymal lineages including myofibroblast and fibroblast (Sarugaser et al. 2009). A comparison of in vitro passaged first trimester-derived

(FTM) umbilical cord cells isolated with collagenase (representing all cells of the developing umbilical cord) and term HUCPVC showed that a similar proportion of cells isolated from each of these sources expresses the pericyte-associated markers CD146, NG2, and PDGFRβ and multipotency-associated marker SSEA-4 in culture. However, Oct4A and Sox17 expression was restricted to FTM HUCPVC, and these cells also showed increased proliferative capacity and in vitro differentiation capacity toward both mesenchymal and all three germ cell lineages when compared to term HUCPVC (Hong et al. 2013; Szaraz et al. 2016; Shlush et al. 2017). In in vitro (Iqbal et al. 2017) and in vivo (under review Iqbal et al.) angiogenesis assays, FTM HUCPVC showed increased levels of CD146 and PDGFRβ expressions and increased angiogenic activity and pericyte-like interactions with endothelial cells of developing vasculature when compared to term HUCPVC or bone marrow MSC (Fig. 12.5b, c). Altogether, the results discussed here and in the previous section, and the fact that vascular maturation and perivascular/connective tissue growth are significant in the first and second trimester, suggest that a more primitive source of pericyte/MSC resides in the early gestation umbilical cord and might function as progenitors for myofibroblasts and vascular smooth muscle cells in the cord. This could have important implications for their use in regenerative medicine. In keeping with the idea that ontologically younger sources of MSC have increased plasticity, it was also shown that term HUCPVCs have higher proliferative capacity and multi-differentiation potential than bone marrow-derived MSC and that this multi-differentiation potential is maintained in CD146-sorted HUCPVC (Baksh et al. 2007).

Indirect evidence for a source of pericytes or pericyte progenitors in the umbilical cord arterial perivascular region arises from the human aortic ring (hAR) assay. In this angiogenic assay, angiogenic sprouts from a denuded artery form capillary-like structures in vitro where NG2-positive pericyte-like cells associate with endothelial cell tube formations (Seano et al. 2013). Covas et al. showed that MSCs isolated from the subendothelial layer of the umbilical cord vein (UCV) or umbilical cord arteries (UCA) and passaged three to five times in culture were CD146+ and showed considerable morphological, immunophenotypical, and ultrastructural similarity to CD146-sorted human fetal or adult retinal pericytes. This included the observation of large nuclei with prominent nucleoli, abundant ER, and mitochondria that are characteristic of metabolically active cells. In addition, gene expression analysis of many cell types showed closest clustering between UCV-derived MSC, bone marrow MSC, and human retinal pericytes (Covas et al. 2008). WJ cells were not compared to perivascular cells in this study.

At least two studies systematically compared MSC isolated from various compartments of the cord (Ishige et al. 2009; Subramanian et al. 2015). Ishige et al. compared the proliferative, CFU-F, immunophenotype, and in vitro tri-lineage differentiation potential of cells isolated from umbilical cord Wharton's jelly (UCWJ), umbilical vein perivascular area (UCV), and umbilical artery perivascular area (UCA) using an explant outgrowth method. While all three cell populations showed similar immunophenotype with regard to the usual MSC and non-MSC markers, cells derived from the UCV exhibited the highest colony-forming unit frequency

(CFU-F) and the slowest proliferation rate. All three cell populations were able to differentiate toward the chondrogenic and adipogenic lineages; however, UCWJ showed reduced effectiveness toward osteogenic differentiation in this study. Surprisingly, UCA cells showed increasing alkaline phosphatase activity after a week in culture, even in the absence of osteogenic induction conditions. Although the tri-lineage differentiation capacity was not demonstrated in clonal experiments, these results suggest that MSC-like cells can be harvested from all compartments of the umbilical cord but that the CD146+ population in UCV/UCA may provide a better source.

Subramanian et al. used previously established enzyme digestion and explant protocols to isolate cells from mixed cords (MC) and three compartments including the amnion (AM), subamnion (SA), Wharton's jelly (WJ), and perivascular zone (PV). They further demonstrated that, although WJ gave the highest cell numbers in primary culture, cells derived from each of the regions including AM and SA, and passaged 1–10 times in culture, were positive for signature MSC markers. WJ cells showed increased expression levels (mean fluorescent intensity) for CD29, CD44, and CD73 and for CD24 and CD108, markers thought to distinguish stem cell MSCs from non-stem cell mesenchymal cells in bone marrow and adipose MSC cultures (Wetzig et al. 2013). Interestingly, in this study, greater than 90% of the cells isolated from all regions were positive for pericyte-associated marker CD146, with increased expression levels in WJ and PV. WJ and PV cells also showed increased expression of another pericyte marker CD271, while PDGFRβ (CD140b), CD49d (integrin alpha 4), and CD40 (a marker associated with non-stem cell MSC or contaminating cells) were detected in increased proportions of SA-, AM-, and MC-derived cells (approximately 40%) and at higher levels when compared to WJ and PV. Early passaged cells showed similar telomerase activity, but WJ cells showed increased levels by passage 10. While all compartments showed detectable mRNA expression of pluripotency-associated genes such as OCT4A, NANOG, and SOX2 by qRT-PCR (cycle threshold values were not specified), levels appeared significantly reduced in SA and AM when compared to WJ. Fibroblast-associated genes showed the opposite pattern. Finally, while cultures derived from all UC compartments could undergo tri-lineage differentiation, here the authors concluded that the WJ, followed by PV, had increased levels of osteogenic and adipogenic differentiation. Although in vitro cell phenotypes are somewhat incongruous with the description of cell phenotypes in situ, this study and many described here illustrate a major issue in umbilical cord MSC research. There is very little consistency with regard to cell isolation methods, definition, and selection of umbilical cord compartments, and there are no standardized culture conditions, immunophenotyping, and differentiation assays. While it is difficult to make definitive conclusions as to the source of MSC and pericyte-like cells in the umbilical cord, this study suggests that multiple subpopulation of MSC/pericytes may exist in the umbilical cord and umbilical cord-derived cell cultures (Fig. 12.4).

12.5 Pericytes/Perivascular Cells and Umbilical Cord Pathologies and Gestational Complications

While structural deficits in the umbilical cord have been described and associated with obstetrical complications involving the umbilical cord, the involvement of pericytes and perivascular mesenchymal cell progenitors in such pathologies or as an endogenous source of cells that could participate in potential repair mechanisms in vivo has not been explored. Common umbilical cord pathologies including abnormally short or long cords, abnormal number of blood vessels or vessel structure, and reduced stromal tissue are reviewed in detail in other textbooks (Cullen 1916; Benirschke and Kaufmann 1995).

Reduced volume of the stromal tissue or Wharton's jelly (WJ) around the vein and/or absence of WJ around the umbilical cord arteries (Fig. 12.6), the latter of which is very rare and has never been observed with regard to the umbilical vein, is associated with poor fetal outcomes, intrauterine growth restriction, preterm delivery, and perinatal death. This is presumably because a lack of WJ protection leaves the vessels susceptible to compression and torsion, leading to impaired placental and fetal blood flow (Cole and Israfil-Bayli 2016). While the cellular and molecular mechanisms for such anomalies are unclear, it has been suggested that progenitors in the UC may play a role in the degeneration or malformation of the WJ around the vessels as a result of improper early umbilical cord development and mesenchymal

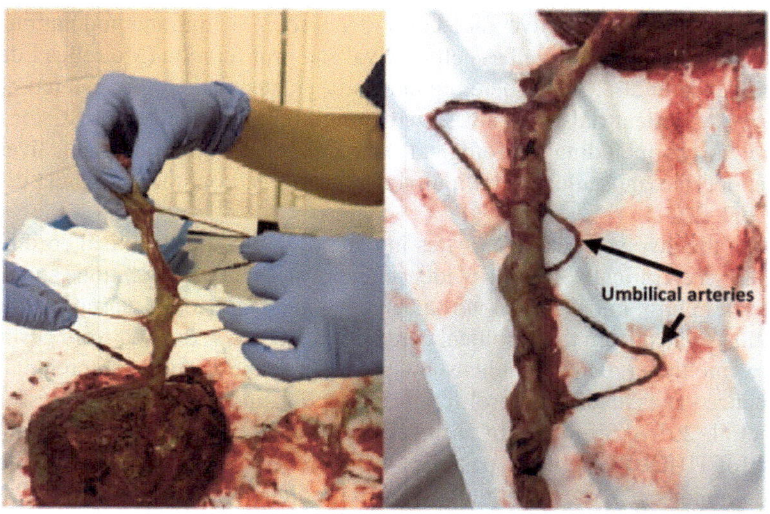

Fig. 12.6 Image of umbilical cord pathology showing absence of Wharton's jelly around arteries. Lack of sufficient stromal tissue renders cord axially disintegrated. Umbilical arteries are detached and form loops along the cord—risking entanglement and insufficient blood flow. Reprinted with permission from Cole, J. and F. Israfil-Bayli (2016). "Wharton's jelly: The significance of absence." *J Obstet Gynaecol* **36**(4): 500–501. www.tandfonline.com

structures (Kulkarni et al. 2007). Progenitors in the umbilical cord, which likely include pericyte-like cells described in previous sections and which potentially give rise to myofibroblasts, may therefore play a role in protecting against umbilical cord and fetal pathologies by fueling the matrix that ensures adequate placental blood flow to the fetus.

Pericytes or pericyte progenitors in the umbilical cord may play additional roles in protecting against pathologies. MSCs derived from WJ have been shown to have tumoricidal properties, and, because of this, it has been suggested that WJ cells act as a natural defense against the migration of cancer cells between fetus and mothers with ovarian, placental, or mammary cancers (Yang and Chao 2013). As discussed in the previous section, MSC/pericytes with in vitro and in vivo multipotential differentiation capacity and paracrine properties can be isolated from either the perivascular or WJ compartments of the umbilical cord and expanded in culture. Over the last 15 years, MSCs derived from the umbilical cord as well as those from other sources such as bone marrow, adipose tissue, and other fetal sources have gained increasing interest in the field of regenerative and immunomodulatory medicine. MSCs, including those derived from the umbilical cord, have been studied as cell therapy candidates in many areas of regenerative medicine (Guadix et al. 2017; Samsonraj et al. 2017). MSCs were initially isolated and characterized from the bone marrow (BM) (Friedenstein et al. 1970), but it is now well established that various types of MSC can be isolated from vascularized tissues throughout the body (Crisan et al. 2008; Corselli et al. 2012). MSCs are well characterized as multipotent cells that can be expanded in culture and are capable of differentiation toward bone, cartilage, adipose, fibrous, and muscle tissue. They are best known for their capacity to secrete bioactive molecules that promote immunomodulation, anti-fibrosis, angiogenesis, as well as cell survival and proliferation (Meirelles Lda et al. 2009; Shende et al. 2018). There are currently greater than 100 registered clinical trials (www.clinicaltrials.gov) investigating umbilical cord-derived MSC for multiple indications including neurological, hematological, immunological, liver, cardiac, endocrine, musculoskeletal, pulmonary, skin, and ophthalmological diseases (Can et al. 2017).

While the involvement of pericytes in pathologies of the umbilical cord has not been shown empirically, there is evidence that obstetrical complications and pathologies can, in turn, impact the in vitro phenotype and behavior of mesenchymal stromal cells isolated from the umbilical cord. UC-MSCs (WJ) derived from gestational diabetes mellitus (GDM) patients show premature senescence and upregulation of senescence genes, decreased proliferative potential, and mitochondrial dysfunction (Kim et al. 2015). Similarly, in an independent study, decreased cell yield of and growth of HUCPVC in early passages (but recovery in later passages under low glucose conditions) were reported, along with altered paracrine-mediated wound healing properties in vitro (An et al. 2017). These findings were not surprising given that exposure to hyperglycemia causes a production and reactive oxygen species and inflammation which, at least in other tissues, induces the accelerated loss of pericytes in capillaries (Manea et al. 2015). Adverse effects of GDM were also observed on the growth, angiogenic, and anticancer properties of Wharton's jelly MSC (Wajid et al. 2015).

Similarly, changes in gene and protein expression of Wharton's jelly MSC have been described in samples derived from preeclamptic patients (Bankowski et al. 1996, 2004; Jurewicz et al. 2014; Joerger-Messerli et al. 2015). Bankowski described "aging" of Wharton's jelly in preeclampsia (PE), but it is unclear whether changes in perivascular and Wharton's jelly cells reflect an adaptation to PE or participate in the development and symptoms of PE. In addition, a thicker smooth muscle cell layer has been reported in PE when compared to normal gestation umbilical cord arteries. This may reflect a functional adaptation of the umbilical cord arteries to altered hemodynamic conditions in PE (Junek et al. 2000) which may well involve regulation of pericyte or vSMC progenitors. Altogether, these studies suggest that obstetrical complications may impact the quality and therapeutic efficacy of umbilical cord-derived MSC.

12.6 Future Trends and Directions

Our understanding of the role of umbilical cord pericyte-like cells in the development and maintenance of the cord, umbilical cord pathologies, and as a source of MSC is incomplete. There are currently over 100 early phase clinical trials underway to test the safety and efficacy of umbilical cord-derived MSC for the treatment of various medical indications (Can et al. 2017), yet there is still no consensus on methods to isolate and expand umbilical cord MSC, or subpopulations of MSC, with the greatest quality and efficacy to treat each of these indications. Our understanding of pericyte-like cells in the umbilical cord could be improved by systematic histological analyses of pericyte-associated markers in all regions of normal and abnormal cords during development and at term, and by use of inducible pericyte-associated gene knockout mouse models and cell fate tracing tools. To our knowledge, this has not been done but could increase our understanding of the contribution of pericyte-like cells to umbilical cord development and pathologies. The emergence of single-cell sequencing technologies will enable researchers to complement previous histological findings by fully characterizing the phenotypes and intercellular interactions of the pericyte-like cell populations that reside in the umbilical cord throughout gestation, both in normal pregnancies and those with obstetrical complications.

Acknowledgments The authors would like to thank Farwah Iqbal (Librach lab) for providing unpublished data included in Fig. 12.5 and Denis Gallagher (CReATe Fertility Centre) for his careful review and edits of the manuscript. The authors would like to acknowledge the authors of previously published books and scientific articles for contributions to figures. Some images found in Figs. 12.2, 12.3, and 12.6 were obtained with permission as described in figure legends.

References

An B, Kim E, Song H, Ha KS, Han ET, Park WS, Ahn TG, Yang SR, Na S, Hong SH (2017) Gestational diabetes affects the growth and functions of perivascular stem cells. Mol Cells 40(6):434–439

Andreeva ER, Pugach IM, Gordon D, Orekhov AN (1998) Continuous subendothelial network formed by pericyte-like cells in human vascular bed. Tissue Cell 30(1):127–135

Baksh D, Yao R, Tuan RS (2007) Comparison of proliferative and multilineage differentiation potential of human mesenchymal stem cells derived from umbilical cord and bone marrow. Stem Cells 25(6):1384–1392

Bankowski E, Sobolewski K, Romanowicz L, Chyczewski L, Jaworski S (1996) Collagen and glycosaminoglycans of Wharton's jelly and their alterations in EPH-gestosis. Eur J Obstet Gynecol Reprod Biol 66(2):109–117

Bankowski E, Sobolewski K, Palka J, Jaworski S (2004) Decreased expression of the insulin-like growth factor-I-binding protein-1 (IGFBP-1) phosphoisoform in pre-eclamptic Wharton's jelly and its role in the regulation of collagen biosynthesis. Clin Chem Lab Med 42(2):175–181

Benirschke K, Kaufmann P (1995) Pathology of the human placenta. Springer, New York

Blanco MV, Vega HR, Giuliano R, Grana DR, Azzato F, Lerman J, Milei J (2011) Histomorphometry of umbilical cord blood vessels in preeclampsia. J Clin Hypertens (Greenwich) 13(1):30–34

Caplan AI (2008) All MSCs are pericytes? Cell Stem Cell 3(3):229–230

Can A, Karahuseyinoglu S (2007) Concise review: human umbilical cord stroma with regard to the source of fetus-derived stem cells. Stem Cells 25(11):2886–2895

Can A, Celikkan FT, Cinar O (2017) Umbilical cord mesenchymal stromal cell transplantations: a systemic analysis of clinical trials. Cytotherapy 19(12):1351–1382

Clarke JA (1965) An x-ray microscopic study of the human umbilical arteries. Z Zellforsch Mikrosk Anat 66(2):293–299

Cole J, Israfil-Bayli F (2016) Wharton's jelly: the significance of absence. J Obstet Gynaecol 36(4):500–501

Corrao S, La Rocca G, Lo Iacono M, Corsello T, Farina F, Anzalone R (2013) Umbilical cord revisited: from Wharton's jelly myofibroblasts to mesenchymal stem cells. Histol Histopathol 28(10):1235–1244

Corselli M, Chen CW, Sun B, Yap S, Rubin JP, Peault B (2012) The tunica adventitia of human arteries and veins as a source of mesenchymal stem cells. Stem Cells Dev 21(8):1299–1308

Coskun H, Can A (2015) The assessment of the in vivo to in vitro cellular transition of human umbilical cord multipotent stromal cells. Placenta 36(2):232–239

Covas DT, Panepucci RA, Fontes AM, Silva WA Jr, Orellana MD, Freitas MC, Neder L, Santos AR, Peres LC, Jamur MC, Zago MA (2008) Multipotent mesenchymal stromal cells obtained from diverse human tissues share functional properties and gene-expression profile with CD146+ perivascular cells and fibroblasts. Exp Hematol 36(5):642–654

Crisan M, Yap S, Casteilla L, Chen CW, Corselli M, Park TS, Andriolo G, Sun B, Zheng B, Zhang L, Norotte C, Teng PN, Traas J, Schugar R, Deasy BM, Badylak S, Buhring HJ, Giacobino JP, Lazzari L, Huard J, Peault B (2008) A perivascular origin for mesenchymal stem cells in multiple human organs. Cell Stem Cell 3(3):301–313

Cullen, T. S. (1916). Embryology, anatomy, and diseases of the umbilicus, together with the diseases of the urachus. Philadelphia Saunders

Davies JE, Walker JT, Keating A (2017) Concise review: Wharton's jelly: the rich, but enigmatic, source of mesenchymal stromal cells. Stem Cells Transl Med 6(7):1620–1630

Diaz-Flores L, Gutierrez R, Madrid JF, Varela H, Valladares F, Acosta E, Martin-Vasallo P, Diaz-Flores L Jr (2009) Pericytes. Morphofunction, interactions and pathology in a quiescent and activated mesenchymal cell niche. Histol Histopathol 24(7):909–969

Dominici M, Le Blanc K, Mueller I, Slaper-Cortenbach I, Marini F, Krause D, Deans R, Keating A, Prockop D, Horwitz E (2006) Minimal criteria for defining multipotent mesenchymal stromal cells. The International Society for Cellular Therapy position statement. Cytotherapy 8(4):315–317

Downs KM (1998) The murine allantois. Curr Top Dev Biol 39:1–33
Downs KM, Gifford S, Blahnik M, Gardner RL (1998) Vascularization in the murine allantois occurs by vasculogenesis without accompanying erythropoiesis. Development 125(22):4507–4520
Ellison JP (1971) The nerves of the umbilical cord in man and the rat. Am J Anat 132(1):53–60
Friedenstein AJ, Chailakhjan RK, Lalykina KS (1970) The development of fibroblast colonies in monolayer cultures of Guinea-pig bone marrow and spleen cells. Cell Tissue Kinet 3(4):393–403
Friedman R, Betancur M, Boissel L, Tuncer H, Cetrulo C, Klingemann H (2007) Umbilical cord mesenchymal stem cells: adjuvants for human cell transplantation. Biol Blood Marrow Transplant 13(12):1477–1486
Guadix JA, Zugaza JL, Galvez-Martin P (2017) Characteristics, applications and prospects of mesenchymal stem cells in cell therapy. Med Clin (Barc) 148(9):408–414
Holm A et al (2018) Microvascular mural cell Organotypic heterogeneity and functional plasticity. Trends Cell Biol 28(4):302–316
Hong SH, Maghen L, Kenigsberg S, Teichert AM, Rammeloo AW, Shlush E, Szaraz P, Pereira S, Lulat A, Xiao R, Yie SM, Gauthier-Fisher A, Librach CL (2013) Ontogeny of human umbilical cord perivascular cells: molecular and fate potential changes during gestation. Stem Cells Dev 22(17):2425–2439
Hoyes AD (1969) Ultrastructure of the epithelium of the human umbilical cord. J Anat 105(Pt 1):149–162
Inman KE, Downs KM (2007) The murine allantois: emerging paradigms in development of the mammalian umbilical cord and its relation to the fetus. Genesis 45(5):237–258
Iqbal F, Szaraz P, Librach M, Gauthier-Fisher A, Librach CL (2017) Angiogenic potency evaluation of cell therapy candidates by a novel application of the in vitro aortic ring assay. Stem Cell Res Ther 8(1):184
Ishige I, Nagamura-Inoue T, Honda MJ, Harnprasopwat R, Kido M, Sugimoto M, Nakauchi H, Tojo A (2009) Comparison of mesenchymal stem cells derived from arterial, venous, and Wharton's jelly explants of human umbilical cord. Int J Hematol 90(2):261–269
Joerger-Messerli M, Bruhlmann E, Bessire A, Wagner A, Mueller M, Surbek DV, Schoeberlein A (2015) Preeclampsia enhances neuroglial marker expression in umbilical cord Wharton's jelly-derived mesenchymal stem cells. J Matern Fetal Neonatal Med 28(4):464–469
Junek T, Baum O, Lauter H, Vetter K, Matejevic D, Graf R (2000) Pre-eclampsia associated alterations of the elastic fibre system in umbilical cord vessels. Anat Embryol (Berl) 201(4):291–303
Jurewicz E, Kasacka I, Bankowski E, Filipek A (2014) S100A6 and its extracellular targets in Wharton's jelly of healthy and preeclamptic patients. Placenta 35(6):386–391
Karahuseyinoglu S, Cinar O, Kilic E, Kara F, Akay GG, Demiralp DO, Tukun A, Uckan D, Can A (2007) Biology of stem cells in human umbilical cord stroma: in situ and in vitro surveys. Stem Cells 25(2):319–331
Khati NJ, Enquist EG, Javitt MC (1998) Imaging of the umbilicus and periumbilical region. Radiographics 18(2):413–431
Kim J, Piao Y, Pak YK, Chung D, Han YM, Hong JS, Jun EJ, Shim JY, Choi J, Kim CJ (2015) Umbilical cord mesenchymal stromal cells affected by gestational diabetes mellitus display premature aging and mitochondrial dysfunction. Stem Cells Dev 24(5):575–586
Kulkarni ML, Matadh PS, Ashok C, Pradeep N, Avinash T, Kulkarni AM (2007) Absence of Wharton's jelly around the umbilical arteries. Indian J Pediatr 74(8):787–789
Lim J, Razi ZR, Law J, Nawi AM, Idrus RB, Ng MH (2016) MSCs can be differentially isolated from maternal, middle and fetal segments of the human umbilical cord. Cytotherapy 18(12):1493–1502
Loibl M, Binder A, Herrmann M, Duttenhoefer F, Richards RG, Nerlich M, Alini M, Verrier S (2014) Direct cell-cell contact between mesenchymal stem cells and endothelial progenitor cells induces a pericyte-like phenotype in vitro. Biomed Res Int 2014:395781
Malas MA, Sulak O, Gokcimen A, Sari A (2003) Morphology of umbilical vessels in human fetuses: a quantitative light microscopy study. Eur J Morphol 41(5):167–174

Manea A, Manea SA, Todirita A, Albulescu IC, Raicu M, Sasson S, Simionescu M (2015) High-glucose-increased expression and activation of NADPH oxidase in human vascular smooth muscle cells is mediated by 4-hydroxynonenal-activated PPARalpha and PPARbeta/delta. Cell Tissue Res 361(2):593–604

McElreavey KD, Irvine AI, Ennis KT, McLean WH (1991) Isolation, culture and characterisation of fibroblast-like cells derived from the Wharton's jelly portion of human umbilical cord. Biochem Soc Trans 19(1):29S

Meirelles Lda S, Fontes AM, Covas DT, Caplan AI (2009) Mechanisms involved in the therapeutic properties of mesenchymal stem cells. Cytokine Growth Factor Rev 20(5–6):419–427

Montemurro T, Andriolo G, Montelatici E, Weissmann G, Crisan M, Colnaghi MR, Rebulla P, Mosca F, Peault B, Lazzari L (2011) Differentiation and migration properties of human foetal umbilical cord perivascular cells: potential for lung repair. J Cell Mol Med 15(4):796–808

Nanaev AK, Kohnen G, Milovanov AP, Domogatsky SP, Kaufmann P (1997) Stromal differentiation and architecture of the human umbilical cord. Placenta 18(1):53–64

Parry EW (1970) Some electron microscope observations on the mesenchymal structures of full-term umbilical cord. J Anat 107(Pt 3):505–518

Parry EW, Abramovich DR (1970) Some observations on the surface layer of full-term human umbilical cord epithelium. J Obstet Gynaecol Br Commonw 77(10):878–884

Parry EW, Abramovich DR (1972) The ultrastructure of human umbilical vessel endothelium from early pregnancy to full term. J Anat 111(Pt 1):29–42

Pearson AA, Sauter RW (1970) Nerve contributions to the pelvic plexus and the umbilical cord. Am J Anat 128(4):485–498

Ryu YJ, Seol HS, Cho TJ, Kwon TJ, Jang SJ, Cho J (2013) Comparison of the ultrastructural and immunophenotypic characteristics of human umbilical cord-derived mesenchymal stromal cells and in situ cells in Wharton's jelly. Ultrastruct Pathol 37(3):196–203

Samsonraj RM, Raghunath M, Nurcombe V, Hui JH, van Wijnen AJ, Cool SM (2017) Concise review: multifaceted characterization of human mesenchymal stem cells for use in regenerative medicine. Stem Cells Transl Med 6(12):2173–2185

Sarugaser R, Lickorish D, Baksh D, Hosseini MM, Davies JE (2005) Human umbilical cord perivascular (HUCPV) cells: a source of mesenchymal progenitors. Stem Cells 23(2):220–229

Sarugaser R, Ennis J, Stanford WL, Davies JE (2009) Isolation, propagation, and characterization of human umbilical cord perivascular cells (HUCPVCs). Methods Mol Biol 482:269–279

Schugar RC, Chirieleison SM, Wescoe KE, Schmidt BT, Askew Y, Nance JJ, Evron JM, Peault B, Deasy BM (2009) High harvest yield, high expansion, and phenotype stability of CD146 mesenchymal stromal cells from whole primitive human umbilical cord tissue. J Biomed Biotechnol 2009:789526

Seano G, Chiaverina G, Gagliardi PA, di Blasio L, Sessa R, Bussolino F, Primo L (2013) Modeling human tumor angiogenesis in a three-dimensional culture system. Blood 121(21):e129–e137

Sexton AJ, Turmaine M, Cai WQ, Burnstock G (1996) A study of the ultrastructure of developing human umbilical vessels. J Anat 188(Pt 1):75–85

Sheppard BL, Bishop AJ (1973) Electron microscopical observations on sheep umbilical vessels. Q J Exp Physiol Cogn Med Sci 58(1):39–45

Shende P, Gupta H, Gaud RS (2018) Cytotherapy using stromal cells: current and advance multi-treatment approaches. Biomed Pharmacother 97:38–44

Shlush E, Maghen L, Swanson S, Kenigsberg S, Moskovtsev S, Barretto T, Gauthier-Fisher A, Librach CL (2017) In vitro generation of Sertoli-like and haploid spermatid-like cells from human umbilical cord perivascular cells. Stem Cell Res Ther 8(1):37

Sims DE (2000) Diversity within pericytes. Clin Exp Pharmacol Physiol 27(10):842–846

Singh N, Rao S, Sobti P, Khurana N (2012) Multiple vessels in the umbilical cord: a report of four cases. Indian J Pathol Microbiol 55(4):597–598

Spurway J, Logan P, Pak S (2012) The development, structure and blood flow within the umbilical cord with particular reference to the venous system. Australas J Ultrasound Med 15(3):97–102

Stehbens WE, Wakefield JS, Gilbert-Barness E, Zuccollo JM (2005) Histopathology and ultrastructure of human umbilical blood vessels. Fetal Pediatr Pathol 24(6):297–315

Subramanian A, Fong CY, Biswas A, Bongso A (2015) Comparative characterization of cells from the various compartments of the human umbilical cord shows that the Wharton's jelly compartment provides the best source of clinically utilizable mesenchymal stem cells. PLoS One 10(6):e0127992

Szaraz P, Librach M, Maghen L, Iqbal F, Barretto TA, Kenigsberg S, Gauthier-Fisher A, Librach CL (2016) In vitro differentiation of first trimester human umbilical cord perivascular cells into contracting cardiomyocyte-like cells. Stem Cells Int 2016:7513252

Takechi K, Kuwabara Y, Mizuno M (1993) Ultrastructural and immunohistochemical studies of Wharton's jelly umbilical cord cells. Placenta 14(2):235–245

Wajid N, Naseem R, Anwar SS, Awan SJ, Ali M, Javed S, Ali F (2015) The effect of gestational diabetes on proliferation capacity and viability of human umbilical cord-derived stromal cells. Cell Tissue Bank 16(3):389–397

Wetzig A, Alaiya A, Al-Alwan M, Pradez CB, Pulicat MS, Al-Mazrou A, Shinwari Z, Sleiman GM, Ghebeh H, Al-Humaidan H, Gaafar A, Kanaan I, Adra C (2013) Differential marker expression by cultures rich in mesenchymal stem cells. BMC Cell Biol 14:54

Yang HT, Chao KC (2013) Foetal defence against cancer: a hypothesis. J Cell Mol Med 17(9):1096–1098

Chapter 13
The Pluripotent Microvascular Pericytes Are the Adult Stem Cells Even in the Testis

Michail S. Davidoff

Abstract The pericytes of the testis are part of the omnipresent population of pericytes in the vertebrate body and are the only true pluripotent adult stem cells able to produce structures typical for the tree primitive germ layers: ectoderm, mesoderm, and endoderm. They originate very early in the embryogenesis from the pluripotent epiblast. The pericytes become disseminated through the whole vertebrate organism by the growing and differentiating blood vessels where they remain in specialized periendothelial vascular niches as resting pluripotent adult stem cells for tissue generation, maintenance, repair, and regeneration. The pericytes are also the ancestors of the perivascular multipotent stromal cells (MSCs). The variable appearance of the pericytes and their progeny reflects the plasticity under the influence of their own epigenetic and the local environmental factors of the host organ. In the testis the pericytes are the ancestors of the neuroendocrine Leydig cells. After activation the pericytes start to proliferate, migrate, and build transit-amplifying cells that transdifferentiate into multipotent stromal cells. These represent progenitors for a number of different cell types in an organ. Finally, it becomes evident that the pericytes are a brilliant achievement of the biological nature aiming to supply every organ with an omnipresent population of pluripotent adult stem cells. Their fascinating features are prerequisites for future therapy concepts supporting cell systems of organs.

Keywords Pericytes · Testis · Pericyte origin · Testis microvasculature · Vascular niche · Stem cell migration · Transit amplifying cells · Transdifferentiation · Adult stem cells · Neuroendocrine features · Peritubular cells · Neural crest cells · Periendothelial cells · Perivascular cells · Pericyte plasticity

M. S. Davidoff (✉)
University Medical Center Hamburg-Eppendorf, Hamburg Museum of Medical History, Hamburg, Germany
e-mail: davidoff@uke.de

A. Birbrair (ed.), *Pericyte Biology in Different Organs*, Advances in Experimental Medicine and Biology 1122,
https://doi.org/10.1007/978-3-030-11093-2_13

13.1 Introduction

Until shortly the view existed that single organs or tissues possess its own tissue-specific (typical) adult stem cells (Soncin and Ward 2011; Rolandsson et al. 2014). Presently three main cell populations are supposed to be adult stem cells, namely, the stromal (mesenchymal) stem cells, the neural crest stem cells, and the pericytes.

We already evidenced that pericytes are the ancestors of the neuroendocrine Leydig cells of the testes (Davidoff et al. 1993, 2004, 2009; Davidoff 2017). We applied in vivo the chemical ablation experiment with EDS (ethane dimethanesulfonate) for adult Leydig cell destruction and subsequent full regeneration and succeeded to verify the individual developmental stages, to identify and localize the Leydig stem cells, and to follow up the fate of regenerating Leydig cell progenitors (Kerr et al. 1985, 1986, 1987a, b; Molenaar et al. 1985; Benton et al. 1995; Teerds 1996; Teerds et al. 1999, 2007; Teerds and Rijntjes 2007; Davidoff et al. 2004, 2009).

The most important conclusions obtained in our previous studies are the following:

1. At the beginning we provided evidence that the Leydig cells are neuroendocrine cells (Davidoff et al. 1993).
2. We performed an in vivo experiment. This is important because as known any in vitro manipulation of the stem cells, inclusive isolation, and putting them in a cell culture lead to changes of their original status and move away the in vivo features, as this is the case with the artificial embryonic stem cells (Rossant 2001; Smith 2001; Buehr and Smith 2003; Zwaka and Thomson 2005).
3. The stem cells (ancestors) of the testicular Leydig cells are the pericytes and the smooth muscle cells of the microvasculature. Both cell types share the same lineage.
4. The pericytes are distributed during embryogenesis throughout the organism by the developing cardiovascular system that is the first functioning organ system during the embryonal development.
5. The pericytes remain the whole life within the authentic microvascular niches as reserve stem cell population for tissue and organ repair and regeneration.
6. The pericytes are pluripotent cells, such as the epiblast (De-Miguel et al. 2010). They possess ectodermal, endodermal, and mesodermal lineage progeny which underlines their plasticity and heterogeneity.
7. The testis pericytes behave as authentic stem cells (Davidoff et al. 2004, 2009). After activation they start to proliferate, self-renew, and migrate out from the vascular niche to the perivascular space and the interstitium in form of transit-amplifying (intermediate) cells which transdifferentiate and generate multipotent stromal (mesenchymal) cells, resp., immature committed cells. The resulting cells, under the influence of the existing local factors, differentiate further as typical for an organ or cell culture mature cell phenotype (Leydig cells, smooth muscle cells, peritubular myoid cells, macrophages, fibroblasts, germ cells—DeFalco et al. 2015; Yoshida et al. 2007) characteristic for the

organ or tissue in which they are situated. This illustrates the high plasticity of and how the pericytes adopt their phenotype when transplanted in another place or organ and provides evidence that the pericytes are true stem cells.

8. The impossibility to observe the fast transdifferentiation of the transit-amplifying stem cells caused misrecognition of some structures and misinterpretation of the results obtained in a lot of previous studies.
9. The pericytes are the ancestors of the multipotent adult stromal cells and the only authentic adult stem cell (ASC) phenotype of the vertebrate organism. Thus, the migrating transit-amplifying pericyte descendants represent the multipotent mesodermal (until shortly mesenchymal stem) stromal cells.
10. Definitively, the applied ethane dimethanesulfonate (EDS) experiment provides evidence that the testis pericytes are authentic (adult) stem cells which are pluripotent and are able to produce structures characteristic for the tree embryologic germ layers, namely, ectoderm, mesoderm, and endoderm. Thus, the pericytes of the testis are not organ-specific adult stem cells, but rather they are part of the common for the whole organism omnipresent pericyte population. The neuroendocrine qualities of the Leydig cells allowed the presumption that pericytes may derive from the neural crest cells or from a common stem cell population segregated early in embryogenesis with progeny for both the neural crest and the mesoderm lineage.
11. The pericytes are the adult stem cells of the vertebrate organism inclusive in the testis.

This chapter synthesizes current knowledge on the features of the microvascular pericytes in the testes. Part of this chapter is the continuation of our earlier work (Davidoff et al. 2004, 2009; Friedrich et al. 2012). We will mainly use the term pericytes while it's proven that the pericytes and the vascular smooth muscle cells share a common lineage (Etchevers et al. 2001). Because the pericytes of the testis belong to the whole body pericyte population, we try to present evidence that the testis pericytes are actually member of the omnipresent (pan-organ stem cells; Péault et al. 2007) pluripotent adult stem cell family of the vertebrate organism, partly by comparing the data from the testis with these of other organs, especially the pericytes of the nervous system. The testicular pericytes possess all features of the pericytes in other organs. Their heterogeneity and differences depend on the factors existing in the microenvironment of the host organ and the stage of differentiation and activity. This is why if transplanted into another organ, the pericytes differentiate into typical for the host organ cells (Dar et al. 2012; Tang et al. 2008 b; Bhagavati 2008; Dellavalle et al. 2011; Bouacida et al. 2012). This is a typical feature of the pericytes as adult stem cells and explains their great plasticity and ability to transdifferentiate (Bjornson et al. 1999; Brazelton et al. 2000; Abedin et al. 2004—MSCs in vascular walls; Millner et al. 2008; Diaz-Flores et al. 2006, 2009—in the nervous system; Wang et al. 2012; Jiang et al. 2003; Hermann et al. 2016). For example, in a series of publications, Birbrair et al. (2013a, b, 2014) and Birbrair and Delbono 2015) established even in the skeletal musculature the existence of two major pericyte subpopulations: Type I and Type II, which possess differing features.

Probably this fact could be explained with the different maturation stages of the pericytes under study.

During the last decades, evidence was provided which leads to changes of some views and interpretations on the pericytes in the testis. These comprise the origin of the pericytes, the relationship between pericytes and mesenchymal stem cells, and their contribution in the development of the testicular structures as well as their behavior during some experimental and pathological conditions (see for details, pericytes and stem cells in the testis, Davidoff et al. 2004, 2009; Ge et al. 2006; Chikhovskaya et al. 2012, 2014; Chen et al. 2010, 2017; Stanley et al. 2012; Kilcoyne et al. 2014; Makala et al. 2015; Pericytes in other organs, predominantly in the nervous system, Dore-Duffy 2008; Le Douarin et al. 2008; Krueger and Bechmann 2010; Dore-Duffy and Cleary 2011; Sá-Pereira et al. 2012; Kurtz 2016).

13.1.1 Short Repetition of the Testis Structure and Nomenclature

The testes and epididymides are paired organs located in the scrotal sac of the external male genital system (Fig. 13.1). The testis consists of 200–400 lobules containing seminiferous tubules and the interstitium situated between them (Brennan and Capel 2004). The testicular interstitium contains connective tissue (fibroblasts, tissue-resident macrophages, and compartmentalizing co-cells), Leydig cells, nerve fibers, blood vessels, and lymphatics (Fawcett et al. 1973; Holstein and Davidoff

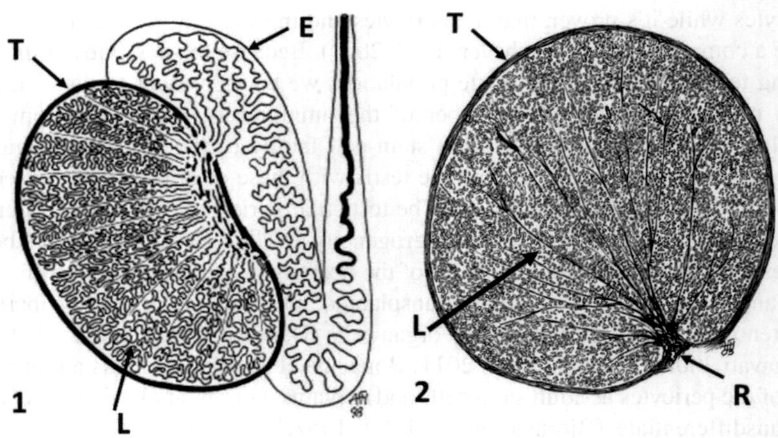

Fig. 13.1 (*1*) Semi-schematic drawing of longitudinal sectioned structural components of the testis and epididymis (E). Arrangement of the seminiferous tubules in the human testis and of the excurrent ductular system of the epididymis. Tunica albuginea (T), lobulus (L) of seminiferous tubule in a space bordered by two testis septa. × 2.5. (*2*) Cross section of a seminiferous tubule of a fertile man 32 years of age. Drawing of a semithin section showing the tunica albuginea (T), a lobulus (L), and the rete testis (R), × 300 (from Holstein et al. 2003)

1997; Holstein 1999; Holstein et al. 2003; DeFalco et al. 2014, 2015; Bhushan and Meinhardt 2017). The seminiferous tubules are enclosed by a layer peritubular myofibroblast cells (PMCs) and extracellular matrix produced by the cooperation between the Sertoli cells and the testicular peritubular cells (Holstein et al. 1996; Skinner et al. 1985) and are compartmentalized in lobules by testicular septa (Figs. 13.1 and 13.2). The testis cords and extracellular matrix are covered by a thick protective cap [a connective tissue testicular capsule named tunica albuginea (Leeson and Cookson 1974)] to build an oval-shaped organ (Figs. 13.1 and 13.2). The tunica albuginea of the human testis contains abundantly contractile elements and is also distinguished by extraordinarily high concentrations of cyclic GMP (cGMP)-dependent protein kinase I, known to mediate cGMP-dependent relaxation (Leeson and Cookson 1974; Holstein et al. 1996; Middendorff et al. 2002).These contractile elements are of importance for the movement of the non-motile spermatozoa from the testis to the epididymis.

The testes have important functions in the origin, maintenance, and functions of the male organism. They produce the sperm and the male hormones (androgens) in vertebrates (Russell and de França 1995; Svingen and Koopman 2013). As mentioned, the testis cords comprise coiled seminiferous tubules with germinal epithelium and peritubular tissue (lamina propria). The germinal epithelium consists of cells that include different developmental stages of germ cells, namely, spermatogonia, primary and secondary spermatocytes, and spermatids. These are located within invaginations of Sertoli cells (Fig. 13.3).The spermatogenesis takes part within the seminiferous tubules (Holstein 1999; Holstein et al. 2003), and the steroid hormones are produced by the neuroendocrine Leydig cells (see Davidoff et al. 2009)

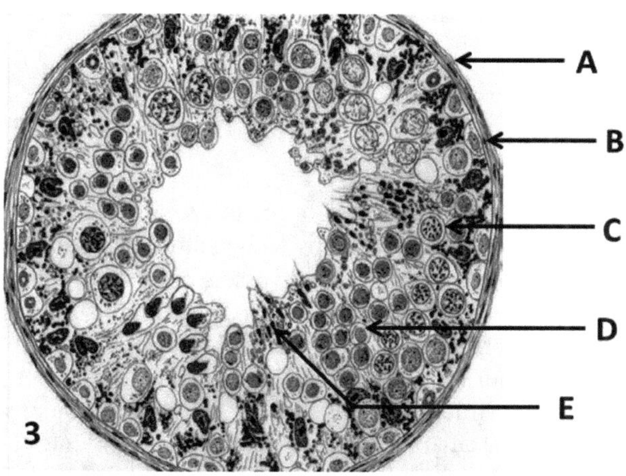

Fig. 13.2 (*3*) Cross section of a seminiferous tubule of a fertile man 32 years of age. Drawing of a semithin section. (**a**) Peritubular tissue (lamina propria of seminiferous tubule). (**b**) Spermatogonium. (**c**) Primary spermatocyte. (**d**) Immature spermatids. (**e**) Mature spermatids ×300 (from Holstein et al. 2003)

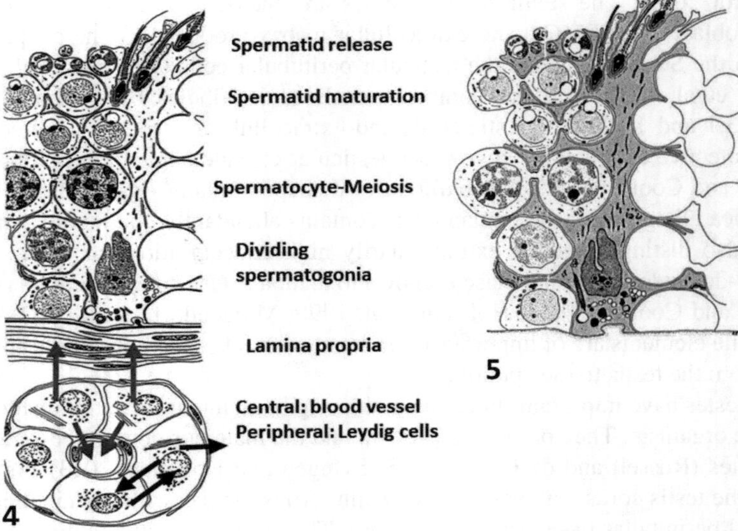

Fig. 13.3 (*4* and *5*) Schematic drawing of a testis segment presenting on the bottom a section of the germinal epithelium with basal lamina propria and below a cluster of Leydig cells surrounding a capillary. Arrows indicate different influences of secreted hormones and growth factors. On the top the main processes of spermatogenesis are shown. (*5*) Demonstrates an enlarged drawing of the top part in (*4*). Visible are different stages of the germ cells development. The germ cells are situated in deep invaginations of a Sertoli cell (from Holstein et al. 2003)

which are situated within the interstitium (Holstein and Davidoff 1997) surrounding the blood vessels and the seminiferous tubules (Fig. 13.3).

13.1.2 Testis Vasculature

It must be emphasized that the pericytes of the testis, similar to the pericytes of all vertebrate organs, show close relationships with the microvasculature. The association of stem cell-like progenitors and blood vessels occurs early in the embryogenesis of the testis (cf. Brennan and Capel 2004). The relationship reaches a level in which the microcirculatory (endothelial) cells direct the testis cord formation and determine their course during embryonic development (Combes et al. 2009). The pericytes themselves are structural component of the microvascular wall.

The testis is a well-vascularized organ that contains (1) "intertubular" arteries and veins running parallel to the tubules and (2) their ramifications building a microcirculatory network consisting of transversally or obliquely running "peritubular" small arterioles, precapillaries, capillaries, postcapillary venules, and small veins (Fig. 13.4; see Müller 1957; Weerasooriya and Yamamoto 1985; Suzuki and Nagano 1986; Murakami et al. 1989; Yoshida et al. 2007; Svingen and Koopman 2013). In

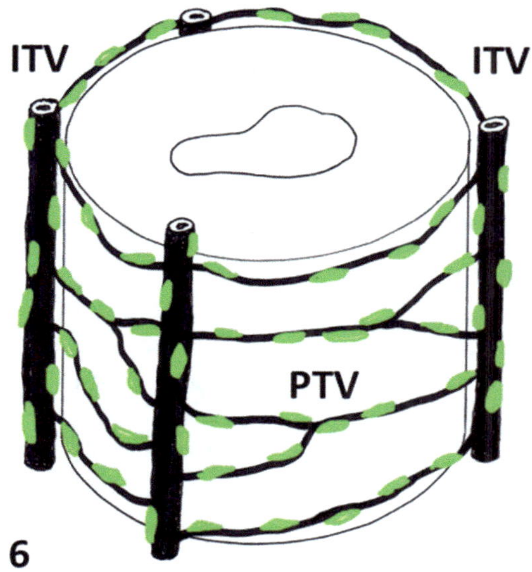

Fig. 13.4 (*6*) Schematic presentation showing the relationship between the testicular blood vessels and the seminiferous tubule. *ITV* intertubular vessels, *PTV* peritubular vessels. The oval blood vessel components illustrate the pericytes. Note the close relationship between the blood vessels/pericytes and the tubule wall and that parts of the tubule surface are devoid of blood vessels (from Davidoff 2017)

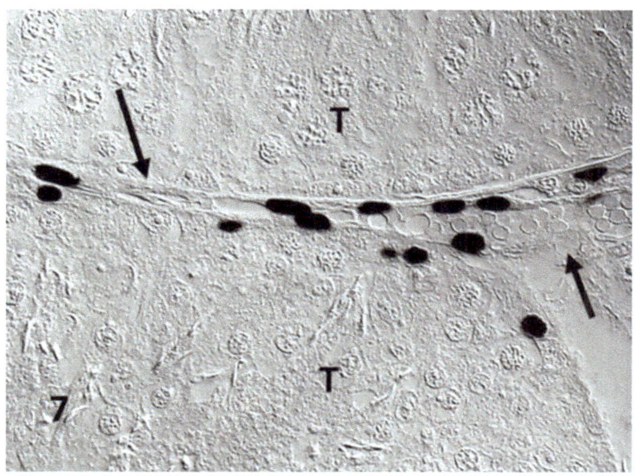

Fig. 13.5 (*7*) Visualization of Bromodeoxyuridine (BrdU) in rat pericyte nuclei 2 days after EDS injection. The capillary is designated by the arrows. Note the close relationship between the vessel/pericytes and the tubules (T) walls. In the lower tubule, single spermatogonia also show BrdU immunoreactivity. ×550 (from Davidoff 2017)

the wall of the microvasculature, numerous pericytes and smooth muscle cells (periendothelial cells, mural cells) are included (Fig. 13.4). The pericytes and the vascular smooth muscle cells share a common lineage (Etchevers et al. 2001). The peritubular microvessels and the seminiferous tubules are in close spatial contact (Fig. 13.5). This fact is the reason that numerous scientists determine inaccurately the lamina propria of the seminiferous tubules as the site of origin of the testicular Leydig cells. These cells are in contact with the endothelial cells and are covered by their common basal membrane that build spaces called vascular niches (Fig. 13.6, electron micrograph; see Yoshida et al. 2007; Scadden 2006; Nikolova et al. 2007). In the human testis, occasionally single capillaries run within the lamina propria and

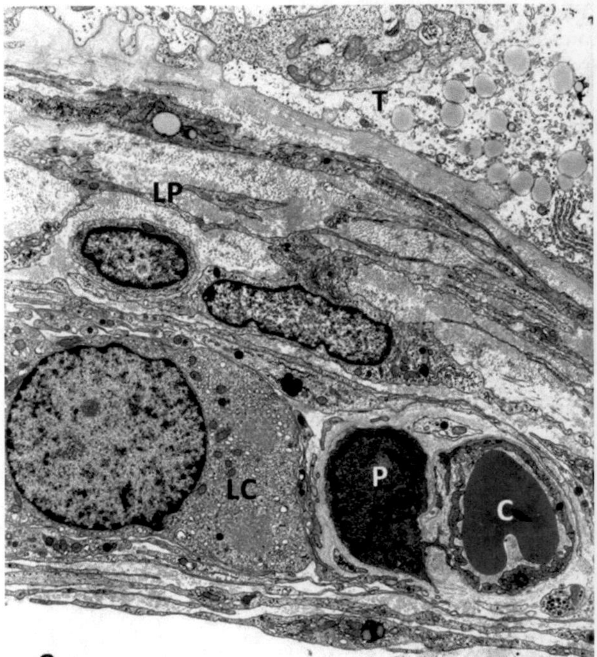

8

Fig. 13.6 (*8*) This is an electron microscopic picture of a human peritubular segment. At the top the basal part of the tubule is seen (T), followed by the multilayered lamina propria (LP) and at the outside surface a part of the interstitium with a capillary (C), periendothelial pericyte (P) in the vascular niche, surrounded by the bright basal lamina, and a Leydig cell (LC). ×4800 (from Davidoff et al. 2009)

are surrounded by processes of the myofibroblasts. The intramural microvessels within the lamina propria possess a partly fenestrated endothelial segment and a continuous basal lamina in which pericytes are included (Ergün et al. 1994, 1996).

13.2 The Pericytes

13.2.1 Short Overview on the History of the Pericytes

The history of the pericyte discovery is presented to some extent contradictory in the literature (Ribatti et al. 2011; Armulik et al. 2011; Attwell et al. 2016).

It was Eberth (1871) who in the small capillaries of the frog eye hyaloid membrane discovered and described stellate "adventitial cells" with wall encircling processes, whereas in the larger capillaries, arteries and veins of the human brain, spinal cord, and retina, these cells he called "outer vessel epithelium" or "perithelium."

Two years later Rouget (1873, 1879) provided a more detailed description (he observed heavily ramified cells that with their processes build perivascular rings at the capillary walls) and observed the gradual transition of these cells into the vascular smooth muscle cells. Rouget expressed the presumption that these "perivascular cells" are contractile structures. In fact, he recognized two types of perivascular cells, namely, amoeboid migratory cells and fusiform contractile cells.

Later Vimtrup (1922) offer a more detailed description of the perivascular cell features and renamed the "adventitial cells" in "Rouget cells" with the argumentation that the first discoverer of these cells was Rouget.

In this respect, Sims (1986) wrote: "Rouget (1873, 1879) is often credited with discovering the pericyte, although Bensley and Vimtrup (1928) acknowledge that Eberth (1871) had previously described a capillary adventitial cell. In 1922, Vimtrup described pericytes in somewhat greater detail. Subsequently, Krogh (1919, 1929), who was one of the original investigators of capillary recruitment (Palm et al. 1983), employed the name "Rouget cell" to describe what had been previously referred to as adventitial cell" (p. 153).

That is not a precise description, because Krogh (1922) writes that he has took over the name "Rouget cells" from Vimtrup. He wrote: "As there can be no doubt that the richly ramified muscle cells on the capillary wall are the same as those originally found by Rouget in the hyaloid membrane, Vimtrup has named them after the first discoverer, and we shall speak of them henceforward as Rouget cells" (p. 54).

It was Zimmermann (1923) who renamed the Rouget cells in pericytes including their transitional forms to smooth muscle fibers. He describes tree pericyte types, namely, precapillary pericytes, capillary pericytes, and postcapillary pericytes. Zimmermann (1923) emphasized that the cells described by Eberth (1871)with minor exceptions are other cells than these described by Rouget (1873). This is not absolutely correct because as later detected, the pericyte population is heterogeneous and can be observed in variable morphological forms depending on the stage of their development and activity (Attwell et al. 2016). Thus, Eberth (1871) was the first who described and depicted these cells that he designated "adventitial cells." According to his description, these adventitial cells are anastomosing stellate cells that with their ramifications wrap the wall of the blood vessels.

In the following years, different single and combined terms were used: "Rouget-Mayer cells" in the human skin (Günther 1917), "adventitial (Rouget) cells" (Clark and Clark 1925a), "Rouget cells" (Dogiel 1899; Mayer 1902; Krogh 1922; Florey and Carleton 1926; Benninghoff 1926; Bensley and Vimtrup 1928; Clark and Clark 1925b; Cervós-Navarro 1963), "Rouget cells (Pericytes)" (Michels 1936), as well as "mural cells," "periendothelial cells," "histiocytes," and "deep cells" (Zimmermann 1923; Hirschi and D'Amore 1996; Bergers and Song 2005; Dore-Duffy 2008; Bergers 2008; Krueger and Bechmann 2010; Armulik et al. 2011; Zouani et al. 2013; van Dijk et al. 2015; see Collett and Canfield 2005, for review).

13.2.2 *The Origin of the Pericytes*

Early studies refer to different sources of the pericyte origin such as the mesenchymal stem cells (Huhtaniemi and Pelliniemi 1992; Gondos 1980; Sbanti et al. 2007; Dore-Duffy 2008) or, in relation to their neuroectodermal properties, the neural crest stem cells (NCSCs; Davidoff et al. 2004, 2009; Calloni et al. 2007; Le Douarin et al. 2008; Dore-Duffy 2008) as well as the peritubular myoid cells (Ariyaratne et al. 2000). There is also a proposal that all pluripotent stem cells of the organism belong to the germ lineage (Ratajczak et al. 2007). However, Brons et al. (2007) provide evidence that the epiblast stem cells are not derivatives of the primordial germ cells. Also the human testis-derived multipotent stem cells do not arise from germ cells (Chikhovskaya et al. 2012, 2014). Gonzalez et al. (2009) termed cells isolated from human testis as gonadal stem cell (GSC) which express pluripotent stem cell markers Oct4, Nanog, and SSEA-4. These GSCs have growth kinetics and marker expression similar to MSCs and differentiate into mesodermal lineage cells. In the testis were found also embryonic cell-like cells (Mizrak et al. 2010). Most likely this cell variability reflects the transit-amplifying cells of the pericyte ancestors in different stages of their maturation and differentiation (Nakagawa et al. 2007).

Concerning the origin of the pericytes, there is actually no consensus beside that they originate from stem cells. However, it became strongly evident that the most possible pericyte ancestor is the early pluripotent embryonal epiblast. In the following some arguments for this conclusion will be presented.

13.2.2.1 Relationship Between the Pericytes and the Multipotent Stromal Cells

The pericytes are the ancestors of the multipotent stromal cells and the only true adult stem cell phenotype of the vertebrate organism.

One main problem is that numerous authors do not make difference between pluripotent pericytes and multipotent stromal cells (MSCs) (Özen et al. 2012; Gökçinar-Yagei et al. 2015; Guimarães-Camboa et al. 2017; but see Cano et al. 2017; see Wong et al. 2015). As Skinner et al. (1985) noted, pericytes are often mixed up with adjacent cell types of the vascular wall, the perivascular space, and the juxtavascular parenchyma. This is important because the most experimentally isolated and cultured mesenchymal stem cells consist of a mixture of both: pericytes and adventitial stem/progenitor cells.

As mentioned, the pericytes have a periendothelial position, whereas the MSCs are situated perivascularly (Rolandsson et al. 2014). Numerous studies postulate that pericytes and multipotent stromal cells (MSCs; until now dubbed mesenchymal stem cells) are the same cells or that all mesenchymal stem cells are pericytes (Bianco et al. 2008; Crisan et al. 2008a; Caplan 2008; Diaz-Flores et al. 2009; Collas 2010; da Silva Meirelles et al. 2013). This view is partially correct, but there is now strong evidence that the pericytes are the ancestors of stromal cells (Caplan 2008; Crisan et al. 2008a; Ratajczak et al. 2008; Lindner et al. 2010; Collas 2010; Zimmerlin et al. 2010, 2012; Murray et al. 2014; Trost et al. 2016), because especially the earlier MSCs—descendants share close similarities with their pericyte ancestors (Ochs et al. 2013).

Pericytes are highly undifferentiated pluripotent cells (De-Miguel et al. 2010; Dar et al. 2012), whereas the MSCs as more differentiated are multipotent. For example, cultured bone marrow-derived pericytes showed stem cell properties, with higher plasticity (Kucia et al. 2005) and immaturity state than standard MSCs (Bouacida et al. 2012).

Furthermore, a lot of authors interpret the existence or the absence of one or two marker substances as enough to define or deny a different lineage of cells, which are in fact pericyte daughter cells in various stages of differentiation. There are results showing that the MSCs (APCs, adventitial progenitor cells) in cell culture conditions possess the capability to dedifferentiate into pericytes: "centripetal" relationship (Pacini and Petrini 2014; Rowley and Johnson 2014; Chen et al. 2017). The reason for such events is the great plasticity of the stem cells appearing especially during experimental or pathological conditions.

As mentioned, pericytes are also the ancestors of the adult perivascular multipotent mesodermal cells that represent the migrating transit-amplifying descendants of the periendothelial pericytes. Crisan et al. (2008b) found that pericytes and vascular smooth muscle cells express mesenchymal stem cells (MSCs) markers in human tissues and behave as MSCs in vivo as well as after heterogeneous transplantation. The pericytes contribute to tissue-specific lineages in vivo through the transdifferentiation of their transit-amplifying mesodermal progeny which is primarily located perivascularly (Sbanti et al. 2007; Dellavalle et al. 2007; Tang et al. 2008; Traktuev et al. 2008; Chen et al. 2015) .

Outside the vascular niche, the progeny of the pericytes behave as multipotent stromal stem cells. These cells undergo genetic changes (Stanley et al. 2011) and obtain different morphology and function in comparison with their ancestors. This is an important event that caused numerous investigators to be not able to recognize the fast transdifferentiation of the pericytes and to define the transit-amplifying cells not as progenitor cells but as stem cells. This fact leads to the contradictory interpretations of a number of results in the literature and the determination of the lineage relationship between the ancestors and the progeny of a cell type (Davidoff et al. 2009; Davidoff 2017). Especially in the testis, the pluripotent pericytes are Leydig stem cells, and their migrating progeny in the interstitium are the multipotent stromal cells or mesodermal stromal cells with spindle-shaped form. It is also possible that the peritubular myoid cells (PMC) of the lamina propria of the seminiferous tubules represent differentiating multipotent PMC progenitor cells. The same is true for the testis tissue pericyte-macrophages (Thomas 1999).

There is also strong evidence that adipocytes arise from progenitor cells (mural cells, pericytes) of the adipose vasculature (Kahn 2008; Tang et al. 2008). Moreover, the pericytes have the most potent adipogenic potential (Zimmerlin et al. 2010). In addition, the combination of endothelial progenitor cells with adipose stem cells in ischemic tissues shows a distinct higher neovascularization capacity in comparison to a single stem cell application (Traktuev et al. 2008).

Wang et al. (2012) were able to perform induced differentiation of human adipose-derived stem cells into Leydig cells.

13.2.2.2 The Pericytes and the Peritubular Myofibroblast Cells

There is strong evidence that during normal development and after EDS treatment (between days 20 and 30), new Leydig cells were observed in peritubular position as sheaths surrounding the seminiferous tubules (Davidoff et al. 2009; Fig. 13.7). A number of authors presume that these Leydig cells arise from the peritubular myoid stem cells (Maekawa et al. 1996; Ariyaratne et al. 2000; Ge et al. 2006; Stanley et al. 2012). However, some of the authors designate these cells as Leydig cell progenitors (PDGF-A positive) which is more accurate (Fecteau et al. 2006). Studies concerning the testis development describe that the peritubular myoid cells are interstitial stromal cells that are attracted by the Sertoli cells, both building the basal membrane and extracellular matrix which stabilize the differentiating seminiferous tubules (Skinner et al. 1985; Svingen and Koopman 2013). We presume that the peritubular myoid cells are also pericyte progeny (see Hall 2006; Diaz-Flores et al. 2012). It is well known that the activated pericytes generate myofibroblasts under different physiological and pathological circumstances (Ribatti and Vacca 2008; Kennedy et al. 2014; Hosaka et al. 2016; Rowley and Johnson 2014). After the seminiferous tubules become created, developing Leydig cells in form of transit-amplifying cells become attracted, to build an additional fetal-type Leydig cell sheath around the tubules. These Leydig cells provide the tubules and the peritubular myoid cells with hormones necessary for the final differentiation of the tubular structures.

EDS treatment of adult rats also leads to damage of the seminiferous epithelium.

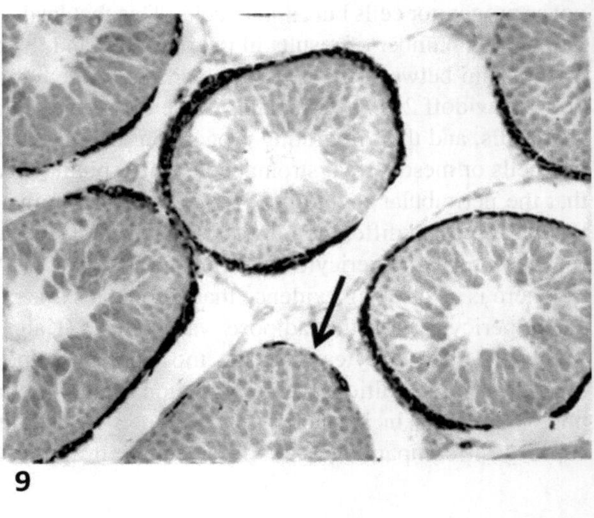

9

Fig. 13.7 (*9*) Immunoreactivity for Cytochrome P450 cholesterol side chain cleavage enzyme (CytP450scc) in young rat Leydig cells on day 21 after EDS treatment. Note the massive peritubular accumulation. Over some tubules fragments a lower number or no Leydig cell are seen (arrow), probably at sites where the spermatogenesis in the tubules became restored. ×450 (from Davidoff et al. 2004)

The fact that early Leydig cells accumulate around tubules with impaired spermatogenesis suggests that Sertoli cell- or germ cell-derived factors promote the increase in number and activity of adult-type Leydig cells (Kerr and Sharpe 1985; Paniagua et al. 1988; Sharpe 1994; Teerds 1996). It is known that Leydig cell numbers can change in adult species depending on the stages of spermatogenesis (Paniagua et al. 1988). In addition, there is strong evidence for a narrow relationship between Leydig cells and Sertoli cells with respect to the regulation of their number and functional activity (Sharpe 1994).

The close relationship between the blood vessels and the seminiferous tubules provides the best conditions for exchange and influence of substances. Thus, the migrating pericyte progeny move along the peritubular vessels and remain in contact with the tubular surface after being attracted and linked by the Sertoli cells and the peritubular myofibroblasts. In this respect, the Sertoli cells, the peritubular myoid cells, and the Leydig progenitor cells exchange signals that regulate these events. The most important among them are the PDGF proteins and their receptors, the androgen receptors, and some other factors. For example, Fecteau et al. (2006) report that in the rat, fetus PDGF-A expression corresponds with the rapid proliferation (Basciani et al. 2002—fetal and adult human testis) and the testosterone expression of the Leydig cells. PDGFs were demonstrated in rat Sertoli cells which attract αSMA-expressing peritubular myoid cells (PMCs) in close proximity to the seminiferous tubules (Gnessi et al. 1995, 2000). Mariani et al. (2002) found that PDGF--B/α-ß interactions may be involved in the proliferation and migration of peritubular myoid cell (PMC) precursors from the intertubular space to the peritubular. Ergün et al. (1994, 1999) provide evidence that endothelin produced in Leydig and Sertoli cells can act in a paracrine manner via ET-B receptors in the human testicular microvasculature and the peritubular myofibroblast. In addition, Ergün et al. (1996) found in the lamina propria of the human testis intramural vessels which are partly fenestrated and possess pericytes. This seems to be partially true for the rat testis in which accumulation of peritubular regenerating Leydig cells in capillaries closely running peritubularly has been observed (Fig. 13.6; Davidoff et al. 2004, 2009). Earlier study by Ge et al. (2006) provide evidence that the testicular pericytes are PDGFR-α and PDGFR-ß positive and that at 1 week of postnatal development spindle-shaped cells in the testicular interstitium, positive for PDGFR-α, differentiate into adult Leydig cells which become immunoreactive for all PDGFs and their receptors.

Concerning the androgen receptors (AR), it was shown that testicular AR mRNA exists in Leydig cells, pericytes, peritubular myoid cells, and Sertoli cells (Shan et al. 1995). The Leydig cells are maximally sensitive to androgen during puberty, which is consistent with the hypothesis that androgens facilitate their differentiation. As mentioned, Kilcoyne et al. (2014) found that the interstitial Leydig cell progenitors express androgen receptors (Shan and Hardy 1992). In addition, it was established that the androgen action via testicular peritubular myoid cells (PMC) is essential for normal testis function, spermatogenesis, and fertility in males (Welsh et al. 2009). Moreover, the androgen action via testicular arteriole smooth muscle cells is important for Leydig cell differentiation, function, vasomotion, and fluid dynamics (Welsh et al. 2010, 2011). There is no convincing proof that the PMCs are the Leydig stem cells.

In conclusion, the peritubular myofibroblasts may be also derived by the pericyte transit-amplifying cells. In combination with the Sertoli cells, the PMC becomes peritubular cells that are important for the differentiation of the seminiferous tubules and the steroidogenesis of the newly arising Leydig cells as well as for the fetal microcirculation. Important significance for these processes seems to have PDGFs and their receptors, the androgen receptors (ARs) and the COUP-TFII, and its receptor H3K27me3 (Haider et al. 2007; Kilcoyne et al. 2014). The PTM-AR signaling is important for normal development, ultrastructure, and function of the adult LCs (Shan and Hardy 1992; Shan et al. 1995, 1997; Welsh et al. 2009, 2010; Takamoto 2005; Shibata et al. 2001– in human adrenals). COUP-TFII plays important role in progenitor Leydig cells formation and early testis organogenesis. In addition, it plays a major role in differentiation but not maintenance of Leydig cells (You et al. 2005).

In addition, new findings established that testicular macrophages play essential roles in normal testis homeostasis and fetal testicular development (Bhushan and Meinhardt 2017). Testicular macrophages are closely involved in regulating steroidogenesis of Leydig cells and in maintaining homeostasis within the testis (Goluža et al. 2014). Concerning the macrophages it has to be clarified what kind of macrophages (pericyte-macrophages; originating from pericytes, Ling and Wong 1993; Balabanov et al. 1996; Thomas 1999; Mori and Leblond 1969; Barón and Gallego 1972) or resident tissue macrophages (mesodermal cells from the yolk sac) are responsible for these events (DeFalco et al. 2014).

The EDS destruction of the newly formed peritubular Leydig cells leads to significant lowering of their number, but the new building of interstitial Leydig cells remains not affected. As Stanley et al. (2012) found that after EDS treatment, the adult testis peritubular Leydig cells originate from stem cells that are not macrophages, pericytes, and smooth muscle cells of vessels. At the same time, they established that the removal of newly differentiated, testosterone-producing LCs from the tubule surfaces did not result in the depletion of the precursor cells. Rather, new Leydig cells were able to form repeatedly. It is also interesting to note that when progenitor Leydig cells acquire LH receptors, they are located partially away from the peritubular region of the testis. In addition, they are definitely rounder in shape, in contrast to their original spindle-shape configuration. These findings suggest that in cells committed to the Leydig cell lineage, acquisition of LH receptors occurs during a stage of development later than that at which they first gain the steroidogenic enzymes. Accordingly, these findings suggest that the function of LH in Leydig cell differentiation occurs subsequent to formation of the progenitor cells (i.e., transformation of progenitors to immature and then mature Leydig cells; Stanley et al. 2012). According to our hypothesis, these results confirm that the progenitor cells of these new Leydig cells are transit-amplifying pericyte progeny which after activation migrate toward the tubules surface. The pericyte ancestors become activated through the lowering of the testosterone production and eventually the hypoxia arising under these conditions. The authors were unable to notice the fast transdifferentiation process to found the important relationship between the pericyte stem cells and their transit-amplifying cells, the Leydig cell progenitors. As mentioned above, the transit-amplifying cells in their further differentiation undergo changes and start to produce factors such PDGFR-α. We would like to underline that during testis development, the new arising Leydig cells remain probably the whole life in direct contact with the testis vasculature through their long branches (Figs. 13.8 and 13.9).

There is strong evidence that the pericyte progeny change the biochemical composition during its commitment to immature and further differentiation to adult,

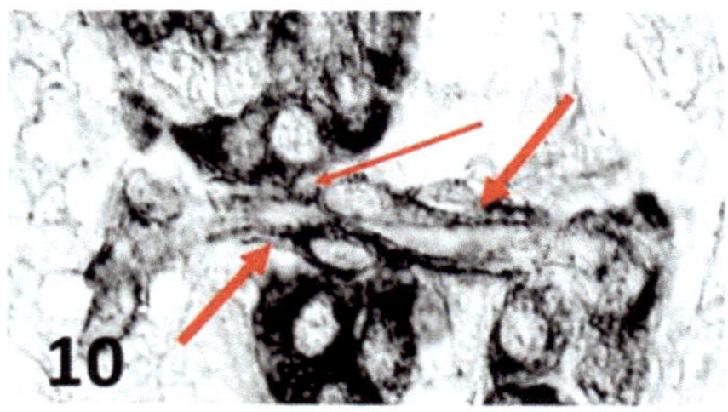

Fig. 13.8 (*10*) Nestin immunoreactivity in activated pericytes 14 days after EDS application. Note the spindle-shaped form of the pericytes on the capillary wall (thick arrows). The migrating pericytes form aggregates in the vicinity of the vessels. Some pericytes have still contact with the vessel wall (thin arrow). ×950 (from Davidoff et al. 2004)

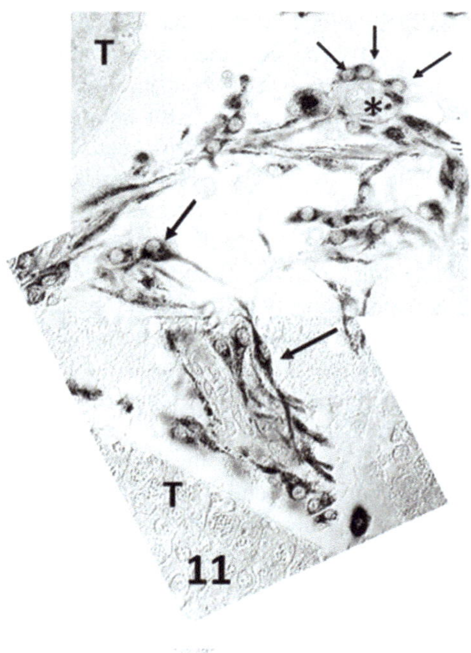

Fig. 13.9 (*11*) Photo montage of Rat testis interstitium. Immunoreactivity for CytP450scc on day 60 of postnatal development. Numerous positive young (immature) Leydig cells possess spindle-shaped form (large arrows). Their outgrowths show contacts with the vessel walls or cover the walls over long distances. The bodies of cross-sectioned Leydig cells in contact with the vessel (asterisk) are round or oval (small arrows). *T* tubules. ×380 (from Davidoff 2017)

mature Leydig cells (Stanley et al. 2011). The leading event is the process of fast transdifferentiation (Tosh and Slack 2002, p. 128; Shen et al. 2004a, b; Tavazoie et al. 2008) in which the early progeny of the pericytes become transformed into transit-amplifying cells, immature and mature Leydig cells that lose their similarities with the stem cell ancestors. This process affects also the potency of the stem/ progenitor cells. The process of transdifferentiation of the transit-amplifying stem cell progeny, under the existing environmental factors, has led to misrecognition of some structures and to misinterpretation of the results obtained in a lot of previous studies (de Souza et al. 2016—MSCs produce pericytes?). For example, Klein et al. (2011, 2014) and Chan-Ling (1997) show that CD44+ multipotent stem cells differentiate into pericytes and smooth muscle cells. In addition Klein et al. (2014) show that Nestin+ tissue-resident multipotent stem cells stimulate tumor progression by differentiating into pericytes and smooth muscle cells. On the other hand, Stanley et al. (2012) informed that the peritubular cells that are immunoreactive for PDGFR-α and negative for 3-ßHSD are stem Leydig cells. In similar cells 6 days after EDS treatment, Kilcoyne et al. (2014) found COUP-TFII immunoreactivity. In other studies the PDGFR-α-positive spindle-shaped interstitial cells of 1 week postpartum rats were called putative SLCs (Ge et al. 2006) or correctly designated as peritubular progeny of Leydig stem cells (Landreh et al. 2013, 2014) in 8, 20, 40, and 69 dpp rats.

It seems very likely that one of the earliest changes of the pericyte progeny in the testis is the expression of the PDGFR-α (see Brennan et al. 2003; Stanley et al. 2012), followed by the expression of other factors such as Coup—TFII, CD51, CD90, 3ß-HSD, and additional enzymes involved in the process of steroidogenesis. These cells are correctly to be defined as progenitor Leydig cells (Chen et al. 2017; Ye et al. 2017) because these are certainly migrating transit-amplifying cells in process of maturation. However, in numerous publications they are qualified as adult Leydig stem cells produced, for example, from COUP-TFII-expressing, undifferentiated, spindle-shaped interstitial cells, that do not express adult Leydig cell markers (see Kilcoyne et al. 2014). However, the authors of these studies neglect the fact that COUP-FMII is actively involved in the formation of the microcirculation of an organ, and that between the stem cells and the first differentiating Leydig cells, no genetic similarity was found. The reason for this is that the differentiated Leydig cells appear after their transdifferentiation from the pluripotent pericytes and via multipotent transit-amplifying cells that further develop to differentiate into adult cells. This means that the transit-amplifying cells differentiate as progenitors of the LCs and are descendants of the pericytes. The transit-amplifying cells develop also as mesenchymal stem cells in the different organs. Thus, the described cells are Leydig cell progenitors, the progeny of the stem cell pericyte that represent the epiblast pluripotent cells in the mature organism.

It is important to note that Kilcoyne et al. (2014) established that the Coup-TFII cells express also androgen receptors. This result provides additional evidence that the cells under study are differentiating progenitor cells and not stem cells. Furthermore, the spindle shape of the interstitial cells is not a solid argument, while pericytes as well as their progeny, the transit-amplifying and immature (young)

Leydig cells have the same shape (Figs. 13.8 and 13.9). As mentioned above, the COUP-TFII protein appears to some extent later than the PDGFR-α, which is the first marker that appear in the pericyte progeny.

13.2.2.3 Pericytes and the Neural Crest

The pericytes and the Leydig cells are not progeny of the neural crest (Breau et al. 2008; Weston et al. 2004; Weston and Thiery 2015; Gilbert 2000; Lee et al. 2013a, b).

The neural crest cells (NCCs) are transient appearance that originate at the neural plate border and reside within the elevating neural folds and dorsal neural tube after its closure (Shyamala et al. 2015). Because of its neuroectodermal (Davidoff et al. 2009) and multipotent qualities (neural and mesodermal progeny), the neural crest was designated by a number of scientists as "the fourth germ layer" (Hall 1998, 2000; Gilbert 2000; Le Douarin et al. 2008; Shyamala et al. 2015; Bronner and Simões-Costa 2016). In this respect Davidoff et al. (2009) presumed that the pericytes may derive from neural crest cells or from a common stem cell population segregated early in embryogenesis with progeny for both the neural crest and the mesoderm lineage (mesectoderm, mesenchymal-neural).

However, it becomes evident that the neural crest stem cells (NCSCs) originate later from the ectoderm than the early from the epiblast occurring pericytes (Fig. 13.10). Already Kastschenko (1888) announced that many cells of the neural crest in the head of the Salachian embryo do not take part in the formation of the nervous system but give rise to reticular mesoderm or mesenchyme. New results also show that the authentic NCSCs do not possess mesenchymal properties and are not able to produce cells with mesenchymal features (Hill and Watson, 1958; Nichols, 1981; Weston et al. 2004; Breau et al. 2008; Lee et al. 2013a, b; Kennedy et al. 2014). As Donoghue et al. (2008) reported, an epithelial-to-mesenchymal transition undergoes a second distinct population of cells within the neural fold. These cells have mesectodermal (ectomesenchyme) lineage potency and are termed metablast (non-neural epithelial domain) that in the neural fold (NF) differentiate in the vicinity of the authentic NCSCs and are responsible for the regulation of the migration and fate of the authentic cranial neural crest-derived cells (Weston et al. 2004; Weston and Thiery 2015).The mouse cranial NF epithelium is heterogeneous, and a sharp boundary exists between the lateral non-neural epithelium and the thickened neural epithelium in the dorsomedial ridge (see also Thiery et al. 1984; Newgreen et al. 1997; Denham et al. 2015). These observations raise the possibility that at least some PDGFR-α ectomesenchyme originates from the lateral non-neural domain of the neural fold epithelium. Previously Lee et al. (2013a, b) and Nichols (1981, 1986) presumed that mesenchymal cells emerge precociously from an epithelial neural fold domain resembling the primitive streak in the early embryonic epiblast, and therefore Weston et al. (2004) propose the name "metablast" for this non-neural epithelial domain to indicate that it is the site of a delayed local delamination of mesenchyme similar to involution of mesoderm during gastrulation. Breau et al. (2008) further propose the hypothesis that neural crest and ectomesenchyme

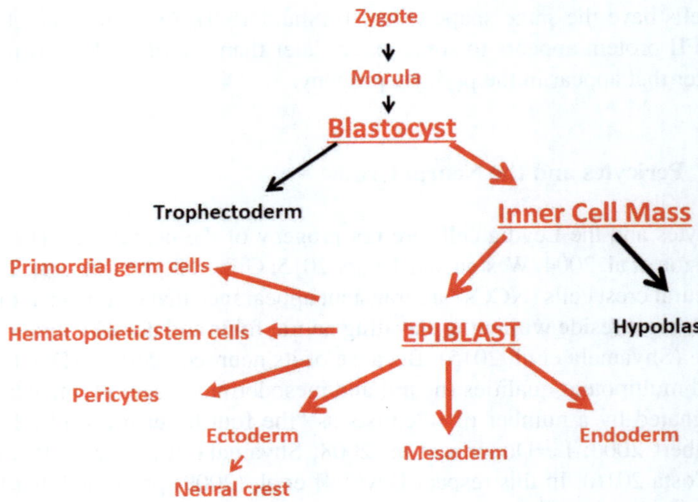

12

Fig. 13.10 (*12*) Represents the very early stages of the zygote and blastocyst differentiation. Note that the first cell populations derived from the epiblast are the primordial germ cells, the hematopoietic stem cells, and the pericytes, whereas the neural crest originates a bit later from the one of the primitive embryonic germ layers, the ectoderm (epiblast/ectoderm)

are developmentally distinct progenitor populations and that at least some ectomesenchyme is metablast-derived rather than neural crest-derived tissue.

To explain some events with the pericytes and keeping in mind that the Leydig cells possess neural properties, we used information concerning the situation in the central nervous system. As mentioned above there is a great similarity between the biology of the developing and adult Leydig cells and the progenitors/immature/mature nerve and glial cells. Also in the nervous system until recently, the brain vasculature/pericytes were neglected. However, the first evidence was provided by Palmer et al. (2000) who found that adult neurogenesis occurs within an angiogenic niche (see Louissaint et al. 2002; Shen et al. 2004a, b). Later it becomes evident that subventricular in the telencephalon, a very well-developed vascular plexus exists (Tavazoie et al. 2008, Shen et al. 2008).

According to Itoh et al. (2011), pericytes and astrocytes play important role for the maintenance of the cerebral microvasculature. In this respect, it seems very probable that the brain microvascular pericytes possess the capability to transdifferentiate into radial astroglia (Alvarez-Buylla et al. 2002; Filippov et al. 2003; Hill et al. 2004; Sild and Rithazer 2011). Thus the conclusion can be made that the pericytes of the subventricular plexus are the neural stem cells (Doetsch 2003) which generate the B progenitors (GFAP+ astrocytes; radial glia), from which the C (transit amplifying) cells and the A (neuroblast) cells that are surrounded by an glial tunnel that lead them toward the olfactory bulb (see Gonzalez-Perez 2012; Jiang et al. 2003; Morshead 2004) differentiate. This hypothesis is also supported by the observation that Nestin and NG2-positive cells are precursors of neurons, oligodendrocytes, and astrocytes in the gray matter (Wei et al. 2002; Zhu et al. 2008). Similar conditions were reported for the subgranular zone of the hippocampal dentate gyrus (Filippov et al. 2003; Bischofberger and Schmidt-Hieber 2006; Hara et al. 2010).

The similarity between the NCSCs and the pericytes is an unequivocal fact (Schulze et al. 1987; Angelova and Davidoff 1989; Angelova et al. 1991; Chiwakata et al. 1991; Davidoff et al. 1993, 1996, 2001; Middendorff et al. 1993; Benton et al. 1995). This fact allowed us and others to determine the Leydig cells as neuroendocrine cells and lead to the presumption that the pericytes, resp., the Leydig cells are of neural crest origin (see Davidoff et al. 2004, 2009; Diaz-Flores et al. 2009). However, we were convinced by new results that the reason for this similarity is the close embryologic timing in arising of these cell types (Fig. 13.8).

Consequently, the neural crest cells arise at the border of the neural plate and the epidermis at the time of neural tube closure (Le Douarin and Dupin 2003; Barembaum and Bronner-Fraser 2005; Shyamala et al. 2015). As new data show, in the embryonic ectoderm of the neural folds, two different cell populations coexist: authentic neural crest cells and mesectodermal stem cells (Weston et al. 2004; Breau et al. 2008; Weston and Thiery 2015).The neural crest cells, derived from the epiblast/ectoderm, possess some restricted multipotent characteristics and represent a migratory cell population that generates diverse cell types during vertebrate development (Trainor et al. 2003; Trainor 2005; Crane and Trainor 2006). In comparison, the pericytes originate something earlier direct from the epiblast in contrast to the neural crest stem cells that arise a bit later from the primary germ layer, the ectoderm, which is also a derivate of the epiblast (epiblast/ectoderm; Fig. 13.8). In this respect Kojima et al. (2014) reveal that the epiblast cells possess different transcriptomes in comparison with the epiblast/ectoderm cells. Thus, the NCSCs arise indirectly from the epiblast, and as recent results show they are not pluripotent but multipotent and are able to produce different cell types with neural qualities, but not mesenchymal cells (Brennan et al. 2003; Ortega et al. 2004a; Yao and Barsoum 2007; Breau et al. 2008; Kennedy et al. 2014).

There is also evidence that the neural crest cells cannot produce pericytes (Brennan et al. 2003; Joseph 2004; Weston et al. 2004; Breau et al. 2008; Griswold and Behringer 2009; Makala et al. 2015). Therefore the only possibility that remains is that the pericytes are the progeny of the epiblast, whereas the neural crest is the daughter structure of the a bit later arising epiblast/ectoderm (De-Miguel et al. 2009).

It has to be noted that the neural crest cells arise and migrate after the origin of the developing vasculature and use the blood vessels and the associated extracellular matrix as leading structures (Spence and Poole 1994). Accordingly, the pericytes are disseminated before the migration of the neural crest stem cells and are not produced by them. Thus, the pericytes, resp., the Leydig cells, are not neural crest derivatives.

13.2.2.4 The Pericyte Progenitors Arise Very Early During Embryogenesis from the Epiblast

The fact that the mesenchymal stem cells are progeny of the pericytes and the authentic neural crest cells and the peritubular myofibroblasts of the lamina propria of the seminiferous tubules are not able to produce mesectoderm allows the conclusion that only the embryonic epiblast can be ancestor of the pericytes. This is supported also by the results which provide evidence that only the epiblast stem cells are pluripotent, whereas the other two cell types are multipotent.

It becomes evident that the pericytes represent the only true pluripotent adult stem cell in the mature vertebrate organism. In accordance, De-Miguel et al. (2009, 2010) and Nichols and Smith (2012) showed that pluripotency is confined to the epiblast, a transient structure that persists for only few days. Only the pericytes (as epiblast progeny) are able in vivo and in vitro to produce structures related to the three main germ layers: ectoderm, mesoderm, and endoderm. The other candidates, namely, the multipotent stromal cells and the neural crest stem cells are multipotent cells (Chikhovskaya et al. 2012, 2014). They are partially committed and with limited potency. In this respect Dore-Duffy (2008) wrote: "We propose that once the pericyte has migrated into the perivascular space it makes functional and phenotypic adjustments or adaptations that may include differentiation along multiple lineages and that this is the basis for their pluripotentiality."

Thus, in the vascular niche, the pericytes remain as resting pluripotent stem cells. After their activation, proliferation, and migration, they become transit-amplifying cells that transdifferentiate to give different multipotent progenitors. With the continuation of the differentiation, the progenitor cells lose their multipotency and become oligopotent and unipotent.

There are numerous publications which presume that the adult stem cells are of multiple origins and correspondingly show different chemical content (markers) and other features (see Davidoff et al. 2009; Davidoff 2017). However, these differences reflect most probably the great plasticity of the pluripotent adult stem cells depending on their position in the niches, stage of differentiation, and their behavior under the influence of various local biophysical and biochemical factors.

As noted, recent studies provide evidence that the true adult stem cells must be the progeny of the epiblast. The epiblast is the product of the inner cell mass at the early blastocyst stage of embryogenesis (Kaufman 1992; Kucia et al. 2006; Ratajczak et al. 2008, 2012a, b; Montiel-Eulefi et al. 2009), and the SSEA-1-positive mesenchymal stem cells are thought to be the most primitive progenitors in the murine bone marrow mesenchymal compartment (Anjos-Afonso and Bonnet 2007; Ratajczak et al. 2008). There is strong evidence that the epiblast is also the ancestor of the primordial germ cells (PGCs; McLaren 2003; Brons et al. 2007; Montiel-Eulefi et al. 2009; Ratajczak et al. 2007; De-Miguel et al. 2009) and the hematopoietic stem cells (HSCs). Brons et al. (2007) emphasized that the epiblast stem cells are not derivatives of the PGCs. Yoshimizu et al. (2001) established that during gastrulation (De-Miguel et al. 2009, 2010), the epiblast cells undergo epithelial-to-mesenchymal transition and delaminate to become the loose mesenchyme of the primitive streak. Epiblast stem cells become also included as pericytes within specific vascular niches (Nikolova et al. 2007; Scadden 2006) of the developing and growing blood vessels with which they become disseminated through the whole body. As already mentioned the cardiovascular system is the first functional organ system in the vertebrate embryo (Flamme et al. 1997). This is important to note while it evidenced that pericytes as stem cells are present very early during the embryo development in the emerging tissues and organs where they can participate in the vascularization, origin, and organization of the corresponding structures and in addition remain as omnipresent adult stem cells in the vascular niches for tissue maintenance, repair, and regeneration (Dore-Duffy 2008).

The pluripotency of the pericytes is the reason for both the mesenchymal and neural qualities of the Leydig cells of the testis. As mentioned above the pericytes are not produced by the neural crest stem cells. By means of the pericyte ancestors, they get their own neuroendocrine quality. This is also true for the determination that the pericytes originate from mesoderm. The brain pericytes are stem cells immunoreactive for αSMA, γ-glutamyl transpeptidase, alkaline phosphatase, nestin, vimentin, and PDGFR-ß (Fisher 2009). These markers are also expressed in the testicular pericytes.

13.3 Conclusion

It becomes evident that the pericytes are periendothelial pluripotent stem cells which can differentiate into cells, tissues, and organs of all three main embryonic germ layers. They arise in the early embryogenesis from the pluripotent epiblast, become disseminated with the differentiating vasculature of the embryo, and stored in true vascular niches as reserve pluripotent adult stem cells for tissue generation, maintenance, repair, and regeneration of the structures in the vertebrate organism. The pericytes are really the adult stem cells and the ancestors of a number of variable progenitor cells. Pericytes are authentic stem cells only until the moment they leave the vascular niche.

After the formation of the seminiferous tubules, the peritubular immature Leydig cells reduce in number or disappear. In some publications this phenomenon is explained with a back migration of these cells toward the interstitial space of the testis where they accumulate in close vicinity of the blood vessels. However, a more plausible explanation is that at this time of the fetal development, fewer hormones are needed for the composition, maintenance, and functioning of the seminiferous tubules which lead to the disappearance of the unnecessary cells. This event resembles the involution of the fetal Leydig cells during the testis development (Orth and Weisz 1980). As consequence the interstitial perivascular cells increase in number and become enlarged. They remain in contact with the blood vessel wall and supply hormones for the male design of the organism. This phenomenon is well known for other organs, especially for the nervous system. During the creation and regeneration of the nervous system for insurance, an excess number of nerve cells get formed. With it, only cells that succeeded to make contacts with a target structure remain alive. The other cells die and disappear.

The described features of the testicular pericytes provide enough evidence that these cells belong to the omnipresent pericyte population of a vertebrate organism. There is a great variability of the pericyte descendants comprising these cells in different migration position, developmental stages, and contacts with structures in their environment. All this explained the great plasticity of the whole pericyte population of an organism and the difficulties in the recognition, lineage determination, and definition of the pericyte progeny (Hermann et al. 2016).

The pericytes are a brilliant achievement of the biological nature aiming to supply every organ with an omnipresent population of pluripotent adult stem cells.

References

Abedin M, Tintut Y, Demer LL (2004) Mesenchymal stem cells and the artery wall. Circ Res 95:671–676

Alvarez-Buylla A, Seri B, Doetsch F (2002) Identification of neural stem cells in the adult vertebrate brain. Brain Res Bull 57:751–758

Angelova P, Davidoff MS (1989) Immunocytochemical demonstration of substance P in Hamster Leydig cells during ontogenesis. Z Mikrosk Anat Forsch 103:560–566

Angelova P, Davidoff MS, Baleva K, Staykova M (1991) Substance P and neuron-specific enolase-like immunoreactivity of rodent Leydig cells in tissue section and cell culture. Acta Histochem 91:131–139

Anjos-Afonso F, Bonnet D (2007) Nonhematopoietic/endothelial SSEA-1+ cells define the most primitive progenitors in the adult murine bone marrow mesenchymal compartment. Blood 109:1298–1306

Ariyaratne HB, Mendis-Handagama SMLC, Hales DB, Mason JI (2000) Studies on the onset of Leydig precursor cell differentiation in the prepubertal rat testis. Biol Reprod 63:165–171

Armulik A, Genové G, Betsholtz C (2011) Pericytes: developmental, physiological, and pathological perspectives, problems, and promises. Dev Cell 21:193–215

Attwell D, Mishra A, Hall CN, O'Farrell FM, Dalkara T (2016) What is a pericyte? J Cerebr Blood Flow Metab 36:451–455

Balabanov R, Washington R, Wagnerova J, Dore-Duffy P (1996) CNS microvascular pericytes express macrophage-like function, cell surface integrin αM, and macrophage marker ED-2. Microvasc Res 52:127–142

Barón M, Gallego A (1972) The relation of the microglia with the pericytes in the cat cerebral cortex. Z Zellforsch Mihosk Anat 128:42–57

Basciani S, Mariani S, Arizzi M, Ulisse S, Rucci N, Jannini EA, Rocca CD, Manicone A, Carani C, Spera G, Gnessi L (2002) Expression of platelet-derived growth factor-A (PDGF-A), PDGF-B, and PDGF receptor- a and -s during human testicular development and disease. J Clin Endocrinol Metab 87:2310–2319

Benninghoff A (1926) Über die Formenreihe der glatten Muskulatur und die Bedeutung der Rouget'schen Zellen an den Kapillaren. Z Zellf Mikr Anat 4:126–170

Bensley RR, Vimtrup BJ (1928) On the nature of the Rouget cells of capillaries. Anat Rec 39:37–55

Benton L, Shan L-X, Hardy MP (1995) Differentiation of adult Leydig cells. J Steroid Biochem Mol Biol 53:61–68

Barembaum M, Bronner-Fraser M (2005) Early steps in neural crest specification. Seminars in Cell & Developmental Biology; 16(6): 642-646

Bergers G (2008) Pericytes, the mural cells of the microvascular system. In: Figg WD, Folkman J (eds) Angiogenesis. An integrative approach from science to medicine, Chapter 4. Springer, New York, pp 45–53
Bergers G, Song S (2005) The role of pericytes in blood-vessel formation and maintenance. Neuro Oncol 7:452–464
Bhagavati S (2008) Stem cell based therapy for skeletal muscle diseases. Curr Stem Cell Res Ther 3(3):219–228
Bhushan S, Meinhardt A (2017) The macrophages in testis function. J Repr Immun 119:107–112
Bianco P, Robey PG, Simmons PJ (2008) Mesenchymal stem cells: revisiting history, concepts, and assays. Cell Stem Cells 2:313–319
Birbrair A, Delbono O (2015) Pericytes are essential for skeletal muscle formation. Stem Cell Rev Rep 11:547–548
Birbrair A, Zhank T, Wang ZM, Messi ML, Enikolopov GN, Mintz A, Delbono O (2013a) Role of pericytes in skeletal muscle regeneration and fat accumulatioin. Stem Cells Dev 22:2298–22314
Birbrair A, Zhang T, Wang Z-M, Messi ML, Enikolopov GN, Mintz A, Delbono O (2013b) Skeletal muscle neural progenitor cells exhibit properties of NG2-glia. Exp Cell Res 319:45–63
Birbrair A, Zhang T, Files DC, Mannava S, Smith T, Wang Z-M, Messi ML, Mintz A, Delbono O (2014) Type-1 pericytes accumulate after tissue injury and produce collagen in an organ-dependent manner. Stem Cell Res Ther 5:122
Bischofberger J, Schmidt-Hieber C (2006) Adulte Neurogenese im Hippocampus. e-Neuroforum 3:212–221
Bjornson CRR, Rietze RL, Reynolds BA, Magli MC, Vescovi A (1999) Turning brain into blood: adult neural stem cells adopt a hematopoietic fate in vivo. Science 283:534–537
Bouacida A, Rosset P, Trichet V, Guilloton F, Espagnolle N, Cordonier T, Heymann D, Layrolle P, Sensébé L, Deschaseaux F (2012) Pericyte-like progenitors show high immaturity and engraftment potential as compared with mesenchymal stem cells. PLoS One 7(11):e486548
Brazelton TR, Rossi FMV, Keshet GI, Blau HM (2000) From marrow to brain: expression of neuronal phenotypes in adult mice. Science 290:1775–1779
Breau MA, Pietri T, Stemmler MP, Thiery P, Weston JA (2008) A nonneural epithelial domain of embryonic cranial neural folds gives rise to ectomesenchyme. Proc Natl Acad Sci U S A 105:7750–7755
Brennan J, Capel B (2004) One tissue, two fates: molecular genetic events that underlie testis versus ovary development. Nat Rev Gen 5(7):509–521
Brennan J, Tilmann C, Capel B (2003) Pdgfr-α mediates testis cord organization and fetal Leydig cell development in the XY gonad. Genes Dev 17:800–810
Bronner ME, Simões-Costa M (2016) The neural crest migrating into the 21st century. Curr Top Dev Biol 116:115–134
Brons IGM, Smithers LE, Trotter MWB, Rugg-Gunn P, Sun B, Lopes SMCS, Howlett SK, Clarkson A, Ahrlund-Richter L, Pedersen RA, Vallier L (2007) Derivation of pluripotent epblast stem cells from mammalian embryos. Nature 448:191–196
Buehr M, Smith A (2003) Genesis of embryonic stem cells. Philos Trans R Soc Lond B Biol Sci 358:1397–1402
Calloni GW, Glavieux-Pardanaud C, Le Douarin NM, Dupin E (2007) Sonic hedgehog promotes the development of multipotent neural crest progenitors endowed with both mesenchymal and neural potentials. Proc Natl Acad Sci U S A 104:19879–19884
Cano E, Gebala V, Gerhardt H (2017) Pericytes or mesenchymal stem cells: is that the question? Cell Stem Cell 20(3):296–297
Caplan AI (2008) All MSCs are pericytes? Cell Stem Cell 3:229–230
Cervós-Navarro J (1963) Electronenmicroscopische Befunden an den Capillaren der Hirnrinde. Arch Psych Ztschr ges Neurologie 204:484–504
Chan-Ling T (1997) Glial, vascular, and neuronal cytogenesis in whole-mounted cat retina. Microsc Res Tech 36:1–16
Chen H, Stanley E, Jin S, Zirkin BR (2010) Stem Leydig cells: From fetal to aged animals. Birth Defects Res C Embryo Today 90:272–283

Chen WCW, Baily JE, Corselli M, Diaz ME, Sun B, Xiang G, Gray GA, Huard J, Péault B (2015) Human myocardial pericytes: multipotent mesodermal precursors exhibiting cardiac specificity. Stem Cells 33:557–573
Chen H, Wang Y, Ge R, Zirkin BR (2017) Leydig stem cells: identification, proliferation and differentiation. Mol Cell Endocrinol 445:65–71
Chikhovskaya JV, Jonker MJ, Meissner A, Breit TM, Repping S, van Pelt AMM (2012) Human testis-derived embryonic stem cell-like cells are not pluripotent, but possess potential of mesenchymal progenitors. Hum Reprod 27(1):210–221
Chikhovskaya JV, van Daalen SKM, Kover CM, Repping S, van Pelt AMM (2014) Mesenchymal origin of multipotent human testis-derived stem cells in human testicular cell cultures. Mol Human Reprod 20:155–167
Chiwakata C, Brackmann B, Hunt N, Davidoff M, Schulze W, Ivell R (1991) Tachykinin (Substance P) gene expression in Leydig cells of the human and mouse testis. Endocrinology 128:2441–2448
Clark ER, Clark EL (1925a) The development of adventitial (Rouget) cells on the blood capillaries of amphibian larvae. Amer J Anat 35:239–264
Clark ER, Clark EL (1925b) The relation of Rouget cells to capillary contractility. Amer J Anat 35: 265-282
Collas P (2010) Programming differentiation potential in mesenchymal stem cells. Epigenetics 5(6):476–482
Collett GDM, Canfield AE (2005) Angiogenesis and pericytes in the initiation of ectopic calcification. Circ Res 96:930–938
Combes AN, Wilhelm D, Davidson T, Dejana E, Harley V, Sinclair A, Koopman P (2009) Endothelial cell migration directs testis cord formation. Dev Biol 326:112–120
Crane JF, Trainor PA (2006) Neural Crest stem and progenitor cells. Ann Rev Cell Dev Biol 22 (1):267-286
Crisan M, Deasy B, Gavina M, Zheng B, Huard J, Lazzari L, Péault B (2008a) Purification and long-term culture of multipotent progenitor cells affiliated with the walls of human blood vessels: myoendothelial cells and pericytes. Methods Cell Biol 86:295–309
Crisan M, Yap S, Casteilla L, Chen C-W, Corselli M, Park TS, Andriolo G, Sun B, Zheng B, Zhang L, Norotte C, Teng P-N, Trass J, Schugar R, Deasy BM, Badylak S, Bühring H-J, Giacobino J-P, Lazzari L, Huard J, Péault B (2008b) A perivascular origin for mesenchymal stem cells in multiple human organs. Cell Stem Cell 3:301–313
da Silva Meirelles L, Caplan AI, Nardi NB (2013) Pericytes as the source of mesenchymal stem cells. In: Goldenberg RC d S, de Carvalho ACC (eds) Resident stem cells and regenerative therapy. Elsevier, Amsterdam, pp 233–250. https://doi.org/10.1016/B978-0-12-416012-5.00012-8
Dar A, Domev H, Ben-Yosef O, Tzukerman M, Zeevi-Levin N, Novak A, Germanguz I, Amit M, Itskovitz-Eldor J (2012) Multipotent vasculogenic pericytes from human pluripotent stem cells promote recovery of murine ischemic limb. Circulation 125:87–99
Davidoff MS (2017) The Leydig cells of the testis originate from the microvascular pericytes. Biomed Rev 28:5–25
Davidoff MS, Schulze W, Middendorff R, Holstein A-F (1993) The Leydig cell of the human testis—a new member of the diffuse neuroendocrine system. Cell Tiss Res 271:429–439
Davidoff MS, Middendorff R, Holstein AF (1996) Dual nature of Leydig cells of the human testis. Biomed Rev 6:11–41
Davidoff MS, Middendorff R, Koeva Y, Pusch W, Jezek D, Muller D (2001) Glial cell line-derived neurotrophic factor (GDNF) and its receptors GFR a −1 and GFR a −2 in the human testis. Ital J Anat Embryol 106(Suppl 2):p173–p180
Davidoff MS, Middendorff R, Enikolopov G, Rietmacher D, Holstein AF, Müller D (2004) Progenitor cells of the testosterone-producing Leydig cells revealed. J Cell Biol 167: 935–944
Davidoff MS, Middendorff R, Müller D, Holstein AF (2009) The Neuroendocrine Leydig cells and their stem cell progenitors, the pericytes. In: Sutovsky P, Clascá F, Kmiec Z, Korf H-W, Singh

B, Timmermans J-P, Schmeisser MJ (eds) Advances in anatomy, embryology and cell biology, vol 205. Springer, New York, pp 1–154

de Souza LEB, Malta TM, Kashima Hadat S, Covas DT (2016) Mesenchymal stem cells and pericytes: to what extent are they related? Stem Cells Dev 25(24):1843–1852

DeFalco T, Bhattacharya I, Williams AV, Sams DM, Capel B (2014) Yolk-sac-derived macrophages regulate fetal testis vascularization and morphogenesis. Proc Natl Acad Sci U S A 111:E2384–E2393

DeFalco T, Potter SJ, Williams AV, Waller B, Kan MJ, Capel B (2015) Macrophages contribute to the spermatogonial niche in the adult testis. Cell Rep 12:1107–1119

Dellavalle A, Sampoalesi M, Tonlorenzi R, Tagliafico E, Sacchetti B, Perani L, Innocenzi A, Galvez BG, Messina G, Morosetti R, Li S, Belicchi M, Peretti G, Chamberlain IS, Wright WE, Torrente Y, Ferrari S, Bianco P, Cossu G (2007) Pericytes of human skeletal muscle are myogenic precursors distinct from satellite cells. Nat Cell Biol 9:255–267

Dellavalle A, Maroli G, Covarello D, Azzoni E, Innocenti A, Perani L, Antonioni S, Sambasivan R, Brunelli S, Tajbakhsh S, Cossu G (2011) Pericytes resident in postnatal skeletal muscle differentiate into muscle fibres and generate satellite cells. Nat Commun 2:499. https://doi.org/10.1038/ncomms1508

De-Miguel MP, Arnalich-Montiel F, Lopez-Iglesias P, Blasquez-Martinez A, Nistal M (2009) Epiblast-derived stem cells in embryonic and adult tissues. Int J Dev Biol 53:1529–1540

De-Miguel MP, Fuentis-Julian S, Alcaina Y (2010) Pluripotent stem cells: origin, maintenance and induction. Stem Cell Rev Rep 6:633–649

Denham M, Hasegawa K, Menheniott T, Rollo B, Zhang D, Hough S, Alshawa A, Febbraro F, Ighanian S, Leung J, Elliott DA, Newgreen DF, Pera MF, Dottori M (2015) Multipotent caudal neural progenitors derived from humans pluripotent stem cells that give rise to lineages of the central and peripheral nervous system. Stem Cells 33(6):1759–1770

Diaz-Flores L Jr, Madrid JF, Guitérrez R (2006) Adult stem and transit-amplifying cell location. Histol Histopathol 21:995–1027

Diaz-Flores L, Gutiérrez R, Madrid JF, Varela H, Valladares F, Acosta E, Martin-Vasallo P, Diaz-Flores L Jr (2009) Pericytes. Morphofunction, interactions and pathology in a quiescent and activated mesenchymal cell niche. Histol Histopathol 24:909–969

Diaz-Flores L, Gutiérrez R, Garcia MP, Diaz-Flores L Jr, Valladares F, Madrid JF (2012) Ultrastructure of myopericytoma: a continuum of transitional phenotypes of myopericytes. Ultrastr Pathol 36:189–194

Doetsch F (2003) The glial identity of neural stem cells. Nat Neurosci 6:1127–1134

Dogiel AS (1899) Über den Bau der Ganglien in den Geflechten des Darms und der Gallenblase des Menschen und der Säugetiere. Arch Anat Physiol Anat Abt A 3–4:100–158

Donoghue PCJ, Graham A, Kelsh RN (2008) The origin and evolution of the neural crest. BioEssays 30:530–541

Dore-Duffy P (2008) Pericytes: pluripotent cells of the blood brain barrier. Curr Pharm Des 14(16):1581–1593

Dore-Duffy P, Cleary K (2011) Morphology and properties of pericytes. Meth Mol Biol 686 (Part 1):49–68

Eberth CJ (1871) Von den Blutgefässen. Capitel VIII. In: Stricker S (ed) Handbuch der Lehre von den Geweben des Menschen und der Tiere, Bd 1. Verlag von Wilhelm Engelmann, Leipzig, pp 191–213

Ergün S, Stingl J, Holstein AF (1994) Microvasculature of the human testis in correlation to Leydig cells and seminiferous tubules. Andrologia 26(5):255–262

Ergün S, Davidoff M, Holstein AF (1996) Capillaries in the lamina propria of human seminiferous tubules are partly fenestrated. Cell Tissue Res 286(1):93–102

Ergün S, Harneit S, Paust HJ, Mukhopadhyay AK, Holstein AF (1999) Endothelin and endothelin receptors A and B in the human testis. Anat Embryol 199:207–214

Etchevers HC, Vincent C, Le Douarin NM, Couly GF (2001) The cephalic neural crest provides pericytes and smooth muscle cells to all blood vessels of the face and forebrain. Development 128:1059–1068

Fawcett DW, Neaves WB, Flores MN (1973) Comparative observations on intertubular lymphatics and the organization of the interstitial tissue of the mammalian testis. Biol Reprod 9:500–532
Fecteau KA, Markonjich L, Mason JI, Mendis-Handagama SMLC (2006) Detection of platelet-derived growth factor-α (PDGF-A) protein in cells of Leydig lineage in the postnatal rat testis. Histol Histopathol 21:1295–1302
Filippov V, Kronenberg G, Pivneva T, Reuter K, Steiner B, Wang L-P, Yamaguchi M, Kettenmann H, Kempermann G (2003) Subpopulation of nestin-expressing progenitor cells in the adult murine hippocampus shows electrophysiological and morphological characteristics of astrocytes. Mol Cell Neurosci 23:373–382
Fisher M, (2009) Pericyte signaling in the neurovascular unit. Stroke 40 (3, Suppl 1):S13-S15
Flamme I, Frölich T, Risau W (1997) Molecular mechanisms of vasculogenesis and embryonic angiogenesis. J Cell Physiol 173:206–210
Florey HW, Carleton HM (1926) Rouget cells and their function. Proc R Soc Lond B 100:23–31
Friedrich R, Holstein AF, Middendorff R, Davidoff MS (2012) Vascular wall cells contribute to tumorigenesis in cutaneous neurofibromas of patients with Neurofibromatosis type 1. A comparative histological, ultrastructural and immunohistochemical study. Anticancer Res 32:2139–2158
Ge R-S, Dong Q, Sottas CM, Papadopoulos V, Zirkin BR, Hardy MP (2006) In search of rat stem Leydig cells: identification, isolation, and lineage-specific development. Proc Natl Acad Sci U S A 103:2719–2724
Gilbert SF (2000) The Neural Crest. Developmental Biology, 6th edn. Sinauer Associates, Sunderland, MA. Available from: https://www.ncbi.nlm.nih.gov/books/NBK10065/
Griswold SL, Behringer RR (2009) Fetal Leydig cell origin and development. Sexual Development 3 (1):1–15
Gnessi L, Emidi A, Jannini EA, Carosa E, Maroder M, Arizzi M, Ulisse S, Spera G (1995) Testicular development involves the spatiotemporal control of PDGFs and PDGF receptors gene expression and action. J Cell Biol 131:1105–1121
Gnessi L, Basciani S, Mariani S, Arizzi M, Spera G, Wang C, Bondjers C, Karlsson L, Betsholtz C (2000) Leydig cell loss and spermatogenic arrest in platelet-derived growth factor (PDGF)-A-deficient mice. J Cell Biol 149(5):1019–1025
Gökçinar-Yagei B, Uçkan-Çetinkaya D, Çelebi-Saltik B (2015) Pericytes: Properties, functions and applications in tissue engineering. Stem Cell Rev Rep 11:549–559
Goluža T, Boscanin A, Cvetko J, Kozina V, Kosović M, Bernat MM, Kasum M, Kaštelan Ž, Ježek D (2014) Macrophages and Leydig cells in testicular biopsies of azoospermic men. Biomed Res Int 2014:828697, 14 pages
Gondos B (1980) Development and differentiation of the testis and male reproductive tract. In: Steinberger A, Steinberger E (eds) Testicular development structure and function. Raven, New York, pp 3–20
Gonzalez R, Griparic L, Vargas V, Burgee K, Santacruz P, Anderson R, Schiewe M, Silva F, Patel A (2009) A putative mesenchymal stem cells population isolated from adult human testes. Biochem Biophys Res Commun 385:570–575
Gonzalez-Perez O (2012) Neural stem cells in the adult human brain. Biol Biomed Rep 2(1):59–69
Guimarães-Camboa N, Cattaneo P, Sun Y, Moore-Morris T, Gu Y, Dalton ND, Rockenstein E, Masliah E, Peterson KL, Stallcup WB, Chen J, Evans SM (2017) Pericytes of multiple organs do not behave as mesenchymal stem cells in vivo. Cell Stem Cell 20(3):345–359
Günther H (1917) Die mechanische Erregbarkeit der Hautmuskeln und Hautgefäße. In: Kraus F et al (eds) Ergebnisse der Inneren Medizin und Kinderheilkunde, vol 15. Julius Springer, Berlin, pp 620–714
Haider SG, Servos K, Tran N (2007) Structural and histological analysis of Leydig cell steroidogenic function. In: Payne AH, Hardy MP (eds) Contemporary endocrinology: the Leydig cell in health and disease. Humana Press, Totowa, NJ, pp 33–45
Hall BK (1998) Germ layers and the germ-layer theory revisited: Primary and secondary germ layers, neural crest as a fourth germ layer, homology, demise of the germ-layer theory. Evol Biol 30:121–186

Hall BK (2000) The neural crest as a fourth germ layer and vertebrates as quadroblastic not triploblastic. Evol Dev 2:3–5
Hall AP (2006) Review of the pericyte during angiogenesis and its role in cancer and diabetic retinopathy. Toxicol Pathol 34:763–775
Hara Y, Nomura T, Yoshizaki K, Frisén J, Osumi N (2010) Impaired hippocampal neurogenesis and vascular formation in ephrin-A5-deficient mice. Stem Cells 28:974–983
Hermann M, Bara JJ, Sprecher CM, Menzel U, Jalowiec JM, Osinga R, Scherberich A, Alini M, Verrier S (2016) Pericyte plasticity—comparative investigation of the angiogenic and multilineage potential of pericytes from different human tissues. Eur Cells Mater 31:236–249
Hill JP, Watson KM (1958) The early development of the brain in marsupials preliminary communication. J Anat 92(Pt4):493–497
Hill WD, Hess DC, Martin-Studdard A, Carotheres JJ, Zheng J, Hale D, Maeda M, Fagan SC, Carroll JE, Conway SJ (2004) SDF-1 (CXCL12) is upregulated in the ischemic penumbra following stroke: association with bone marrow cell homing to injury. J Neuropathol Exp Neurol 63:84–96
Hirschi KK, D'Amore PA (1996) Pericytes in the microvasculature. Cardiovasc Res 32:687–698
Holstein AF (1999) Spermatogenese beim Menschen: Grundlagenforschung und Klinik. Ann Anat 181:427–436
Holstein AF, Davidoff MS (1997) Organization of the intertubular tissue of the human testis. In: Motta PM (ed) Recent advances of cells, tissues and organs. Antonio Delfino Editore, Rome, pp 569–577
Holstein AF, Maekawa M, Nagano T, Davidoff MS (1996) Myofibroblasts in the lamina propria of human seminiferous tubules are dynamic structures of heterogeneous phenotype. Arch Histol Cytol 59:109–125
Holstein AF, Schulze W, Davidoff M (2003) Understanding spermatogenesis is a prerequisite for treatment. Reprod Biol Endocrinol 1:107
Hosaka K, Yang Y, Sekia T, Fischera C, Dubeya O, Fredlundc E, Hartman J, Religa P, Morikawa H, Ishii Y, Sasahara M, Larsson O, Cossu G, Cao R, Lim S, Cao Y (2016) Pericyte–fibroblast transition promotes tumor growth and metastasis. Proc Natl Acad Sci U S A 113(38): E5618–E5627
Huhtaniemi I, Pelliniemi LJ (1992) Fetal Leydig cells: cellular origin, morphology, life span, and special functional features. Proc Soc Exp Biol Med 201:125–140
Itoh Y, Toriumi H, Yamada S, Hoshino H, Suzuki N (2011) Astrocytes and pericytes cooperatively maintain a capillary-like structure composed of endothelial cells on gel matrix. Brain Res 1406:74–83
Jiang Y, Hernderson D, Blackstad M, Chen A, Miller RF, Verfaillie CM (2003) Neuroectodermal differentiation from mouse multipotent adult progenitor cells. Proc Natl Acad Sci U S A 100:11854–11860
Joseph NM (2004) Neural crest stem cells undergo multilineage differentiation in developing peripheral nerves to generate endoneurial fibroblasts in addition to Schwann cells. Development 131 (22):5599-5612
Kahn CR (2008) Can we nip obesity in its vascular bud? Science 322:542–543
Kastschenko N (1888) Zur Entwicklungsgeschichte des Salachierembryos. Anatomischer Anzeiger III (1–32) 16:445–467
Kaufman M (1992) The atlas of mouse development. Academic, London, p 512
Kennedy E, Mooney CJ, Hakimjavadi R, Fitzpatrick E, Guha S, Collins LE, Loscher CE, Morrow D, Redmond EM, Cahill PA (2014) Adult vascular smooth muscle cells in culture express neural stem cell markers typical of resident multipotent vascular stem cells. Cell Tissue Res 358:203–216
Kerr JB, Sharpe RM (1985) Stimulatory effect of follicle-stimulating hormone on rat Leydig cells. A morphometric and ultrastructural study. Cell Tissue Res 239:405–415
Kerr JB, Donachie K, Rommerts FFG (1985) Selective destruction and regeneration of rat Leydig cells in vivo. Cell Tiss Res 242:145–156

Kerr JB, Bartlett JMS, Donachie K (1986) Acute response of testicular interstitial tissue in rats to the cytotoxic drug ethane dimethanesulphonate. An ultrastructural and hormonal study. Cell Tissue Res 243:405–414
Kerr JB, Knell CM, Abbott M, Donachie K (1987a) Ultrastructural analysis of the effect of ethane dimethanesulphonate on the testis of the rat, guinea pig, hamster and mouse. Cell Tissue Res 249:451–457
Kerr JB, Bartlett JMS, Donachie K, Sharpe RM (1987b) Origin of regenerating Leydig cells in the testis of the adult rat. An ultrastructural, morphometric and hormonal assay study. Cell Tissue Res 249:367–377
Kilcoyne KR, Smith LB, Atanassova N, Macpherson S, McKinell C, van den Driesche S, Jobling MS, Ghambers TJG, De Gendt K, Verhoeven G, O'Hara L, Platts S, de Franca LR, Lara NLM, Anderson RA, Sharpe RM (2014) Fetal programming of adult Leydig cell function by androgenic effects on stem/progenitor cells. Proc Natl Acad Sci U S A 111:E1924–E1932
Klein D, Weißhardt P, Kleff V, Jastrow H, Jakob HG, Ergün S (2011) Vascular wall-resident CD44+ multipotent stem cells give rise to pericytes and smooth muscle cells and contribute to new vessel maturation. PLoS One 6:e20540
Klein D, Meissner N, Kleff V, Jastrow H, Yamaguchi M, Ergün S, Jendrossek V (2014) Nestin(+) tissue-resident multipotent stem cells contribute to tumor progression by differentiating into pericytes and smooth muscle cells resulting in blood vessel remodeling. Front Oncol 4:Article 169
Kojima Y, Kaufman-Francis K, Studdert JB, Steiner KA, Power MD, Loebel DAF, Jones V, Hor A, de Alencastro G, Logan GJ, Teber ET, Tam OH, Stutz MD, Alexander IE, Pickett HA, Tam PPL, (2014) The transcriptional and functional properties of mouse epiblast stem cells resemble the anterior primitive streak. Cell Stem Cell 14 (1):107–120
Krogh A (1919) The number and distribution of capillaries in muscles with calculations of the oxygen pressure head necessary for supplying the tissue. J Physiol 52:409–415
Krogh A (1922) The anatomy and physiology of the capillaries. New Haven Yale University Press, London, pp 1–304
Krogh A (1929) Anatomie und Physiologie der Capillaren. p.362, Springer, Berlin-Heidelberg.
Krueger M, Bechmann I (2010) CNS pericytes: concepts, misconceptions, and a way out. Glia 58:1–10
Kucia M, Reca R, Jala VR, Dawn B, Ratajczak J, Ratajczak MZ (2005) Bone marrow as a home of heterogeneous populations of nonhematopoietic stem cells. Leukemia 19:1118–1127
Kucia M, Machalinski B, Ratajczak MZ (2006) The developmental deposition of epiblast/germ cell-line derived cells in various organs as a hypothetical explanation of stem cell plasticity? Acta Neurobiol Exp 66:331–341
Kurtz A (2016) Commentary for "human kidney pericytes produce renin". Kidney Int 90:1153–1154
Landreh L, Stukenborg J-B, Söder O, Svechnikov K (2013) Phenotype and steroidogenic potential of PDGFα-positive rat neonatal peritubular cells. Mol Cell Endocrinol 372:96–104
Landreh L, Spinner K, Schubert K, Häkkinen MR, Auriola S, Poutanern M, Söder O, Svechnikov K, Mayerhofer A (2014) Human testicular peritubular cells host putative stem Leydig cells with steroidogenic capacity. J Clin Endocrinol Metab 99(7):E1227–E1235
Le Douarin NM, Dupin E (2003) Multipotentiality of the neural crest. Current Opinion in Genetics & Development 13 (5):529–536
Le Douarin NM, Callani GW, Dupin E (2008) The stem cells of the neural crest. Cell Cycle 7:1013–1019
Lee RTH, Knapik EW, Thiery JP, Carney TJ (2013a) An exclusively mesodermal origin of fin mesenchyme demonstrates that zebrafish trunk neural crest does not generate ectomesenchyme. Development 140:2923–2932
Lee RTH, Negai H, Nakaya Y, Sheng G, Trainor PA, Weston JA (2013b) Cell delamination in the mesencephalic neural fold and its implication for the origin of ectomesenchyme. Development 140:4890–4902
Leeson TS, Cookson FB (1974) The mammalian testicular capsule and its muscle elements. J Morph 144:237–254

Lindner U, Kramer J, Rohwedel J, Schlenke P (2010) Mesenchymal stem or stromal cells: toward a better understanding of their biology? Transfus Med Hemother 37:75–83
Ling E-A, Wong W-C (1993) The origin and nature of ramified and amoeboid microglia: a historical review and current concepts. Glia 7:9–18
Louissaint A Jr, Rao S, Leventhal C, Goldman SA (2002) Coordinated interaction of neurogenesis and angiogenesis in the adult songbird brain. Neuron 34(6):945–960
Maekawa M, Kamimura K, Nagano T (1996) Peritubular myoid cells in the testis: their structure and function. Arch Histol Cytol 59:1–13
Makala H, Pothana L, Sonam S, Malla A, Goel S (2015) Regeneration of Leydig cells in ectopically autografted adult mouse testis. Reproduction 149:259–268
Mariani S, Basciani S, Arizzi M, Spera G, Gnessi L (2002) PDGF and the testis. Trends Endocrinol Metab 13(1):11–17
Mayer S (1902) Die Muskularisierung der capillaren Blutgefäße. Nachweis des anatomischen Substrats ihrer Kontraktilität. Anat Anz, Jena 21:442–455
McLaren A (2003) Primordial germ cells in the mouse. Dev Biol 262:1–15
Michels NA (1936) The structure of capillaries and the un-myogenic character of Rouget cells (Pericytes) in the omentum of rabbits and the web of living frogs. Anat Rec 65:99–125
Middendorff R , Davidoff MS , Holstein AF (1993) Neuroendocrine marker substances in human Leydig cells – changes by disturbances of testicular function. Andrologia; 25:257–262
Middendorff R, Müller D, Mewe M, Mukhopadhyay AK, Holstein A-F, Davidoff MS (2002) The tunica albuginea of the human testis is characterized by complex contraction and relaxation activities regulated by cyclic GMP. J Clin Endocrinol Metab 87:3486–3499
Millner R, Hung S, Erokwu B, Dore-Duffy P, LaManna JC, del Zoppo GJ (2008) Increased expression of fibronectin and the α5ß1integrin angiogenic cerebral blood vessels of mice subject to hypobaric hypoxia. Mol Cell Neurosci 38:43–52
Mizrak SC, Chikhovskaya JV, Sadri-Ardekani H, van Daalen S, Korver CM, Hovingh SE, Roepers-Gajadien HL, Raya A, Fluiter K, de Reijke TM, de la Rosette JJMCH, Knegt AC, Belmonte JC, van der Veen F, de Rooij DG, Repping S, van Pelt AMM (2010) Embryonic stem cell-like cells derived from adult human testis. Hum Reprod 25:158–167
Molenaar R, de Rooij DG, Rommerts FFG, Reuvers PJ, van der Molen HJ (1985) Specific destruction of Leydig cells in mature rats after in vivo administration of Ethane dimethyl sulfonate. Biol Reprod 33:1213–1222
Montiel-Eulefi E, Sánchez R, Rojas M, Bustos-Obregon E (2009) Epiblast embryo stem cells give origin to adult pluripotent cell populations: primordial germ cell, pericytic and haematopoietic stem cells. A review. Int J Morphol 27:1325–1333
Mori S, Leblond CP (1969) Identification of microglia in light and electron microscopy. J Comp Neurol 135:57–79
Morshead CM (2004) Adult neural stem cells: attempting to solve the identity crisis. Dev Neurosci 26:93–199
Müller I (1957) Kanälchen—und Capillararchitektonik des Rattenhodens. Z Zellforsch 45:522–537
Murakami T, Uno Y, Ohtsuka A, Taguchi T (1989) The blood vascular architecture of the rat testis: a scanning electron microscopic study of corrosion casts followed by light microscopy of tissue sections. Arch Histol Cytol 52:151–172
Murray IR, West CC, Hardy WR, James AW, Park TS, Nguyen A, Tawonsawatruk T, Lazzari L, Soo C, Péault B (2014) Natural history of mesenchymal stem cells, from vessel walls to culture vessels. Cell Mol Life Sci 71(8):1353–1374
Nakagawa T, Nabeshima Y, Yoshida S (2007) Functional identification of the actual and potential stem cell compartments in mouse spermatogenesis. Dev Cell 12(2):195–206
Newgreen DF, Kerr RS, Minichiello J, Warren N (1997) Changes in cell adhesion and extracellular matrix molecules in spontaneous spinal neural tube defects in avian embryos. Teratology 55:195–207
Nichols DH (1981) Neural crest formation in the head of the mouse embryo as observed using a new histological technique. J Embryol Exp Morph 64:105–120

Nichols DH (1986) Formation and distribution of neural crest mesenchyme to the first pharyngeal arch region of the mouse embryo. Am J Anat 176:221–231
Nichols J, Smith A (2012) Pluripotency in the embryo and in culture. Cold Spring Harbor Perspectives in Biology 4 (8):a008128-a008128
Nikolova G, Strilic B, Lammert E (2007) The vascular niche and its basement membrane. Trends Cell Biol 17:19–25
Ochs K, Sahm F, Opitz CA, Lanz TV, Oezen I, Couraud P-O, von Deimling A, Wick W, Platten M (2013) Immature mesenchymal stem cell-like pericytes as mediators of immunosuppression in human malignant glioma. J Neuroimmunol 265:106–116
Ortega HH, Lorente JA, Salvetti NR (2004a) Immunohistochemical study of intermediate filaments and neuroendocrine marker expression in Leydig cells of laboratory rodents. Anat Histol Embryol 33(5):309–315
Orth J, Weisz J (1980) Development of Δ5-3ß-hydroxysteroid dehydrogenase and glucose-6-phosphatase activity in Leydig cells of the fetal rat testis: a quantitative cytochemical study. Biol Reprod 22:1201–1209
Özen I, Boix J, Paul G (2012) Perivascular mesenchymal stem cells in the adult human brain: a future target for neuroregeneration? Clin Transl Med 1(1):30
Pacini S, Petrini I (2014) Are MSCs angiogenic cells? New insights on human nestin-positive bone-marrow-derived multipotent cells. Front Cell Dev Biol 2:Article 20. https://doi.org/10.3389/fcell.2014.00020
Palm T, Nielsen SL, Lassen NA (1983) Vascular recruitment in forearm muscles during exercise. Clin Physiol 3:445–451
Palmer TD, Willhoite AR, Gage FH (2000) Vascular niche for adult hippocampal neurogenesis. J Comp Neurol 425:479–494
Paniagua R, Rodriguez MC, Nistal M, Fraile B, Regadera J, Amat P (1988) Changes in surface area and number of Leydig cells in relation to the 6 stages of the cycle of the human seminiferous epithelium. Anat Embryol 178(5):423–427
Péault B, Rudnicki M, Torrente Y, Cossu G, Trmblay JP, Partridge T, Gussoni E, Kunkel LM, Huard J (2007) Stem and progenitor cells in skeletal muscle development, maintenance, and therapy. Mol Ther 15:867–877
Ratajczak MZ, Machalinski B, Wojakowski W, Ratajczak J, Kucia M (2007) A hypothesis for embryonic origin of pluripotent Oct-4+ stem cells in adult bone marrow and other tissues. Leukemia 21:860–867
Ratajczak MZ, Zuba-Surma EK, Wysoczynski M, Ratajczak J, Kucia M (2008) Very small embryonic-like stem cells: characterization, developmental origin, and biological significance. Exp Hematol 36:742–751
Ratajczak MZ, Shin D-M, Liu R, Mierzeiewska K, Ratajczak J, Kucia M, Zuba-Surma EK (2012a) Very small embryonic/epiblast-like stem cells (VSELs) and their potential role in aging and organ rejuvenation—an update and comparison to other small stem cells isolated from adult tissues. Aging 4:235–246
Ratajczak MZ, Zuba-Surma E, Kucia M, Poniewierska A, Suszynska M, Ratajchak J (2012b) Pluripotent and multipotent stem cells in adult tissues. Adv Med Sci 57:1–17
Ribatti D, Vacca A (2008) Overview of angiogenesis during tumor growth, Chapter 14. In: Figg WD, Folkman J (eds) Angiogenesis. An integrative approach from science to medicine. Springer, New York, NY, pp 161–167
Ribatti D, Nico B, Crivellato E (2011) The role of pericytes in angiogenesis. Int J Dev Biol 55:261–268
Rolandsson S, Andersson Sjöland A, Brune JC, Li H, Kassem M, Martens F, Westergren A, Eriksson L, Hanson L, Skog I, Bjermer L, Scheding S, Westergren-Thorson G (2014) Primary mesenchymal stem cells in human transplanted lungs are CD90/CD105 perivascularly located tissue-resident cells. BMJ Open Resp Res 1:e000027. https://doi.org/10.1136/bmjresp-2014-00002
Rossant J (2001) Stem cells from the mammalian blastocyst. Stem Cells 19:477–482
Rouget CMB (1873) Memoiré sur le développment, la structure et les propriétés physiologiques des capillaires sanguins et lymphatiques. Arch Physiol 5:603–661

Rouget CMB (1879) Sur la contractilite des capillaires sanguins. Compt Rend Acad Sci Paris 88:916–918
Rowley JE, Johnson JR (2014) Pericytes in chronic lung disease. Int Arch Allergy Immunol 164:178–188
Russell LD, de França LR (1995) Building a testis. Tissue Cell 27:129–147
Sá-Pereira I, Brites D, Brito MA (2012) Neurovascular unit: a focus on pericytes. Mol Neurobiol 45:327–347
Sbanti RM, Li W-J, Nesti LJ, Wang Y, Tuan RS (2007) Adult mesenchymal stem cells: biological properties, characteristics, and application in maxillofacial surgery. J Oral Maxillofac Surg 65:1640–1647
Scadden DT (2006) The stem-cell niche as an entity of action. Nature 44:1075–1079
Schulze W, Davidoff MS, Holstein A-F (1987) Are Leydig cells of neural origin? Substance P-like immunoreactivity in human testicular tissue. Acta Endocrinol 115:373–377
Shan L-X, Hardy MP (1992) Developmental changes in levels of luteinizing hormone receptor and androgen receptor in rat Leydig cells. Endocrinology 131:1107–1114
Shan L-X, Zhu L-J, Bardin CW, Hardy MP (1995) Quantitative analysis of androgen receptor messenger ribonucleic acid in developing Leydig cells and Sertoli cells by in situ hybridization. Endocrinology 136:3856–3862
Shan L-X, Bardinb CW, Hardy MP (1997) Immunohistochemical analysis of androgen effects on androgen receptor expression in developing Leydig and Sertoli cells. Endocrinology 138:1259–1266
Sharpe RM (1994) Regulation of spermatogenesis. In: Knobil E, Neil JD (eds) The physiology of reproduction, vol 1, 2nd edn. Raven Press, New York, pp 1363–1434
Shen C-N, Burke ZD, Tosh D (2004a) Transdifferentiation, metaplasia and tissue regeneration. Organogenesis 1:36–44
Shen Q, Goderie SK, Jin L, Karanth N, Sun Y, Abramova N, Vincent P, Pumiglia K, Temple S (2004b) Endothelial cells stimulate self-renewal and expand neurogenesis of neural stem cells. Science 304:1338–1340
Shen Q, Wang W, Kokovay E, Lin G, Chang S-M, Goderie SK, Roysam B, Temple S (2008) Adult SVZ stem cells lie in a vascular niche: a quantitative analysis of niche-cell interactions. Cell Stem Cell 3:289–300
Shibata H, Ikeda Y, Mukai T, K-i M, Kurihara I, Ando T, Suzuki T, Kobayashi S, Murai M, Saito I, Saruta T (2001) Expression profiles of COUP-TF, DAX-1, and SF-1 in human adrenal gland and adrenocortical tumors: possible implications in steroidogenesis. Mol Gen Metabol 74:206–216
Shyamala K, Yanduri S, Girish HC, Murgod S (2015) Neural crest: the fourth germ layer. J Oral Maxillofac Pathol 19(2):221–229
Sild M, Rithazer ES (2011) Radial glía: progenitor, pathway, and partner. Neuroscience 17(3):288–302
Sims DE (1986) The pericyte—a review. Tiss Cell 18:153–174
Skinner MK, Tung PS, Fritz IB (1985) Cooperativity between Sertoli cells and testicular peritubular cells in the production and deposition of extracellular matrix components. J Cell Biol 100:1941–1947
Smith AG (2001) Embryo-derived stem cells: of mice and men. Annu Rev Cell Dev Biol 17:435–462
Soncin F, Ward CM (2011) The function of E-cadherin in stem cell pluripotency and self- renewal. Genes 2:229–259
Spence SG, Poole TJ (1994) Developing blood vessels and associated extracellular matrix as substrates for neural crest migration in Japanese quail, Coturnix coturnix japonica. Int J Dev Biol 38:85–98
Stanley EL, Johanston DS, Fan J, Papadopoulos V, Chen H, Ge RS, Zirkin BR, Jelinsky SA (2011) Stem Leydig cell differentiation: gene expression during development of the adult rat population of Leydig cells. Biol Reprod 85:1161–1166
Stanley E, Lin C-Y, Jin S, Liu J, Sottas CM, Ge R, Zirkin BR, Chen H (2012) Identification, prtoliferation, and differentiation of adult Leydig stem cells. Endocrinology 153:5002–5010

Suzuki F, Nagano T (1986) Microvasculature of the human testis and excurrent duct system. Resin-casting and scanning electron-microscopic studies. Cell Tissue Res 243:79–89
Svingen T, Koopman P (2013) Building the mammalian testis: origins, differentiation, and assembly of the component cell populations. Genes Dev 27:2409–2426
Takamoto N (2005) COUP-TFII is essential for radial and anteroposterior patterning of the stomach. Development 132 (9):2179–2189
Tang W, Zeve D, Suh JM, Bosnakovski D, Kyba M, Hammer RE, Tallquist MD, Graff JM (2008) White fat progenitor cells reside in the adipose vasculature. Science 322:583–586
Tavazoie M, Van der Veke L, Silva-Varga V, Louissaint M, Colonna L, Zaidi B, Garcia-Verdugo JM, Doetsch F (2008) A specialized vascular niche for adult neural stem cells. Cell Stem Cell 3:279–288
Teerds K (1996) Regeneration of Leydig cells after depletion by EDS: a model for postnatal Leydig cell regeneration. In: Payne AH, Hardy MP, Russel LD (eds) The Leydig cell. Cache River, Vienna, pp 203–219
Teerds K, Rijntjes E (2007) Dynamics of Leydig cell regeneration after EDS. A model for postnatal Leydig cell development. In: Payne AH, Hardy MP (eds) Contemporary endocrinology: the Leydig cell in health and disease. Humana Press, Totowa, NJ, pp 91–116
Teerds KJ, de Boer-Brouwer M, Dorrington JH, Balvers M, Ivell R (1999) Identification of markers for precursor and Leydig cell differentiation in the adult rat testis following ethane dimethyl sulphonate administration. Biol Reprod 60:1437–1445
Teerds KJ, Rijntjes E, Veldhuizen-Tsoerkan MB, Rommerts FFG, de Boer-Brouwer M (2007) The development of rat Leydig cell progenitors in vitro: how essential is luteinizing hormone? J Endocrinol 194:579–593
Thiery JP, Delouvee A, Gallin W, Cunningham B, Edelman G (1984) Ontogenetic expression of cell adhesion molecules: L-CAM is found in epithelia derived from the three primary germ layers. Dev Biol 102:61–78
Thomas WE (1999) Brain macrophages: on the role of pericytes and perivascular cells. Brain Res Rev 31:42–57
Tosh D, Slack JMW (2002) How cells change their phenotype. Nat Rev 3:187–194
Trainor PA, Melton KR, Manzanares M (2003) Origins and plasticity of neural crest cells and their roles in jaw and craniofacial evolution. Int J Dev Biol 47: 541–553
Trainor PA (2005) Spezification and pattering of neural crest cells during craniofacial development. Brain Behav Evol 66: 266–280
Traktuev DO, Merferld-Clauss S, Li J, Kolonin M, Arap W, Pasqualini R, Johnstone BH, March KL (2008) A population of multipotent CD-34-positive adipose stromal cells share pericyte and mesenchymal surface markers, reside in a periendothelial location, and stabilize endothelial networks. Circ Res 102:77–85
Trost A, Lange S, Schroedl F, Bruckner D, Motloch KA, Bogner B, Kaser- Eichberger A, Strohmaier C, Runge C, Aigner L, Rivera FJ, Reitsamer HA (2016) Brain and retinal pericytes: origin, function and role. Front Cell Neurosci 10:20
van Dijk CGM, Nieuweboer FE, Pei JY, Xu YJ, Burgisser P, van Mulligen E, el Azzouzi H, Duncker DJ, Verhaar MC, Cheng C (2015) The complex mural cell: pericyte function in health and disease. Int J Cardiol 190:75–89
Vimtrup B (1922) Beiträge zur Anatomie der Capillaren. I. Über contractile Elemente in der Gefäßwand der Blutcapillaren. Anat Embryol 65:150–182
Wang L, Kamath A, Frye J, Iwamoto GA, Chun JL, Berry SE (2012) Aorta-derived mesoangioblasts differentiate into the oligodendrocytes by inhibition of the Rho kinase signaling pathway. Stem Cells Dev 7:1069–1189
Weerasooriya TR, Yamamoto T (1985) Three-dimensional organization of the vasculature of the rat spermatic cord and testis. Cell Tissue Res 241:317–323
Wei LC, Shi M, Chen LW, Cao R, Zhang P, Chan YS (2002) Nestin-containing cells express glial fibrillary acidic protein in the proliferative regions of central nervous system of postnatal developing and adult mice. Brain Res Dev Brain Res 139:9–17
Welsh M, Saunders PT, Atanassova N, Sharpe R, Smith LB (2009) Androgen action via testicular peritubular myoid cells is essential for male fertility. FASEB J 23:4218–4230

Welsh M, Sharpe RM, Moffat L, Atanassova N, Saunders PTK, Kilter S, Bergh A, Smith LB (2010) Androgen action via testicular arteriole smooth muscle cells is important for Leydig cell function, vasomotion and testicular fluid dynamics. PLoS One 5(10):e13632

Welsh M, Moffat L, Belling K, de França LR, Segatelli TM, Saunders PTK, Sharpe RM, Smith LB (2011) Androgen receptor signaling in peritubular myoid cells is essential for normal differentiation and function of adult Leydig cells. Int J Androl 35:25–40

Weston JA, Thiery JP (2015) Pentimento: neural crest and the origin of mesectoderm. Dev Biol 401:37–61

Weston JA, Yoshida H, Robinson V, Nishikawa S, Fraser ST, Nishikawa S (2004) Neural crest and the origin of ectomesenchyme: neural fold heterogeneity suggests an alternative hypothesis. Dev Dyn 229:118–130

Wong SP, Rowley JE, Redpath AN, Tilman JD, Fellous TG, Johnson JR (2015) Pericytes, mesenchymal stem cells and their contributions to tissue repair. Pharmacol Ther 151:107–120. https://doi.org/10.1016/j.pharmthera.2015.03.006

Yao HH-C, Barsoum I (2007) Fetal Leydig cells. Origin, regulation, and function. In: Payne AH, Hardy MP (eds) Contemporary endocrinology: The Leydig Cell in Health and Disease. Humana Press, Totowa, pp. 47-54

Ye L, Li X, Li L, Chen H, Ge R-S (2017) Insights into the development of the adult Leydig cell lineage from stem Leydig cells. Front Physiol 8:430

Yoshida S, Sukeno M, Nabeshima Y-I (2007) A vasculature-associated niche for undifferentiated spermatogonia in the mouse testis. Science 317:1722–1726

Yoshimizu T, Obinata M, Matsui Y (2001) Stage-specific tissue and cell interactions play key roles in mouse germ cell specification. Development 128: 481–490

You LR, Lin F-J, Lee CT, DeMayo FJ, Tsai M-J, Tsai SY (2005) Suppression of notch signaling by COUP-TFII transcription factor regulates vein identity. Nature 435:98–104

Zhu X, Bergles DE, Nishiyama A (2008) NG2 cells generate both oligodendrocytes and gray matter astrocytes. Development 135:145–157

Zimmerlin L, Donnenberg VS, Pfeifer ME, Meyer EM, Péault B, Rubin JP, Donnenberg AD (2010) Stromal vascular progenitors in adult human adipose tissue. Cytometry A 77A:22–30

Zimmerlin L, Donnenberg VS, Donnenberg AD (2012) Pericytes: A universal adult tissue stem cell? Cytometry A 81A: 12-14 doi:10.1002/cyto.a21168

Zimmermann KW (1923) Der feinere Bau der Blutcapillaren. Zeitschr f d ges Anat I Abt 68:29–109

Zouani OF, Lei Y, Durrieu M-C (2013) Pericytes, stem-cell-like cells, but not mesenchymal stem cells are recruited to support microvascular tube stabilization. Small 9:3070–3075

Zwaka TP, Thomson JA (2005) A germ cell origin of embryonic stem cells? Development 132:227–233

Index

A. Birbrair (ed.), *Pericyte Biology in Different Organs*,
Advances in Experimental Medicine and Biology 1122,
https://doi.org/10.1007/978-3-030-11093-2

T

U